FUNDAMENTALS OF QUALITY CONTROL AND IMPROVEMENT

SECOND EDITION

FUNDAMENTALS OF QUALITY CONTROL AND IMPROVEMENT

Amitava Mitra
Auburn University

Prentice Hall, Upper Saddle River, New Jersey 07458

Acquisitions Editor: Tom Tucker
Editor-in-Chief: Natalie Anderson
Assistant Editor: Audrey Regan
Editorial Assistant: Melissa Back
Editorial Director: James Boyd
Production Editor: Marc Oliver
Production Coordinator: Cindy Spreder
Managing Editor: Katherine Evancie
Manufacturing Buyer: Lisa DiMaulo
Senior Manufacturing Supervisor: Paul Smolenski
Manufacturing Manager: Vincent Scelta
Art Director: Jayne Conte
Cover Designer: Bruce Kenselaar
Cover Illustration: Screen capture provided courtesy of Minitab, Inc.
Composition: TSI Graphics, Inc.

A Simon & Schuster Company
Upper Saddle River, New Jersey 07458

Mitra, Amitava.
Fundamentals of quality control and improvement / Amitava Mitra. –
– 2nd ed.
p. cm.
Includes bibliographical references and index.
ISBN 0-13-645086-5
1. Quality control—Statistical methods. I. Title.
TS156.M54 1998
658.5′62′015195—dc21 97-22434
CIP

Prentice-Hall International (UK) Limited, *London*
Prentice-Hall of Australia Pty. Limited, *Sydney*
Prentice-Hall Canada, Inc., *Toronto*
Prentice-Hall Hispanoamericana, S.A., *Mexico*
Prentice-Hall of India Private Limited, *New Delhi*
Prentice-Hall of Japan, Inc., *Tokyo*
Simon & Schuster Asia Pte. Ltd., *Singapore*
Editora Prentice-Hall do Brasil, Ltda., *Rio de Janeiro*

Printed in the United States of America

10 9 8 7 6 5 4 3 2 1

To my parents,
who instilled the importance of
an incessant inquiry for knowledge—
and continue to inspire to this day

Brief Contents

Contents

Appendixes

Preface

This book covers the foundations of modern methods of quality control and improvement that are used in the manufacturing and service industries. Quality is key to surviving tough competition. Consequently, business needs technically competent people who are well-versed in statistical quality control and improvement. This book should serve the needs of students in business and management and students in engineering, technology, and other related disciplines. Professionals will find this book to be a valuable reference in the field.

An outgrowth of many years of teaching, research, and consulting in the field of quality assurance and statistical process control, the methods discussed in this book apply statistical foundations to real-world situations. Mathematical derivations and proofs are kept to a minimum to allow a better flow of material. Although an introductory course in statistics would be useful to a reader of this text, the foundations of statistical tools and techniques discussed in Chapter 4 should enable students without a statistical background to understand the material.

Prominently featured are many real-world examples. For each major concept, at least one example demonstrates its application. Furthermore, case studies at the end of every chapter further student understanding of pertinent issues. These case studies present realistic applications of quality control principles and aid in the mastery of the material.

The book is divided into six parts. Part I, which deals with the philosophy and fundamentals of quality control, consists of three chapters. Chapter 1 is an introduction to quality control and the total quality system. In addition to introducing the reader to the nomenclature associated with quality control and improvement, it provides a framework for the systems approach to quality. Discussions of quality costs and their measurement and of the management of the quality function are presented. Chapter 2 examines philosophies of such leading experts as Deming, Crosby, and Juran. Deming's 14 points for management are analyzed, and the three philosophies are compared. Chapter 3 discusses quality management practices, tools, and standards. Topics such as total quality management, quality function deployment, benchmarking, and the seven tools for quality improvement are presented. The chapter also discusses the criteria for the Malcolm Baldrige National Quality Award in the United States and the International Standards Organization (ISO) 9000 standards.

Part II deals with the statistical foundations of quality control and consists of two chapters. Chapter 4 offers a detailed coverage of statistical concepts and techniques in quality control and improvement. It presents a thorough treatment of inferential statistics. Depending on the students' background, only selected sections of this chapter will need to be covered. Chapter 5 deals with graphical methods of data presentation and quality improvement. Modern tools such as cause-and-effect diagrams, box plots, quantile-quantile plots, matrix plots, and multivariable charts are covered in this chapter.

The field of statistical quality control consists of two areas: statistical process control and acceptance sampling. Part III deals with statistical process control and consists of four chapters. Chapter 6 gives an overview of the principles and use of control charts. A variety of control charts for variables are discussed in detail in Chapter 7. In addition to charts for the mean and range, those for the mean and standard deviation, individual units, cumulative sum, moving average, geometric moving average, and trends are presented. Control charts for attributes are discussed in Chapter 8. Charts such as the *p*-chart, *np*-chart, *c*-chart, *u*-chart, and *U*-chart are presented. The topic of process capability analysis is discussed in Chapter 9. The ability of a process to meet customer specifications is examined in detail. Process capability analysis procedures and process capability indices are also treated in depth. The chapter discusses proper approaches to setting tolerances on assemblies and components. Part III should form a core of material to be covered in most courses.

Part IV deals with acceptance sampling procedures and consists of one chapter. Methods of acceptance of a product based on information from a sample are described. Chapter 10 presents acceptance sampling plans for attributes and variables. In addition to lot-by-lot attribute and variable sampling plans, standardized plans such as ANSI/ASQC Z1.4–1981 for attributes, and ANSI/ASQC Z1.9 for variables are presented. With the emphasis on process control and improvement, sampling plans do not occupy the forefront. Nevertheless, they are included to make the discussion complete.

Part V deals with product and process design and consists of two chapters. With the understanding that quality improvement efforts are generally being moved further upstream, these chapters constitute the backbone of current methodology. Chapter 11 deals with reliability and explores the effects of time on the proper functioning of a product. Design principles by which the reliability of a system may be improved are discussed. Chapter 12 provides the fundamentals of experimental design and the Taguchi method. Different designs, such as the completely randomized design, randomized block design, and Latin square design are presented. Estimation of treatment effects using factorial experiments is included. This chapter also provides a treatment of the Taguchi method for design and quality improvement; the philosophy and fundamentals of this method are discussed. Various sections of Part V could also be included in the core material for a quality control course.

Finally, Part VI deals with applications of quality control and improvement in the service sector. Chapter 13 describes applications of quality control and improvement methods to a variety of service industries such as banking, education, food, government, healthcare services, public utilities, and transportation.

This book may serve as a text for an undergraduate or a graduate course for students in business and management. It may also serve the needs of students in engineering, technology, and other related disciplines. For a one-semester or one-quarter course, Part I, selected portions of Part II (usually parts of Chapter 4 and all of Chapter 5), Part III, and selected portions of Part V and Part VI could be covered. For a two-semester or two-quarter course, all of Part V and Part VI, along with portions from Part IV, could be covered as well.

CHANGES IN THE SECOND EDITION

Several changes have been made in the second edition. First, the management aspects of quality assurance have been strengthened. Chapter 3, which deals with quality management practices, tools, and standards, has been added; it discusses concepts such as total quality management, quality function deployment, benchmarking, quality auditing, vendor selection and certification programs, and the seven tools for quality improvement. Materials on ISO 9000 standards and the Malcolm Baldrige National Quality Award are also included. Second, in the section on statistical process control, additional discussions on the average run length and standardized control charts for variables, which may also be used for short-run production situations, have been introduced. The discussion on process capability in Chapter 9 has been expanded to include the Taguchi capability index, the capability ratio, and the effect of measurement error on capability indices.

Third, some new case studies have been introduced in Chapter 1 and Chapter 2 that deal with employee participation in quality improvement efforts, in Chapter 10, on acceptance sampling, and in Chapter 12, on experimental design and the Taguchi method. Profiles of some Baldrige Award winners have been included in the first three chapters. They demonstrate some of the best-business practices and their performance achievements.

Fourth, the application of computer software has been integrated throughout the text. Examples of using software packages such as EXCEL, SAS, and MINITAB are demonstrated. The MINITAB software has been emphasized in addressing quality control and improvement. Additional exercises have been included in the statistical process control material. Finally, the discussion on acceptance sampling for attributes and variables has been combined into one chapter. Along similar lines, the material on experimental design and the Taguchi method has been combined into one chapter.

ACKNOWLEDGMENTS

Many individuals have contributed to the development of this book, and thanks are due to them. Modern trends in product/process quality through design and improvement, as well as discussions and questions from undergraduate and graduate classes over the years, have shaped this book. Applications encountered in a consulting environment have provided a scenario for examples and exercises. Constructive comments from the reviewers have been quite helpful. Many of the changes in the second edition are based on input from those who have used the book as well as from reviewers. I would like to thank Glenn Milligan of Ohio State University and Tim Krehbiel of Miami University of Ohio for their thorough review of the book. Additional reviewers were Victor Prybutok of the University of North Texas and David E. Booth of Kent State University. Their input has been valuable.

The manuscript preparation center of the College of Business at Auburn University did a remarkable job for which Margie Heard is to be congratulated. I would like to thank Minitab, Inc. (3081 Enterprise Drive, State College, PA 16801-3008) for its assistance in providing software support. My editor, Tom Tucker, is to be commended for his patience and understanding.

Learning is a never-ending process. It takes time and a lot of effort. So does writing and revising a book. That has been my reasoning to my wife, Sujata, and son, Arnab. I believe they understand this—my appreciation to them.

A.M.

PART I: Philosophy and Fundamentals

CHAPTER

1

Introduction to Quality Control and the Total Quality System

Chapter Outline

1-1 INTRODUCTION

September 29, 1988; the countdown was in progress . . . Ten, nine, eight, seven, six, five, four, three, two, one, ignition, and liftoff. The United States Space Shuttle *Discovery* rose majestically from its launch pad and followed its calculated trajectory into the clear blue sky. Dense white fumes from the rocket engines completely covered the surroundings. Cheers from the spectators were belated rather than spontaneous. People were still nervous; memories of the ill-fated *Challenger,* which exploded in January 1986, haunted them. But soon thereafter the cloud of anxiety that seemed to engulf the spectators faded away, and at last the cheers of triumph, accomplishment, and applause drowned everything else. The shuttle was a little white speck in the sky and on its way to a successful mission. America is back as a leader in space programs, and nothing has helped it regain this supremacy more than quality control. Our space program is an example of quality from the total systems point of view—a scenario in which every unit contributing to the program had to develop a commitment to producing a quality product. And this goal is being accomplished. The myth of the staleness of the American management style and the resistance to accepting innovative ideas with an open mind has been destroyed. The United States is becoming, as in the past, a quality-conscious nation.

1-2 EVOLUTION OF QUALITY CONTROL

The quality of goods and services produced has been monitored, either directly or indirectly, since time immemorial. However, using a quantitative base involving statistical principles to control quality is a modern concept.

The ancient Egyptians demonstrated a commitment to quality in the construction of their pyramids. The Greeks set high standards in arts and crafts. The quality of Greek architecture of the fifth century B.C. was so envied that it profoundly affected the subsequent architectural constructions of Rome. Roman-built cities, churches, bridges, and roads inspire us even today.

During the Middle Ages and up to the 1800s, the production of goods and services was predominantly confined to a single individual or a small group of individuals. These small groups were often family-owned businesses, so the responsibility for controlling the quality of a product or service lay with that person or small group. The standard of quality was therefore determined by the individual who was in turn also responsible for producing the item that would conform to those standards. This phase, comprising the time period up to 1900, has been labeled by Feigenbaum (1983) as the *Operator Quality Control* period. The entire product was manufactured by one person or by a very small group of persons. For this reason, the quality of the product could essentially be controlled by a person who was also the operator. The volume of production was limited. The worker felt a sense of accomplishment, which was morale-lifting, and motivated the worker to new heights of excellence. Controlling the quality of the product was thus embedded in the philosophy of the worker because pride in workmanship was widespread.

Starting in the early 1900s and continuing to about 1920, a second phase evolved, called the *Foreman Quality Control* period (Feigenbaum 1983). With the Industrial Revolution came the concept of mass production, which was based on the principle of specialization of labor. An individual was responsible not for the production of the entire product but rather for only a portion of it. One drawback of this approach was the decrease in the workers' sense of accomplishment and pride in their work. However, most tasks were still not very complicated, and workers became skilled at the particular operations that they performed. Individuals who performed similar operations were grouped together. A supervisor who directed that operation now had the task of ensuring that quality was achieved. Foremen or supervisors controlled the quality of the product, and they were also responsible for the operations in their span of control.

The period from about 1920 to 1940 saw the next phase in the evolution of quality control. Feigenbaum (1983) calls this the *Inspection Quality Control* period. Products and processes became more complicated, and production volume increased. As the number of workers reporting to a foreman grew in number, it became impossible for the foreman to keep close watch over individual operations. Inspectors were therefore designated to check the quality of a product after certain operations. Standards were set, and inspectors compared the quality of the produced item with those standards. In the event of discrepancies between a standard and the product, deficient items were set aside from those that met the standard. The nonconforming items were either reworked, if feasible, or were discarded.

During this period, the foundations of statistical aspects of quality control were being developed, although they did not gain wide usage in U.S. industry. In 1924, Walter A. Shewhart of Bell Telephone Laboratories proposed the concept of using statistical charts to control the variables of a product. These came to be known as control charts (sometimes referred to as Shewhart control charts). They play a fundamental role in statistical process control. In the late 1920s, H. F. Dodge and H. G. Romig, also from Bell Telephone Laboratories, pioneered work in the areas of acceptance sampling plans. These plans were to become substitutes for 100 percent inspection.

The 1930s saw the application of acceptance sampling plans in industry, both domestic and abroad. Walter Shewhart continued his efforts to promote the fundamentals of statistical quality control to industry. In 1929, he obtained the sponsorship of the American Society for Testing Materials (ASTM), the American Society of Mechanical Engineers (ASME), the American Statistical Association (ASA), and the Institute of Mathematical Statistics (IMS) in creating the Joint Committee for the Development of Statistical Applications in Engineering and Manufacturing.

Interest in the field of quality control began to gain acceptance in England at this time. The British Standards Institution Standard 600 dealt with applications of statistical methods to industrial standardization and quality control. In the United States, J. Scanlon introduced the *Scanlon Plan,* which dealt with improvement of the overall quality of work life (Feigenbaum 1983). Furthermore, the U.S. Food, Drug, and Cosmetic Act was instituted in 1938 and had jurisdiction over procedures and practices in the areas of processing, manufacturing, and packing.

The next phase in the evolution process occurred between 1940 and 1960 and is termed the *Statistical Quality Control* phase by Feigenbaum (1983). Production requirements escalated during World War II. Since 100 percent inspection was often not feasible, the principles of sampling plans gained acceptance. The American Society for Quality Control (ASQC) was formed in 1946. A set of sampling inspection plans for attributes called MIL–STD–105A was developed by the military in 1950. These plans underwent several modifications in the future, becoming MIL–STD–105B, MIL–STD–105C, MIL–STD–105D, and MIL–STD–105E. Furthermore, in 1957, a set of sampling plans for variables called MIL–STD–414 was also developed by the military.

Use of quality control procedures was, however, nowhere close to the level that it should have been. Even though the fundamental principles had been developed in the United States, U.S. industry was lackadaisical in adopting them and this caused its downfall in the next decade. No other country was in a better position to exploit the benefits of statistical quality control. Nonetheless, U.S. industry did not capitalize on this unique advantage, though other countries did.

Japan, although totally destroyed during World War II, embraced the new philosophy wholeheartedly. When W. Edwards Deming visited Japan and lectured on these new ideas in 1950, Japanese engineers and top management were convinced of the importance of statistical quality control as a means of gaining a competitive edge in the world market. Subsequently, J. M. Juran, another pioneer in quality control, visited Japan in 1954 and further impressed upon them the strategic role that management plays in the achievement of a quality program. The Japanese were quick to realize the profound effects that these principles would have on the future of business, and they made a strong commitment to a massive program of training and education.

Meanwhile, in the United States, developments in the area of sampling plans were taking place. In 1958, the Department of Defense (DOD) developed the *Inspection and Quality Control Handbook H107,* which dealt with single-level continuous sampling procedures and tables for inspection by attributes. Revised in 1959, this book became the *Inspection and Quality Control Handbook H108,* which also covered multilevel continuous sampling procedures, as well as topics in life testing and reliability.

The next phase, *Total Quality Control,* took place during the 1960s (Feigenbaum 1983). An important feature during this phase was the gradual involvement of several departments and management personnel in the quality control process. Previously, most of these activities were dealt with either by people on the shop floor, by the production foreman, or by people from the so-called inspection and quality control department. The commonly held attitude prior to this period was that quality control is the responsibility of the inspection department. Because this concept was so widely accepted, quality came to be considered someone else's "stepchild." Quality in America seemed to fade from the forefront. The 1960s, however, saw some changes in this attitude. People began to realize that each department had an important role in the production of a quality item. The concept of *zero defects,* which centered around achieving productivity through worker involvement, emerged during this time. For critical products and assemblies—for example, missiles and rockets used in the space program by the National Aeronautics and Space Administration (NASA)—this concept proved to be very successful. Along similar lines, the use of **quality circles** was beginning to grow in Japan. The concept of quality circles is based on the participative style of management. It assumes that productivity will improve through an uplift of morale and motivation, which are in turn achieved through consultation and discussion in informal subgroups.

The advent of the 1970s brought what Feigenbaum (1983) calls the *Total Quality Control Organizationwide* phase. This phase involved the participation of everyone in the company, from the operator to the first-line supervisor, manager, vice president, and even the chief executive officer. Quality was associated with every individual. As this notion continued in the 1980s, it was termed by Feigenbaum (1983) the **Total Quality System,** which he defines as follows:

> A quality system is the agreed on companywide and plantwide operating work structure, documented in effective, integrated technical and managerial procedures, for guiding the coordinated actions of the people, the machines, and the information of the company and plant in the best and most practical ways to assure customer quality satisfaction and economical costs of quality.

In Japan, the 1970s marked the expanded use of a graphical tool known as the **cause-and-effect diagram.** This tool was originally introduced in 1943 by K. Ishikawa and is sometimes called an **Ishikawa diagram.** It is also called a **fishbone diagram** because of its resemblance to a fish skeleton. This diagram helps identify possible reasons for a process to go out of control as well as possible effects on the process. This diagram has become an important tool in the use of control charts because it aids in choosing the appropriate action to take in the event of a process being out of control. Also in this decade, G. Taguchi of Japan introduced the concept of quality improvement through statistically designed experiments. Expanded usage of this technique has continued in the 1990s as companies have sought to improve the design phase.

In the 1980s, U.S. advertising campaigns placed quality control in the limelight. Consumers were bombarded with advertisements relating to product quality, and frequent comparisons were made with those of the competitor. These promotional efforts tried to point out certain product characteristics that were superior to that of similar products. Within the industry itself, an awareness of the importance of quality was beginning to evolve at all levels. Top management saw the critical need for the marriage of the quality philosophy to the production of goods and services in all phases, starting with the determination of customer needs and product design and continuing on to product assurance and customer service.

Management gave more than lip service to quality; they adopted training programs in statistical quality control methods for all levels of workers. The idea that education and training for everyone—from hourly wage earners to top management personnel—is a fundamental condition for success had come home.

This lesson was learned the hard way, though. As management looked at the domestic and international market, they found that those corporations that produced quality products had a clear understanding of, and background in, statistical quality control. It was not a fluke that Japan dominated the world market in the 1980s. This decade saw the fruition of a commitment that Japanese industry had made almost 30 years earlier. They had seen the importance of education and training in this area and had systematically developed a business environment in which all personnel in a company, down to the operator level, were immersed in a total quality work culture. The tools and techniques that Japanese industry adopted in the 1950s were not unknown to the United States. In fact, those methods had originated in the United States. But the miserable irony of the situation was that they were not accepted and used by U.S. industry. The success of Japanese industry and our accompanying struggles opened our eyes; only then did we recognize the importance of training and statistical quality control.

As computer use exploded during the 1980s, an abundance of quality control software programs came on the market. The notion of a total quality system increased the emphasis on vendor quality control, product design assurance, product quality audit, and other related areas. Industrial giants such as the Ford Motor Company and General Motors Corporation adopted the quality philosophy and made strides in the implementation of statistical quality control methods. They in turn pressured other companies to use quality control techniques. For example, Ford demanded documentation of statistical process control from its vendors. Thus, smaller companies who had not used statistical quality control methods previously were forced to adopt these methods to maintain their contracts. Requiring evidence of using quality control procedures will likely continue down to the smallest contractor or vendor. The new millennium will see expanded use of quality control measures and increased attention to the customers' needs. The customer will reign supreme as the determiner of acceptable levels of quality. Industry will adjust to this or inevitably lose market share.

1-3 QUALITY

The notion of **quality** has been defined in different ways by various authors. Garvin (1984) divides the definition of quality into five categories—namely, transcendent, product-based, user-based, manufacturing-based, and value-based. Furthermore, he identifies a framework of the following eight attributes that may be used to define quality: performance, features, reliability, conformance, durability, serviceability, aesthetics, and perceived quality. This frequently used definition is attributed to Crosby (1979):

> Quality is conformance to requirements or specifications.

A more general definition proposed by Juran (1974) is as follows:

> Quality is fitness for use.

This text adopts the latter definition and expands it to cover both the manufacturing and service sectors. The service sector accounts for almost 70 percent of our present economy; it is a major constituent that is not to be neglected. Projections indicate that this proportion will expand even further in the future. Hence, quality may be defined as follows:

> The quality of a product or service is the fitness of that product or service for meeting or exceeding its intended use as required by the customer.

So, who is the driving force behind determining the level of quality that should be designed into the product of service? The customer! Therefore, as the needs of customers change, so should the level of quality. If, for example, the customer prefers an automobile that gives adequate service for 15 years, then that is precisely what the notion of a quality product should be. Quality, in this sense, is not something that is held at a constant universal level. In this view, the term *quality* implies different levels of expectations for different groups of consumers. For instance, to some, a quality restaurant may be one that provides extraordinary cuisine served on the finest china with an ambience of soft music. However, to another group of consumers, the characteristics that comprise a quality restaurant may be quite different. It might be excellent food served buffet-style with moderate prices until the early morning hours.

Quality Characteristics

The preceding example demonstrates that one or more elements define the intended quality level of a product or service. These elements are known as **quality characteristics,** which can be categorized in these groupings: **Structural characteristics** include such elements as the length of a part, the weight of a can, the strength of a beam, the viscosity of a fluid, and so on. **Sensory characteristics** include the taste of good food, the smell of a sweet fragrance, and the beauty of a model, among others. **Time-oriented characteristics** include such measures as a warranty, reliability, and maintainability, while **ethical characteristics** include honesty, courtesy, friendliness, and so on.

Variables and Attributes

Quality characteristics fall into two broad classes—namely, **variables** and **attributes.** *Characteristics that are measurable and are expressed on a numerical scale* are called *variables.* The diameter of a bearing expressed in millimeters is a variable, as are the density of a liquid in grams per cubic centimeter and the resistance of a coil in ohms.

Nonconformity and Nonconforming Unit

Prior to defining an attribute, the terms **nonconformity** and a **nonconforming unit** should be defined. A nonconformity is a *quality characteristic that does not meet its stipulated specifications.* Let's say the specifications of the thickness of steel washers are 2 ± 0.1 millimeters (mm). If we have a washer with a thickness of 2.15 mm, then its thickness is a nonconformity. A nonconforming unit has *one or more nonconformities such that the unit is unable to meet the intended standards and is unable to function as required.* An example of a nonconforming unit might be a cast iron pipe that has an internal diameter and a weight that both fail to satisfy specifications, thereby making the unit dysfunctional.

A quality characteristic is said to be an attribute if it is classified as *either conforming or nonconforming to a stipulated specification.* A quality characteristic that cannot be measured on a numerical scale is expressed as an attribute. For example, the smell of a cologne is characterized as either acceptable or not; the color of a fabric is either acceptable or not. However, there are some variables that are treated as attributes because it is simpler to measure them this way or because it is difficult to obtain data on them. Examples in this category are numerous. For instance, the diameter of a bearing is, in theory, a variable. However, if we measure the diameter using a go/no-go gage and classify it as either conforming or nonconforming (with respect to some established specifications), then the characteristic is expressed as an attribute. The reasons for using a go/no-go gage, as opposed to a micrometer, could be economic; that is, the time needed to obtain a measurement using a go/no-go gage may be much shorter and consequently less expensive. Alternatively, an inspector may not have enough time to obtain measurements on a numerical scale using a micrometer, so such a variables classification would not be feasible.

Defect

A **defect** is associated with a quality characteristic that does not meet certain standards. Furthermore, the severity of one or more defects in a product or service may cause it to be unacceptable (or defective). The modern term for a defect is nonconformity, and the term for a defective is a nonconforming item.

The American National Standards Institute (ANSI) and the American Society for Quality Control (ASQC) provide the following definition of a defect as stated in ANSI/ASQC Standard A3 (1987):

> A defect is a departure of a quality characteristic from its intended level or state that occurs with a severity sufficient to cause an associated product or service not to satisfy intended normal or reasonably foreseeable usage requirements.

Standard or Specification

Since the definition of quality involves meeting the requirements of the customer, these requirements need to be documented. A **standard,** or a **specification,** refers to *a precise statement that formalizes the requirements of the customer; it may relate to a product, a process, or a service.* For example, the specifications for an axle might be 2 ± 0.1 centimeters (cm) for the inside diameter, 4 ± 0.2 cm for the outside diameter, and 10 ± 0.5 cm for the length. This means that for an axle to be acceptable to the customer, each of these dimensions must be within the specified values. A definition given by the U.S. National Bureau of Standards (1983) is as follows:

Specification: **A set of conditions and requirements, of specific and limited application, that provide a detailed description of the procedure, process, material, product, or service for use primarily in procurement and manufacturing. Standards may be referenced or included in a specification.**

Additionally, the U.S. National Bureau of Standards (1983) defines a standard as follows:

Standard: **A prescribed set of conditions and requirements, of general or broad application, established by authority or agreement, to be satisfied by a material, product, process, procedure, convention, test method; and/or the physical, functional, performance, or conformance characteristic thereof. A physical embodiment of a unit of measurement (for example, an object such as the standard kilogram or an apparatus such as the cesium beam clock).**

Acceptable bounds on individual quality characteristics (say, 2 ± 0.1 cm for the inside diameter) are usually known as **specification limits,** whereas the document that addresses the requirements of all the quality characteristics is labeled the *standard.*

Quality of Design

Three aspects are usually associated with the definition of quality: *quality of design, quality of conformance,* and *quality of performance.*

Quality of design deals with the stringent conditions that the product or service must minimally possess to satisfy the requirements of the customer. *It implies that the product or service must be designed to meet at least minimally the needs of the consumer.* Generally speaking, the design should be the simplest and least expensive while still meeting the customers' expectations. Quality of design is influenced by such factors as the type of product, cost, profit policy of the firm, demand for product, availability of parts and materials, and product safety. For example, suppose the quality level of the yield strength of steel cables desired by the customer is 100 kg/cm^2. When designing such a cable, the parameters that influence the yield strength would be selected so as to satisfy this requirement at least minimally. In practice, the product is typically overdesigned so that the desired conditions are exceeded. The choice of a safety factor (k) normally accomplishes this purpose. Thus, to design a product with a 25 percent stronger load characteristic over the specified weight, the value of k would equal 1.25, and the product will be designed for a yield strength of $100 \times 1.25 = 125$ kg/cm^2.

In most situations, the effect of an increase in the designed quality level is to increase the cost at an exponential rate. The value of the product, however, increases at a decreasing rate, with the rate of increase approaching zero beyond a certain designed quality level. Figure 1-1 shows the impact of the designed quality level on the cost and value of the product or service. Sometimes, it might be of interest to choose a design quality level b, which maximizes the difference between value and cost, given that the minimal customer requirements a are met. This is done with the idea of maximizing the return on investment. It may be observed from Figure 1-1 that for a designed quality level c, the cost and value are equal. For any level above c (say, d) the cost exceeds the value. This information is important when a suitable design level is being chosen.

Quality of Conformance

Quality of conformance implies that *the manufactured product or the service rendered must meet the standards selected in the design phase.* With respect to the manufacturing sector, this phase is concerned with the degree to which quality is controlled from the procurement of raw material to the shipment of finished goods. It consists of the three broad areas of defect prevention, defect finding, and defect analysis and rectification. As the name suggests, defect prevention deals with the means to deter the occurrence of defects and is usually achieved using statistical process control techniques. Locating defects is conducted through inspection, test, and statistical analysis of data from the process. Finally, the causes behind the presence of defects are investigated, and corrective actions are taken.

Figure 1-2 shows how quality of design, conformance, and performance influence the quality of a product or service. The quality of design has an impact on the quality of conformance. Obviously, one must be able to produce what was designed. Thus, if the design specification for the length of iron pins is 20 ± 0.2 mm, the question that must be addressed is how to design the tools, equipment, and operations such that the manufactured product will meet the design specifications. If such a system of production can be achieved, the conformance phase will be capable of meeting the stringent requirements of the design phase. On the other hand, if such a production system is not feasibly attained (for instance, if the process is only capable of producing pins with a specification of 20 ± 0.36 mm), the design phase is affected. This feedback suggests that the product be redesigned because the current design cannot be produced using the existing capability. Therefore, there should be a constant exchange of information between the design and manufacturing phases so that a feasible design can be achieved.

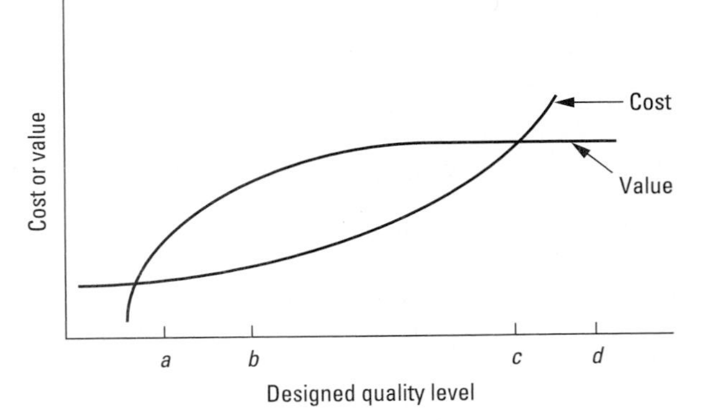

FIGURE 1-1 Cost and value as a function of designed quality.

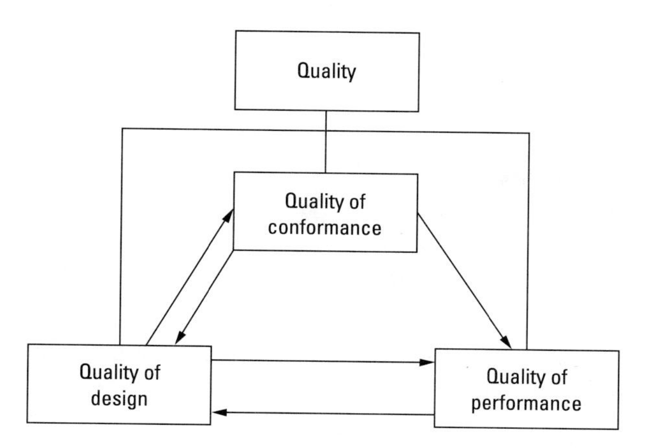

FIGURE 1-2 The three aspects of quality.

Quality of Performance

Quality of performance is concerned with *how well the product functions or service performs when put to use. It measures the degree to which the product or service satisfies the customer.* This is a function of both the quality of design and the quality of conformance. Remember that the final test of product or service acceptance always lies with the customers. Meeting their expectations is the major goal. If a product does not function well enough to meet these expectations, or if a service does not live up to customer standards, then adjustments need to be made in the design or conformance phase. This feedback from the performance to the design phase, as shown in Figure 1-2, may prompt a change in the design because the current design does not produce a product that performs adequately.

1-4 QUALITY CONTROL

Quality control may generally be defined as *a system that is used to maintain a desired level of quality in a product or service.* This task may be achieved through different measures such as planning, design, use of proper equipment and procedures, inspection, and taking corrective action in case a deviation is observed between the product, service, or process output and a specified standard (ASQC 1983; Walsh et al. 1986). This general area may be divided into three main subareas—namely, **off-line quality control, statistical process control,** and **acceptance sampling plans.**

Off-Line Quality Control

Off-line quality control procedures deal with measures to select and choose controllable product and process parameters in such a way that the deviation between the product or process output and the standard will be minimized. Much of this task is accomplished through product and process design. The goal is to come up with a design within the constraints of resources and environmental parameters such that when production takes place, the output meets the standard. Thus, to the extent possible, the product and process parameters are set before production begins. Principles of experimental design and the Taguchi method, discussed in a later chapter, provide information on off-line process control procedures.

Statistical Process Control

> Statistical process control involves comparing the output of a process or a service with a standard and taking remedial actions in case of a discrepancy between the two. It also involves determining whether a process can produce a product that meets desired specifications or requirements.

For example, to control paperwork errors in an administrative department, information might be gathered daily on the number of errors. If the observed number exceeds some specified standard, then on identification of possible causes, action should be taken to reduce the number of errors. This may involve training the administrative staff, simplifying operations if the error is of an arithmetic nature, redesigning the form, or other appropriate measures.

On-line statistical process control means that information is gathered about the product, process, or service while it is functional. When the output differs from a determined norm, corrective action is taken in that operational phase. It is preferable to take corrective actions on a real-time basis for quality control problems. This approach attempts to bring the system to an acceptable state as soon as possible, thus minimizing either the number of unacceptable items produced or the time over which undesirable service is rendered. Chapters 6, 7, 8, and 9 address the background and procedures of on-line statistical process control methods.

One question that may come to mind is: Shouldn't all processes be controlled on an off-line basis? The answer is yes, to the extent possible. The prevailing theme of quality control is that quality has to be designed into the product or service; it cannot be inspected into it. However, in spite of taking off-line quality control measures, there may be a need for on-line quality control, because variation in the manufacturing stage of a product or the delivery stage of a service is inevitable. Therefore, some rectifying measures are needed in this phase. Ideally, a combination of off-line and on-line quality control measures will lead to a desirable level of operation.

Acceptance Sampling Plans

This branch of quality control deals with inspection of the product or service. When 100 percent inspection of all items is not feasible, a decision has to be made on how many items should be sampled or whether the batch should be sampled at all. The information obtained from the sample is used to decide whether to accept or reject the entire batch or lot. In the case of attributes, one parameter is the acceptable number of nonconforming items in the sample. If the observed number of nonconforming items is less than or equal to this number, the batch is accepted. This is known as the acceptance number. In the case of variables, one parameter may be the proportion of items in the sample that are outside the specifications. This proportion would have to be less than or equal to a standard for the lot to be accepted.

A plan that determines the number of items to sample and the acceptance criteria of the lot, based on meeting certain stipulated conditions (such as the risk of rejecting a good lot or accepting a bad lot), is known as *an acceptance sampling plan.*

Let's consider a case of attribute inspection where an item is classified as conforming or not conforming to a specified thickness of 12 ± 0.4 mm. Suppose the items come in batches of 500 units. If an acceptance sampling plan with a sample size of 50 and an acceptance number of 3 is specified, then the interpretation of the plan is as follows. Fifty items will be randomly selected by the inspector from the batch of 500 items. Each of the 50 items will then be inspected (say, with a go/no-go gage) and classified as conforming or not conforming. If the number of nonconforming items in the sample is 3 or less, the entire batch of 500 items is accepted. However, if the number of nonconforming items is greater than 3, the batch is rejected. Alternatively, the rejected batch may be screened; that is, each item is inspected and nonconforming ones are removed.

Acceptance sampling plans for attributes and variables are discussed in Chapter 10.

1-5 QUALITY ASSURANCE

Quality is not just the responsibility of one person in the organization—this is the message. Everyone involved directly or indirectly in the production of an item or in the performance of a service is responsible. Unfortunately, something that is viewed as everyone's responsibility can fall apart in the implementation phase because one person may feel that someone else will follow the appropriate procedures. This behavior can create an ineffective system where the quality assurances exist only on paper. Thus, what is needed is *a system that ensures that all procedures that have been designed and planned are followed.* This is precisely the role and purpose of the quality assurance function.

The objective of the quality assurance function is to have in place a formal system that continually surveys the effectiveness of the quality philosophy of the company. The quality assurance team thus audits the various departments and assists them in meeting their responsibilities for producing a quality product. The ANSI/ASQC Standard A3 (1987) defines quality assurance as follows:

Quality assurance: **All those planned or systematic actions necessary to provide confidence that a product or service will satisfy given needs.**

Quality assurance may be conducted, for example, at the product design level by surveying the procedures used in design. An audit may be carried out to determine the type of information that should be generated in the marketing department for use in designing the product. Is this information representative of the customer's requirements? If one of the customer's key needs in a food wrap is that it withstand a certain amount of force, is that information incorporated in the design? Do the collected data represent that information? How frequently is the data updated? Are the forms and procedures used to calculate the withstanding force adequate and proper? Are the measuring instruments calibrated and accurate? Does the design provide a safety margin? The answers to all of these questions and more will be sought by the quality assurance team. If any discrepancies are found, the quality assurance team will then advise the department in question of the changes that should be adopted. This function acts as a watchdog over the whole system.

1-6 QUALITY CIRCLES AND QUALITY IMPROVEMENT TEAMS

A **quality circle** is typically *an informal group of people that consists of operators, supervisors, managers, and so on, who get together to improve ways to make the product or deliver the service.* The concept behind quality circles is that, in most cases, the persons who are closest to an operation are in a better position to contribute ideas that will lead to an improvement in it. Thus, improvement-seeking ideas do not come only from managers but also from all other personnel who are involved in the particular activity. A quality circle tries to overcome barriers that may exist within the prevailing organizational structure so as to foster an open exchange of ideas.

A quality circle can be an effective productivity improvement tool because it generates new ideas and implements them. Key to its success is its participative style of management. The group members are actively involved in the decision-making process and therefore develop a positive attitude toward creating a better product or service. They identify with the idea of improvement and no longer feel that they are outsiders or that only management may dictate how things are done. Of course, whatever suggestions that a quality circle comes up with will be examined by management for feasibility. Thus, members of the management team must understand the workings and advantages of the proposed action clearly. Only then can they objectively evaluate its feasibility.

Quality circles have been used in Japan since the early 1960s. They have led to numerous improvements in product quality. Toyota, for example, has used this approach to identify critical problems and determine remedial measures. Brainstorming sessions are usually conducted under the guidance of a group leader. In the United States, quality circles were implemented in the early 1970s. Lately, the use of this tool has not been increasing and is nowhere close to the levels seen in Japan.

One possible reason for this difference is the lack of statistical training of U.S. workers. Another reason is the reluctance of U.S. managers to share power with employees. Identification of possible problems and remedial actions requires a statistical background. For almost two decades, starting in the 1950s, Japan has conducted training programs in statistical methods for all of their personnel. U.S. companies have not done this, and they are only now realizing the importance of understanding statistical concepts. However, with a renewed commitment to training *all* people, from operators to management, in these techniques, U.S. companies will find that quality circles are very effective in improving productivity and quality.

The **quality improvement team** is another means of identifying feasible solutions to quality problems. Such teams are typically cross-functional in nature and involve people from various disciplines. It is not uncommon to have a quality improvement team with personnel from design and development, engineering, manufacturing, marketing, and servicing. A key advantage of such a team is that it promotes cross-disciplinary flow of information in real time as it solves the problem. When design changes are made, the feasibility of equipment and tools in meeting the new requirements must be analyzed. It is thus essential for information to flow between design, engineering, and manufacturing. Furthermore, the product must be analyzed from the perspective of meeting customer needs. Do the new design changes satisfy the unmet needs of the customer? What are typical customer complaints on the product? Including personnel from marketing and servicing on these teams assists in answering these questions.

The formation and implementation of quality improvement teams is influenced by several factors. The first deals with selection of team members and its leader. Their knowledge and experience must be relevant to the problem being addressed. People from outside the operational and technical areas can also make meaningful contributions; the objective is to cover a broad base of areas that have an impact. Since the team leader has the primary responsibility for team facilitation and maintenance, he or she should be trained in accomplishing task concerns as well as people concerns, which deal with the needs and motivation of team members.

Team objectives should be clearly defined at the beginning of any quality improvement team project. These enable members to focus on the right problem. The team leader should prepare and distribute an agenda prior to each meeting. Assignments to individual members or subgroups must be clearly identified. Early in the process, the team leader should outline the approach, methods, and techniques to be used in addressing the problem. Team dynamics deals with interactions among members that promote creative thinking and is vital to the success of the project. The team leader plays an important role in creating this climate for creativity. He or she must remove barriers to idea generation and must encourage differing points of view and ideas. All team members should be encouraged to contribute their ideas or to build on others.

Regular feedback on the results and actions taken at meetings is important. It keeps the team on track, helps eliminate personal bias of members, if any, and promotes group effort. Such reviews should ensure that all members have been assigned specific tasks; this should be documented in the minutes. Progress should be reviewed systematically, the objective being to come up with a set of action plans. This review is based on data collected from the process, which is analyzed through basic quality improvement tools (some of which are discussed in Chapters 3 and 5). Based on the results of the analysis, action plans can be proposed. This way, team recommendations will not be based on intuition but on careful analysis.

1-7 BENEFITS OF QUALITY CONTROL

The goal of most companies is to conduct business in such a manner that an acceptable rate of return is obtained by the shareholders. What must be considered in this setting is the short-term goal versus the long-term goal. If the goal is to show a certain rate of return this coming year, this may not be an appropriate strategy, because the benefits of quality control may not be realized immediately. However, from a long-term perspective, a quality control system may lead to a rate of return that is not only better but is also sustainable.

One of the drawbacks of the manner in which many U.S. companies operate is that the output of managers is measured in short time frames. It is difficult for a manager to show an increase of a 5 percent rate of return, say, in the quarter after implementing a quality system. Top management may then doubt the benefits of quality control.

The advantages of a quality control system, however, become obvious in the long run. First and foremost is the improvement in the quality of products and services.

Production improves because a well-defined structure for achieving production goals is present. Second, the system is continually evaluated and modified to meet the changing needs of the customer. Therefore, a mechanism exists to rapidly modify product or process design, manufacture, and service to meet customer requirements so that the company remains competitive. Third, a quality control system improves productivity, which is a goal of every organization. It reduces the production of scrap and rework, thereby increasing the number of usable products. Fourth, such a system reduces costs in the long run. The notion that improved productivity and cost reduction do not go hand in hand is a myth. On the contrary, this is precisely what a quality control system does achieve. With the production of fewer nonconforming items, total costs decrease, which may lead to a reduced selling price and thus increased competitiveness. Fifth, with improved productivity, the lead time for producing parts and subassemblies is reduced, which results in improved delivery dates. Once again, quality control keeps customers satisfied. Meeting their needs on a timely basis helps sustain a good relationship. Last, but not least, a quality control system maintains an "improvement" environment where everyone strives for improved quality and productivity. There is no end to this process—there is always room for improvement. A company that adopts this philosophy and uses a quality control system to help meet this objective is one that will stay competitive.

1-8 RESPONSIBILITY FOR QUALITY

During the industrial revolution of the late nineteenth and early twentieth centuries, the concept of specialized labor was introduced. Prior to this time, the entire product was made by an individual, who was therefore solely responsible for its quality. Mass-production methods were introduced during the Industrial Revolution, and jobs became more specialized as products became more complicated. A supervisor was responsible for quality control, with the focus primarily on inspection. In 1924, when W. A. Shewhart developed the control chart to monitor the product's variables through process control, the emphasis changed from defect detection by inspection to defect prevention. Over the years, defect prevention has become the norm. Thus, no longer is one person or one department responsible for quality; it is everyone's responsibility. The commitment starts with top management and spreads throughout the organization. The quality assurance team must evaluate and oversee each department's responsibility in producing a quality product or service. The following are some specific responsibilities associated with the various departments or units within an organization. Figure 1-3 depicts the responsibilities of the different departments for ensuring quality.

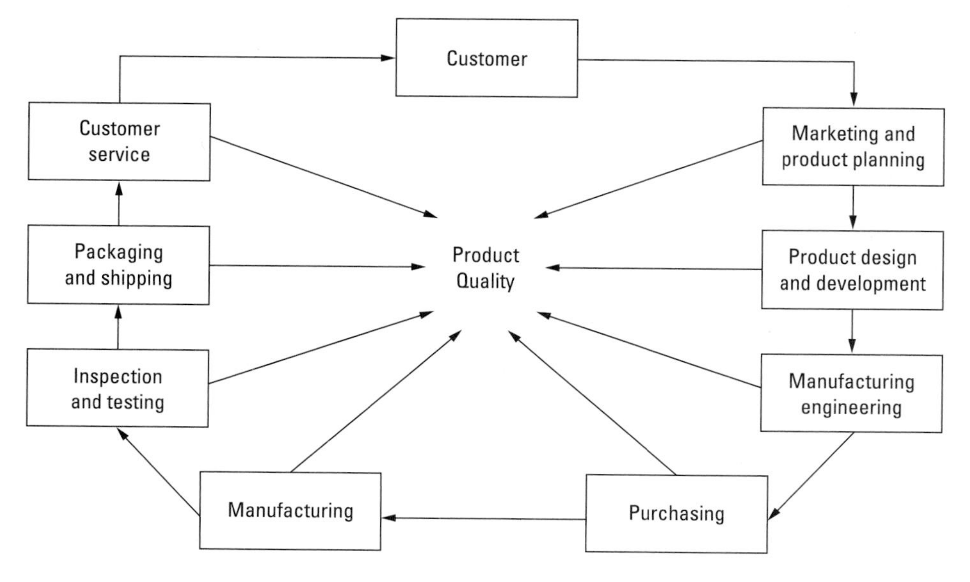

FIGURE 1-3 Responsibility for quality.
(Dale H. Besterfield, *Quality Control*, 3rd ed., © 1990, p. 5. Adapted by permission of Prentice Hall, Inc., Upper Saddle River, New Jersey.)

Marketing and Product Planning

The **marketing and product planning** department determines the needs and requirements of the customer. Additionally, they supply information on the price that the customer is willing to pay. The data obtained here will have an impact on the product quality design. This department obtains information through such means as market surveys conducted via questionnaires, feedback from sales representatives, and customer complaints.

Product Design and Development

The responsibility of the **product design and development** unit is to develop product specifications, determine the raw materials or components to be used, and decide on the performance characteristics of the product. It uses the information on customer needs found by the marketing department as input. Say the marketing department finds that, for a particular fertilizer, the nitrogen content required by the consumer is 15 percent. Product design then comes up with tolerances for a nitrogen content of 15 ± 1 percent. The raw material to be used to produce the fertilizer with this specification is then identified. Remember, quality has to be designed *into* the product, and a good design is one that is feasible given the constraints under which the organization operates. Therefore, a representative from manufacturing should be involved in the design phase.

Manufacturing Engineering

The **manufacturing engineering** department is responsible for determining the details of the manufacturing process. It designs the equipment, work methods and procedures, inspection tools, and sequence of operations. It conducts analyses to determine whether existing manufacturing facilities and resources are capable of producing the product with the quality specified in the design phase. Chapter 9 discusses some methods of analyzing process capability. If production is not feasible, this information must be conveyed to the product design department so that a modified design can be suggested.

Purchasing

The **purchasing** department obtains the raw materials and components for the product. It selects vendors whose products meet certain incoming quality requirements. If there is a lot of variation in the incoming quality of raw materials, an undesirable variability will occur in the outgoing product. Vendor quality control is under the jurisdiction of this department. In selecting vendors, quality, and cost are important factors. Certain companies such as Ford demand that their vendors demonstrate the use and results of statistical quality control procedures in order to maintain their contracts. Vendor selection can also be considered in the product design phase.

Another important aspect in vendor quality control involves developing a long-term relationship with a few high-quality vendors. This practice has been extensively followed in Japan and has several advantages for both parties. For instance, the company does not have to go through the vendor selection process as frequently; this is good because evaluating vendors can be time-consuming and costly. Furthermore, the good relationship enables the company to depend on the vendor when demand for the company's product is uncertain. In peak demand periods, the company can still obtain all of its materials from the same vendor without searching for additional ones. Achieving company goals is easier with one source rather than with many. From the vendor's point of view, mutual trust is extremely desirable. In the case of an occasional poor shipment, the vendor knows that its contract will not be terminated right away. Moreover, because of the long-term relationship, the vendor does not have to worry about frequently renewing contracts. A stable environment therefore exists at the vendor, which helps the vendor maintain a desirable quality level.

Manufacturing

The **manufacturing** unit is responsible for producing a quality product. It must control the operations, process parameters, and operator performance to achieve the desired level of quality. Control charts, discussed in Chapters 6, 7, and 8, are often used for such purposes. Also, for the quality system to work, operators and management must have an adequate background in statistics. Training sessions can achieve this objective.

Inspection and Test

The **inspection** and test unit is responsible for appraising the quality of incoming raw materials and components as well as the quality of the manufactured product or service. It also specifies the type of inspection devices to use and the procedures to follow to measure the quality characteristics of interest. For instance, to measure the thickness of a component, is it better to use a micrometer, an optical sensing device, or a go/no-go gage? Should inspection be done automatically or manually? If inspection is conducted manually, do the inspectors need to be trained? Will the inspection be conducted for all components or only for some? If only for some, with what frequency will the measurements be taken? How will the values be recorded, and what type of analysis will be conducted? These are some of the questions that inspection and testing addresses. The functions of this department may overlap those of the manufacturing department. Say, for an automatic lathe machine, a thickness could be measured at the end of the operation; this task may be done by a member of the manufacturing team.

Inspection is important, but quality *cannot* be inspected into a product or service. Inspection only measures the degree of conformance to a standard in the case of variables. In the case of attributes, inspection merely separates the nonconforming from the conforming. Inspection does not show why the nonconforming units are being produced. Neither does it address what actions should be taken to prevent nonconformities. Methods of statistical process control (Banks 1989; Wadsworth et al. 1986), discussed in Chapters 6, 7, and 8, will take into account some of these issues.

Packaging and Shipping

The packing and shipping unit concerns itself with how the product is packaged and transported to the customer. This function serves to protect the quality of the product during the process of storage and shipment. For instance, in the case of products that are brittle (such as glass mirrors), proper packing methods must be used so that the mirrors are not scratched during transportation and delivery to the customer. Furthermore, unloading and installation should be done carefully. Nothing is more irritating to a customer than opening a box containing the ordered product and finding that it has been damaged during shipment because it was poorly packaged. Since the whole idea of a quality system evolves around satisfying the customer, this unit has an important role in assuring quality.

Customer Service

The **customer service** department is responsible for installation, maintenance, and repair of products. Its purpose is to help the consumer get the most out of the product and to assist the customer when requested. The promptness, politeness, and accuracy of customer service is sometimes all that stands between a satisfied customer and a dissatisfied one. On certain occasions, the customer needs to be trained and made aware of the different functions of a new piece of equipment—say, a drilling machine. This department thus occasionally conducts customer training programs. This unit deals with issues pertaining to **product liability** and warranty. So it also continually seeks feedback from customers on their satisfaction levels. This information is conveyed to the marketing and product planning department; therefore, the customer service unit and the marketing and product planning unit work closely to maximize customer satisfaction and to minimize liability and warranty costs.

1-9 QUALITY AND RELIABILITY

Reliability refers to the ability of a product to function effectively over a certain period of time. Reliability is related to the concept of quality of performance. Since the consumer has the ultimate say on the acceptability of a product or service, the better the performance over a given time frame, the higher the reliability and the greater the degree of customer satisfaction. Achieving desirable standards of reliability requires careful analysis in the product design phase. Analysis of data, obtained on a timely basis during product performance, keeps the design and production parameters updated so that the product may continue to perform in an acceptable manner. Reliability is built in through quality of design.

The product is often overdesigned so that it more than meets the performance requirements over a specified time frame. For example, consider the quality of a highway system where roads are expected to last some minimum time period under certain conditions of use. Conditions of use may include the rate at which the road system is used, the weight of vehicles, and such atmospheric conditions as the proportion of days that the temperature exceeds a certain value. Suppose the performance specifications require the road system to last at least 20 years. In the design phase, to account for the variation in the uncontrollable parameters, the roads might be designed to last 25 years. This performance level may be achieved through properly selected materials and the thickness of the concrete and tar layers.

1-10 TOTAL QUALITY SYSTEM

As previously stated, quality is everyone's responsibility. This means that comprehensive plans should be developed to show the precise responsibilities of the different units, procedures should be defined to check their conformance to the plans, and remedial measures should be suggested in the event of discrepancies between performance and standard. The quality assurance function, as defined earlier, monitors the system. The definition of a quality system, as given by the American National Standards Institute (ANSI)/American Society for Quality Control (ASQC) Standard A3 (1987), is as follows:

***Quality system:* The collective plans, activities, and events that are provided to ensure that a product, process, or service will satisfy given needs.**

The **systems approach** to quality integrates the various functions and responsibilities of the different units and provides a mechanism to ensure that organizational goals are being met through the coordination of the goals of the individual units. The International Organization for Standardization (ISO) has developed standards that describe quality systems. These are discussed in Chapter 3.

ANSI and ASQC have developed generic guidelines for quality systems. This standard, ANSI/ASQC Standard Z-1.15 (1979), defines the following elements of a total quality system.*

1. *Policy, planning, organization, and administration.* A quality policy is essential for developing consistency in the company's goals. Once a plan is developed, an organization must be established to aid in the achievement of the plan. Quality manuals are often created for this purpose. Furthermore, procedures for administering the plans in practice should be detailed, and costs should be identified.

2. *Product design assurance, specification development, and control.* With the customers' requirements in mind, a product design is formulated. Through prototype

---

*ASQC (1979), *ANSI/ASQC Standard Z-1.15-1979.* Reprinted with the permission of ASQC.

development and testing, this design will undergo modifications until it satisfactorily meets all requirements. Tolerances on the product's characteristics will be developed based on careful consideration of the manufacturing capabilities. During the production phase, procedures should be defined to check and control the product characteristics to conform to the standards of design.

3. *Control of purchased materials and component parts.* The production of a quality product is very much influenced by the quality of the raw materials and components used. Procedures must be developed to evaluate the capability and performance of the vendors. Some companies require vendors to demonstrate an adequate use of statistical quality control methods before they can be considered as candidates for selection. Specifications should be set for incoming items and then explained to the vendors. If incoming inspection is to be performed, then the vendors must be notified. Quality, cost, and ability to meet due dates should be considered in choosing vendors. A company should attempt to develop harmonious long-term relationships with its vendors. Corrective actions for controlling purchased materials should be developed if nonconformance occurs and if the materials are to be inspected. The American Society for Quality Control has published a series of materials on this topic, including *Procurement Quality Control: A Handbook of Recommended Practices, How to Evaluate a Supplier's Product,* and *How to Conduct a Supplier Survey.*

4. *Production quality control.* This is a critical aspect of the whole framework for quality. It involves determining process specifications, selecting equipment, training personnel, designing forms and charts, selecting inspection points, and collecting and analyzing process data to determine whether it is under control. It is also important to determine process capability, to conduct experiments on improving the process, and to perform final inspection. It should be emphasized that preventive maintenance and feed-forward structures, in addition to feedback control, are useful in production and process control. In feed-forward control, when changes are made in operations or processes upstream, such information is conveyed to those who deal with subsequent operations. They are also informed on how to conduct concurrent control during the transformation process. Minimizing in-process inventory and using the just-in-time concept, where items are received just when they are needed, can also improve productivity. Proper application of the just-in-time principle is dependent, however, on the timeliness of incoming parts and the reliability of the equipment and resources of the company. Developing accurate forecasts for demand can influence the success of a just-in-time system. Other concerns include designing and developing product-quality audits to detect departures from specified standards before mass production begins. Doing so allows the manufacturer to correct problems before large numbers are produced. Disseminating quality information to key people throughout the organization is another important objective.

5. *User contact and field performance.* A system must exist to collect information from the consumer and to determine the level of performance of the product or service. Poor field performance may necessitate changes in design. How will the information be collected from consumers? Design of the data forms, the frequency with which data is collected, the types of analyses to use, appropriate advertising and promotion schemes, and installation plans and procedures are some of the items to be considered. Feedback on product failure has implications on the type of warranties to be offered and should dictate the procedures for handling product liability issues. Developing policies for installation and servicing and determining guidelines for compliance with these policies are critical to providing customer satisfaction. Careful thought should be given to designing a system that will collect, analyze, and act on consumer feedback.

6. *Corrective action.* Timely corrective action is the key to creating and maintaining a quality system. Problems therefore need to be detected, categorized, and systematically documented. Effective remedial actions also need to be documented so that

results are consistently successful. Furthermore, a structure whose task is to detect problems and take corrective actions needs to be created. Corrective action may be required with such different entities as the incoming materials and components, the vendors, the process and equipment, the operators, and the products. Specific procedures for each of these categories as well as appropriate follow-up must be determined.

7. *Employee selection, training, and motivation.* The employee is the cornerstone in the success of a total quality system. This means every employee—from the temps to the operators to the CEO. People in staff roles contribute significantly too. Errors in administrative paperwork, which delay orders for raw materials and components, are as critical as maladjustments made on the assembly line. Guidelines should be established to select people for particular jobs, a task that may involve job analyses and identification of skills of the available pool of personnel. Job manuals should be developed to train people. In every situation, there must be a clear demonstration by management of their commitment to help select, train, and motivate employees to produce a better-quality product. Recognition of superior effort and the use of motivation programs help reassure employees of the support of management. Some of the common motivational programs are the zero defects, quality circles, and participative quality control programs. The **zero defects program** (ZD) originated in 1961–1962, during which time the Martin Company of Orlando, Florida, delivered a Pershing missile to Cape Canaveral with zero nonconformities. Since then, ZD programs have been used to motivate and challenge operators and supervisors to do their best all the time. ZD programs give operators the authority to make changes in the process when problems occur. However, for these programs to work, employees must first be trained in problem detection and in taking appropriate actions.

8. *Legal requirements—product liability and user safety.* The failure of products within the warranty period and the hazardous effects of malfunction are of grave concern to manufacturers. Liability suits can be an enormous expense, and companies have to budget for warranty costs. Consumer tolerances also change with time. Minor blemishes or cosmetic defects that were once acceptable may not be so because of more stringent demands imposed by the consumer. Furthermore, the U.S. judicial system enforces the rule of **strict liability,** which creates the need to plan for field failures and their possible legal implications. Strict liability refers to the presumption of liability when a given event occurs. For example, if foreign matter is found in food products, strict liability assigns negligence to the manufacturer. Under the rule of strict liability, any enterprise—the manufacturer as well as the seller—must respond immediately to unsatisfactory quality. This response may take place through product service, repair, or replacement. The responsibility extends through the period of the product's use and also includes coverage of environmental effects (say, noise pollution) and safety aspects. There must be clear directions on what safe usage means (for example, a precise description of conditions under which a product should not be used; a hair dryer should not be operated while resting in water). Second, all advertising statements must be supportable by certified data. Therefore, product durability, safety, and environmental concerns must be integrated in the quality system.

9. *Sampling and other statistical techniques.* This segment of the total quality system is composed of the analytical tools and techniques. Most of these procedures are described in the following chapters and constitute the main subject matter of this text. Chapter 4 covers the fundamentals of statistical concepts and techniques used in quality control. Chapter 5 presents some statistical techniques for quality analysis and improvement. The idea of process control through control charts, which is one of the primary quality control tools, is covered in Chapters 6–9. The fundamental principles of control charts are introduced in Chapter 6. Chapter 7 focuses on control charts for variables, while those for attributes are covered in Chapter 8. Statistical methods for determining whether a process is capable of producing items that conform to a standard are described in Chapter 9. These methods involve process capability analysis. The topics of acceptance sampling plans for attributes and variables are found in Chapter 10. Statistical methods dealing with life testing and reliability may be found in

Chapter 11; these techniques concern the performance of a product over a period of time. Designing experiments for use in systematically analyzing and guiding process parameter settings is covered in Chapter 12. Some fundamental concepts of the Taguchi method of off-line quality control are also presented in Chapter 12. Taguchi methods may also be used for on-line quality control, a limited discussion of which is found in Chapter 12. Chapter 13 includes applications of quality control techniques in the service sector. Since more than 70 percent of the gross national product comes from the service sector, techniques for monitoring quality that have primarily been used in the manufacturing sector are also warranted here. Finally, computers play a fundamental role in quality control, and their use will expand even more in the years to come. The applications of computer software for different statistical techniques in quality control are integrated throughout the text.

1-11 QUALITY IMPROVEMENT

Efforts to reduce both the variability of a process and the production of nonconforming items should be ongoing because quality improvement is a *neverending* process. Whereas process control deals with identification and elimination of special causes (those for which an identifiable reason can be determined) that force a system to go out of control (for example, tool wear, operator fatigue, poor raw materials), **quality improvement** relates to the detection and elimination of common causes. **Common causes** are inherent to the system and are always present. Their impact on the output may be uniform relative to that of special causes. An example of a common cause is the variability in a characteristic (say, a diameter) caused by the inherent capability of the particular equipment used (say, a milling machine). This means that, all other factors held constant, the milling machine is unable to produce parts with exactly the same diameter. To reduce the inherent variability of that machine, an alternative might be to install a better or more sophisticated machine. **Special causes** are mainly controllable by the operator, but common causes need the attention of management. Therefore, quality improvement can take place only through the joint effort of the operator and management, with the emphasis primarily on the latter. For instance, a decision to replace the milling machine must be made by management. Eliminating common causes results in an improved process capability, as measured by less variation of the output.

Most quality control experts agree that common causes account for at least 90 percent of the quality problems in an organization. The late W. Edwards Deming, the noted authority on quality, strongly advocated this belief. He concluded that management alone is responsible for common-cause problems and, hence, only management can define and implement remedial actions for these problems. The operator has no control on nonconforming product or service in a majority of the instances. Therefore, if a company is interested in eliminating the root causes of such problems, management must initiate the problem-solving actions.

Quality improvement should be the objective of all companies and individuals. It improves the rate of return or profitability by increased productivity and by cost reduction. It is consistent with the philosophy that a company should continually seek to expand its competitive edge. It supports the principle that no deviation from a standard is acceptable, which is akin to the principle of the loss function developed in the Taguchi methods (Taguchi 1986; Taguchi and Wu 1979). So, even if the product is within the specification limits, an ongoing effort should be made to reduce its variability around the target value.

Let's say that the specifications for the weight of a package of sugar is 5.00 ± 0.05 lb. If the output from the process reveals that all packages weigh between 4.95 and 5.05 lb, then the process is capable and all items will be acceptable. However, not all of the packages weigh exactly 5.00 lb, the target value: that is, there is some variability in the weights of the packages. The Taguchi philosophy states that any deviation from the target value of 5.00 lb is unacceptable, with the loss being proportional to the deviation. Quality improvement is a logical result of this philosophy.

Some methods for quality improvement are discussed in Chapters 3 and 5. These include such graphical techniques as Pareto analysis, histograms, and cause-and-effect or fishbone diagrams. Additional techniques discussed in Chapter 9 deal with process capability analysis. Quality improvement through design may also be achieved through experimental design techniques and the Taguchi method; these are discussed in Chapter 12.

Several authors have identified and categorized the different stages associated with quality improvement. Pall (1987) divides the process into three stages: commitment, consolidation, and maturity (see Figure 1-4).

Commitment Stage

In the commitment stage, management agrees to undertake a quality improvement program. A formal plan and policy are developed, and the organizational structure for implementing the plan is put into place. This stage may be divided into two phases, introduction and initial implementation. The introduction phase may take anywhere from three to six months. Keep in mind that these are rough estimates; the actual time will be influenced by such factors as business climate, type of company, product, process, management perception, and resource availability. During the introduction phase, a policy statement is developed, and plans for creating the infrastructure to implement quality improvement methods are discussed.

In the initial implementation phase, which may take between six months and a year, problems are identified, and causes associated with the production of nonconformities are removed. This phase usually deals with the elimination of special causes, such as a poor quality of raw materials, operator mistakes, or improper tools. The process is thereby brought to a state of statistical control. In Figure 1-4, point *A* represents the percentage of nonconformities produced by the system before this process is implemented. With the introduction of the plan, education and training of personnel, and adequate management support, the percentage drops to *B*. As special causes are detected and eliminated, the quality improves to point *C*.

Consolidation Stage

In the consolidation stage, the objectives are to produce an item that conforms to requirements (quality of conformance) and to start a continual improvement in the efficiency and productivity of the process. Although eliminating special causes helps to attain the first objective, removing or improving common causes is necessary to ac-

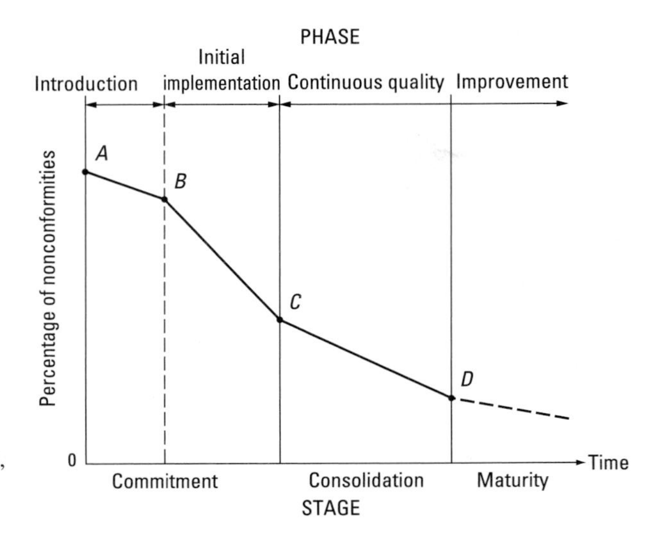

FIGURE 1-4 The three stages of quality improvement. (Gabriel A. Pall, *Quality Process Management*, © 1987, pp. 34, 46. Adapted by permission of Prentice Hall, Inc., Upper Saddle River, New Jersey.)

complish the second one. The latter action improves process capability. A large investment is made in the prevention of defects, which therefore reduces total costs by minimizing the number of items that have to be scrapped or reworked. Education and training are an integral part of this stage. Managing both the intent and the outcome is key to the successful implementation of this stage. As shown in Figure 1-4, quality level improves from point *C* to point *D*, but the rate of improvement is slower than in the initial implementation phase.

Maturity Stage

In this stage, the process is mature; quality improvement and management are a way of life. The process parameters are adjusted to create optimal operating conditions. Very few defects or nonconformities are produced; efforts and costs are focused on defect prevention. Improvements in productivity and quality continue to take place. Every person in the organization believes in the quality philosophy. Employees are enthusiastic about seeking innovative ways to improve the process. They are convinced that quality management is an integral part of the system and are fully aware of its benefits. The rate of improvement, as can be seen from Figure 1-4, slows down. From point *D*, the goal is to asymptotically approach the ideal condition of zero nonconformities while continually improving the process performance.

1-12 QUALITY COSTS

One measure of the performance of the total quality system is the cost associated with it. Careful identification, measurement, and analysis of cost as a function of time aids in tracking the impact of an effective quality control system. Remember that the benefits of a quality system, as measured by the total quality costs, may be realized in the long run rather than in the short run. The full impact of a particular change in the process is usually only felt later.

The American Society for Quality Control (1971) has defined four major categories for **quality costs,** which are discussed here.

Prevention Costs

Prevention costs are incurred in planning, implementing, and maintaining a quality system. They include salaries and developmental costs for product design, process and equipment design, process control techniques (through such means as control charts), information systems design, and all other costs associated with making the product right the first time. Also, costs associated with education and training are included in this category. Other such costs include those associated with defect-cause and removal, process changes, and the cost of a quality audit. Figure 1-5 shows a typical breakdown of the quality costs as a function of time into four categories. As the figure shows, prevention costs increase with the introduction of a quality system and may be a significant proportion of the total quality costs. The rate of increase slows with time. Even though prevention costs increase, they are more than justified by reductions in total quality costs.

Appraisal Costs

Appraisal costs are those associated with measuring, evaluating, or auditing products, components, or purchased materials to determine their degree of conformance to the specified standards. Such costs include dealing with the inspection and test of incoming materials as well as product inspection and testing at various phases of manufacturing and at final acceptance. Other costs in this category include the cost of calibrating and maintaining measuring instruments and equipment and the cost of materials and products consumed in a destructive test or devalued by reliability tests. Appraisal costs typically occur during or after production but before the product is released to the customer. Hence, they are associated with managing the outcome, whereas prevention costs are

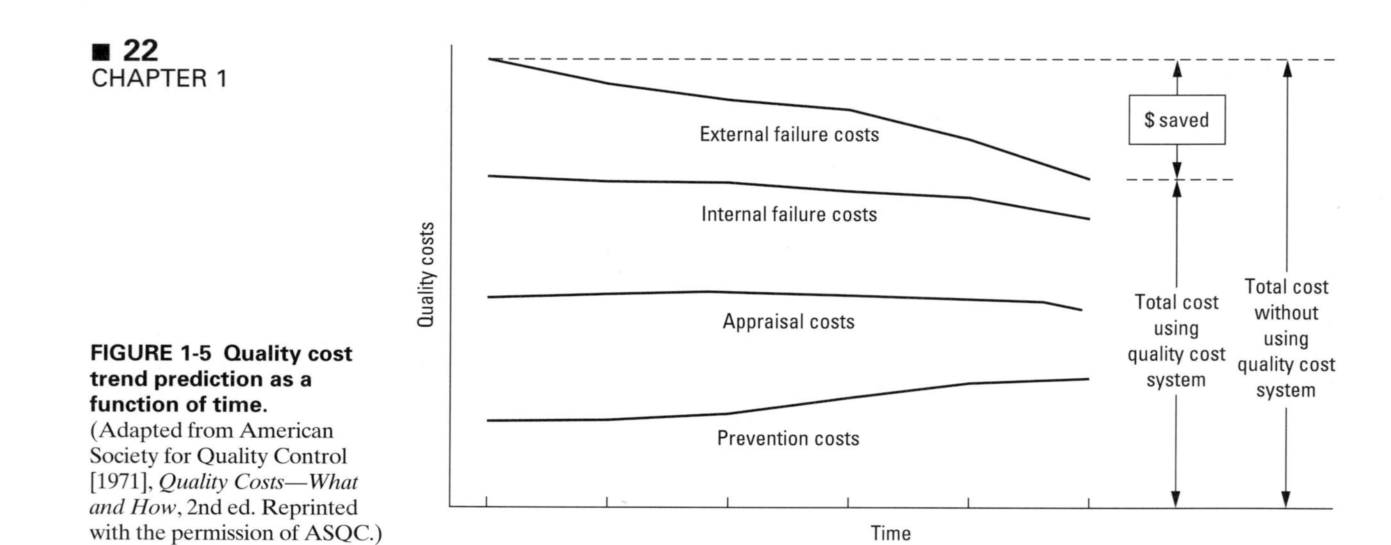

FIGURE 1-5 Quality cost trend prediction as a function of time. (Adapted from American Society for Quality Control [1971], *Quality Costs—What and How*, 2nd ed. Reprinted with the permission of ASQC.)

associated with managing the intent or goal. Figure 1-5 shows that appraisal costs normally decline with time as more nonconformities are prevented from occurring.

Internal Failure Costs

Internal failure costs are incurred when products, components, materials, and services fail to meet quality requirements prior to the transfer of ownership to the customer. These costs would disappear if there were no nonconformities in the product. Internal failure costs include scrap and rework costs for the materials, labor, and overhead associated with production. The cost of correcting nonconforming units, as in rework, can include such additional manufacturing operations as regrinding the outside diameter of an oversized part. If the outside diameter were undersized, it would not be feasible to use it in the finished product, and the part would become scrap. Those costs involved in determining the cause of failure or in reinspecting or retesting reworked products are other examples from this category. The cost of lost production time due to nonconformities must also be considered (for example, if poor quality of raw materials requires retooling of equipment). Furthermore, the revenue lost because a flawed product has to be sold at a lower price—called downgrading costs—is another component. As a total quality system is implemented and becomes effective with time, internal failure costs will decline, as shown in Figure 1-5. Less scrap and rework will result as problems are prevented.

External Failure Costs

External failure costs are incurred when the product does not perform satisfactorily after ownership is transferred to the customer. If no nonconforming units were produced, this cost would vanish. Such costs include those due to customer complaints, which include the costs of investigation and adjustments, and those associated with receipt, handling, repair, and replacement of nonconforming products. Warranty charges (failure of a product within the warranty time) and product liability costs (costs or awards as an outcome of product liability litigation) also fall under this category. As shown in Figure 1-5, a reduction in external failure costs occurs when a quality control system is successfully implemented. A quality system causes a reduction in the internal and external failure costs, which in turn reduces the total quality costs.

Quality Costs Data Requirements

Quality costs should be carefully monitored. Because indirect costs are as important as such direct costs as raw material and labor, well-defined accounting procedures should be set up to determine realistic quality cost estimates. Consider the case where quality

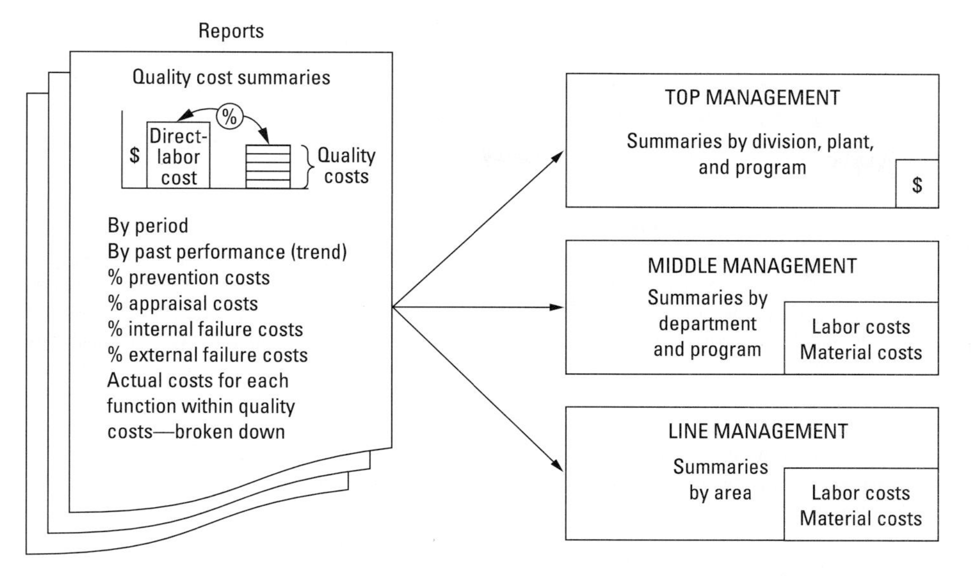

FIGURE 1-6 Quality costs data requirements at different management levels.
(Adapted from American Society for Quality Control [1971], *Quality Costs—What and How*, 2nd ed. Reprinted with the permission of ASQC.)

cost data cross departmental lines. This occurs, for example, when a quality control supervisor in a staff position identifies the reason for scrap or rework, and a lathe operator conducts an extra operation to rework those items. Likewise, should rework or scrap inspire a change in the product design, then the redesign time is assigned to quality costs.

Special forms may be needed to report quality costs data. Figure 1-6 shows the data requirements at different management levels. Data is collected for each product line or project and distributed to each level of management. The needs are somewhat different at each level. Top management may prefer a summary of the total quality costs, broken down into each of the four categories, at the division or plant level. On the other hand, line management or supervisors may want a summary of the direct costs, which include labor and material costs, as it relates to their area.

This means that if a change is made in product or process design, then it is possible for one or more quality cost categories to be affected. The time spent by the design engineer would be allocated, costwise, to prevention cost. On the other hand, if the design calls for new inspection equipment, then that would be allocated to appraisal cost. Thus, a costwise breakdown into the four categories of prevention, appraisal, internal failure, and external failure is incorporated into the accounting system for the different functions performed by management and operators.

1-13 MEASURING QUALITY COSTS

The magnitude of quality costs is important to management because such indices as return on investment are calculated from it. However, for comparing quality costs over time, magnitude may not be the measure to use because conditions often change from one quarter to the next. The number of units produced may change, which affects the direct costs of labor and materials, so the total cost in dollars may not be comparable. To alleviate this situation, a measurement base that accounts for labor hours, manufacturing costs, sales dollars, or units produced could be used to produce an index. These ideas are discussed here.

Labor base index. One commonly used index is the quality costs per direct-labor hour. The information required to compute this index is readily available, since the accounting department collects direct-labor data. This index should be used for short periods because over extended periods, the impact of automation on direct-labor hours may be significant. Another index lists quality costs per direct-labor dollar, thus eliminating the effect of inflation. This index may be most useful for line and middle management.

Cost base index. This index is based on calculating the quality costs per dollar of manufacturing costs. Direct-labor, material, and overhead costs make up manufacturing costs, and the relevant information is readily available from accounting. This index is more stable than the labor base index because it is not significantly affected by price fluctuations or changes in the level of automation. For middle management, this might be an index of importance.

Sales base index. For top management, quality costs per sales dollar may be an attractive index. It is not a good measure for short-term analysis, but for strategic decisions, top management focuses on long-term outlook. Sales lag behind production and are subject to seasonal variations (for example, increased sale of toys during Christmas). These variations have an impact in the short run. However, they smooth out over longer periods of time. Furthermore, changes in selling price also affect this index.

Unit base index. This index calculates the quality costs per unit of production. If the output of different production lines are similar, then this index is valid. Otherwise, if a company produces a variety of products, then the product lines would have to be weighted and a standardized product measure computed. For an organization producing—say, refrigerators, washers, dryers, and electric ranges—it may be difficult to calculate the weights based on a standard product. For example, if 1 electric range is the standard unit, is a refrigerator 1.5 standard units of a product and a washer 0.9 standard unit? The other indexes should be used in such cases.

For all of these indexes, a change in the denominator causes the value of the index to change, even if the quality costs do not change. If the cost of direct labor de-

TABLE 1-1 A Sample Monthly Quality Cost Report

Cost Categories	*Amount*	*Percentage of Total*
Prevention Costs		
Quality planning	15,000	
Quality control engineering	30,000	
Employee training	10,000	
Total prevention costs	55,000	26.31
Appraisal Costs		
Inspection	6,000	
Calibration and maintenance of test equipment	3,000	
Test	2,000	
Vendor control	4,000	
Product audit	5,000	
Total appraisal costs	20,000	9.57
Internal Failure Costs		
Retest and troubleshooting	10,000	
Rework	30,000	
Downgrading expense	3,000	
Scrap	16,000	
Total of internal failure costs	59,000	28.23
External Failure Costs		
Failures—manufacturing	15,000	
Failures—engineering	20,000	
Warranty charges	40,000	
Total external failure costs	75,000	35.89
Total quality costs	209,000	

Bases:		**Ratios:**	
Direct labor	800,000	Internal failure to labor	7.375%
Manufacturing cost	2,000,000	Internal failure to manufacturing	2.950%
Sales	3,800,000	Total quality costs to sales	5.500%

creases, which may happen because of improvement in productivity, the labor base index increases. Such increases should be interpreted cautiously because they can be misconstrued as increased quality costs.

A sample monthly quality cost report is shown in Table 1-1. Monthly costs are depicted for individual elements in each of the major quality cost categories of prevention, appraisal, internal failure, and external failure. Observe that external failure and internal failure are the top two cost categories comprising 35.89 and 28.23 percent, respectively, of the total cost. Comparisons of costs with different bases are also found in Table 1-1. Data on direct-labor cost, manufacturing cost that includes direct-labor, material, and overhead costs, and sales are shown in the table. Internal failure costs, for example, are 7.375 percent of direct-labor costs and 2.950 percent of manufacturing costs. Total quality costs are 5.5 percent of sales. From the information in the table, it seems that management needs to look into measures that will reduce internal and external failure costs, which dominate total quality costs. Perhaps better planning and design, and coordination with manufacturing, will reduce the external failure costs.

1-14 MANAGEMENT OF QUALITY

Management of the total quality system is achieved through the functions of planning, organizing, staffing, directing, and controlling. Figure 1-7 shows these functions as well as the phases in product development and the corresponding actions for each phase.

The major objectives of the management system are to integrate all processes and functional units to meet the goals of the company. Management is responsible for the prevention of nonconformities, and their goal is to design, manufacture, and maintain the product or the service at the least possible cost while still meeting all customer requirements. Presently, companies are finding out that meeting customer requirements is not always enough to remain competitive. Companies have to satisfy their customers to such a degree that they continually repurchase the product. This feat can be achieved only through an organized and thorough evaluation and analysis of performance data integrated into the design phase.

Management Functions	*Product Phase*	*Action to Be Taken*
Plan	Proposal phase	Develop quality policy Plan for quality Set up guidelines for system administration Consider product liability/user safety
Organize	Design/planning phase	Develop an organization structure Design assurance Design change control Develop production quality planning
Staff	Preproduction phase	Select employees Train employees Motivate employees
Direct	Production phase	Monitor purchased materials quality Monitor process quality control Direct final inspection Direct handling/inspection
Control	Production and postproduction phase	Obtain quality information Get data on field performance Take corrective action Conduct statistical quality control Manage quality costs

FIGURE 1-7 Functions of management in the quality system.
(E. G. Schilling [1984], "The Role of Statistics in the Management of Quality," *Quality Progress*, 17[8]: 33.)

Planning for Quality

Planning is a necessary requirement for a successful quality program. In the planning process, the company considers its quality mission and objectives, its key qualities as perceived by the consumer, its market commitments, its available human resources and production facilities, and its financial constraints. Planning for quality assurance takes place in a hierarchical manner. Strategic plans involving broad product-quality decisions are made at the top management level. A strategic plan must be clearly defined; it is often stated in terms of key products or the market share it intends to capture. Strategies should be realistic and achievable. A steel manufacturing company may have a mission of capturing 15 percent of the market share. Plans for new products or improving the quality of existing products should also be considered and developed in conjunction with resource requirements planning and facilities design.

Based on the company's strategy, the different areas of the organization develop goals to support that mission. The rolling mill division of a steel company may set a goal to reduce scrap to 0.5 percent. To determine feasible goals, adequate data must be gathered and analyzed to assess customer demand, competition in the target market, and resource constraints. The next step in the planning process is to develop specific objectives based on the formulated goals. An objective of the rolling mill division may be to produce 1000 tons of acceptable rolled sheets per day. The objectives set at one level in an organization provide boundaries for objectives at a lower level. The strategic plans developed by top management may be used to formulate tactical plans at the middle management level. These plans include the design of quality assurance systems.

Next, the goals and objectives of middle management must be converted to operational plans that will be implemented by all levels of management. This task requires identifying standards in quantitative terms. A high-technology firm making integrated circuits might specify a target or standard such as achieving sales of $20 million annually. Resources, including people, must be allocated in order to meet the strategic objectives. Furthermore, the system of administration that will support quality control has to be set up.

Quality planning may be categorized as administrative plans and product-oriented plans. Administrative plans deal with the operational and business aspects of the company. They begin with the development of an overall quality program and contain a multitude of other activities. First, they require the formation of a quality policy and procedures manual, which guides the entire quality activity. Second, plans must be designed for audits. Third, resource planning (such as personnel requirements for the coming year, new equipment needs, and space requirements) needs to be incorporated.

Product-oriented plans, on the other hand, may cross functional lines such as marketing, product and process design, and production engineering. These would include plans for vendor evaluation, receiving inspection if any, process control, and inspection and test. Planning answers the fundamental questions of what, when, and how it is to be done and who will do it. Chapter 3 discusses some of these aspects of quality management practices.

Organizing for Quality

An organizational structure is created to establish lines of authority and responsibility, to improve communication, and to improve productivity. The major activities involved in organizing for quality consist of defining quality-related activities and the interrelationships between them, assigning responsibility for each of the tasks, and subdividing tasks down to the worker level.

The specific form of the organizational structure is influenced by a variety of factors. An organizational structure must suit the individual company; what is best for it may not necessarily be best for some other company. The size of the organization, the nature of the product (does it deal with general consumer goods, or does it work on special projects that are narrow in scope?) and the management culture (the informal channels of communication in the organization) are factors that mold the formal organizational structure. Authority gives one the right to access the resources needed to carry out a task. Responsibility determines who is to be held accountable for decisions.

Thus, one principle that should be adhered to is that authority must be commensurate with responsibility.

Formal organizational structure is represented through an organizational chart that shows the channels of authority and responsibility. Organizational structures may be categorized as either line, staff, or matrix structures.

In a line structure, the channels are grouped into functional units. Typical functional units are marketing, finance, design and development, and manufacturing. Each functional unit is responsible for quality in its own unit. Thus, the vice president of the marketing department is responsible for quality functions in that department. These functions may include conducting analyses to determine key requirements of customers, designing data collection forms, and implementing a field survey or audit to obtain information on customer needs. The heads of the other functional units have similar quality-related responsibilities in their respective areas.

On paper, the line structure may seem unrelated to the quality function, because no individual unit or person appears to have overall responsibility for quality. Figure 1-8 shows a segment of the line organizational structure. Managers and supervisors are not shown. The blocks with dashed arrows indicate the responsibility in the quality area for each vice president.

In the staff organizational structure, responsibility for quality is formally dealt with by a special department, where the vice president for quality assurance reports to the president. This vice president has overall responsibility in the quality area and also has the authority to make changes as appropriate in functional areas in order to adhere to the formulated quality policy. The vice president has a staff, which consists of the manager of quality assurance and quality improvement, the manager of quality control, and the manager of reliability and maintainability. Under each manager are corresponding supervisors. Figure 1-9 illustrates an organizational structure where quality assurance is a staff function.

In general, the layers of management in the staff structure should be minimal to facilitate efficient flow of communication and flow of information. The span of supervision should be as broad as possible. For small organizations, the span of control is usually broad. The levels of management are few. In such cases, quality assurance managers may have a diverse group of people reporting to them, such as receiving inspection, in-process inspection, final inspection, data collection, process control, shipping, and customer service.

For large organizations, the span of supervision is usually narrower. Tasks are more specialized. However, there are more levels of management, so the efficiency of information flow is affected. The leader in the quality assurance area should be a top-level manager, given the significance and importance of this area.

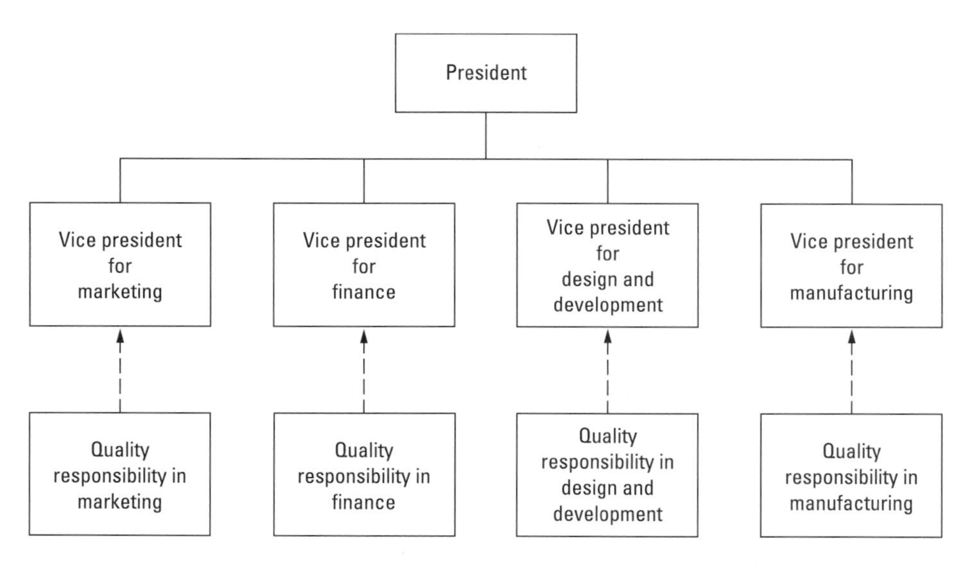

FIGURE 1-8 Line organizational structure.

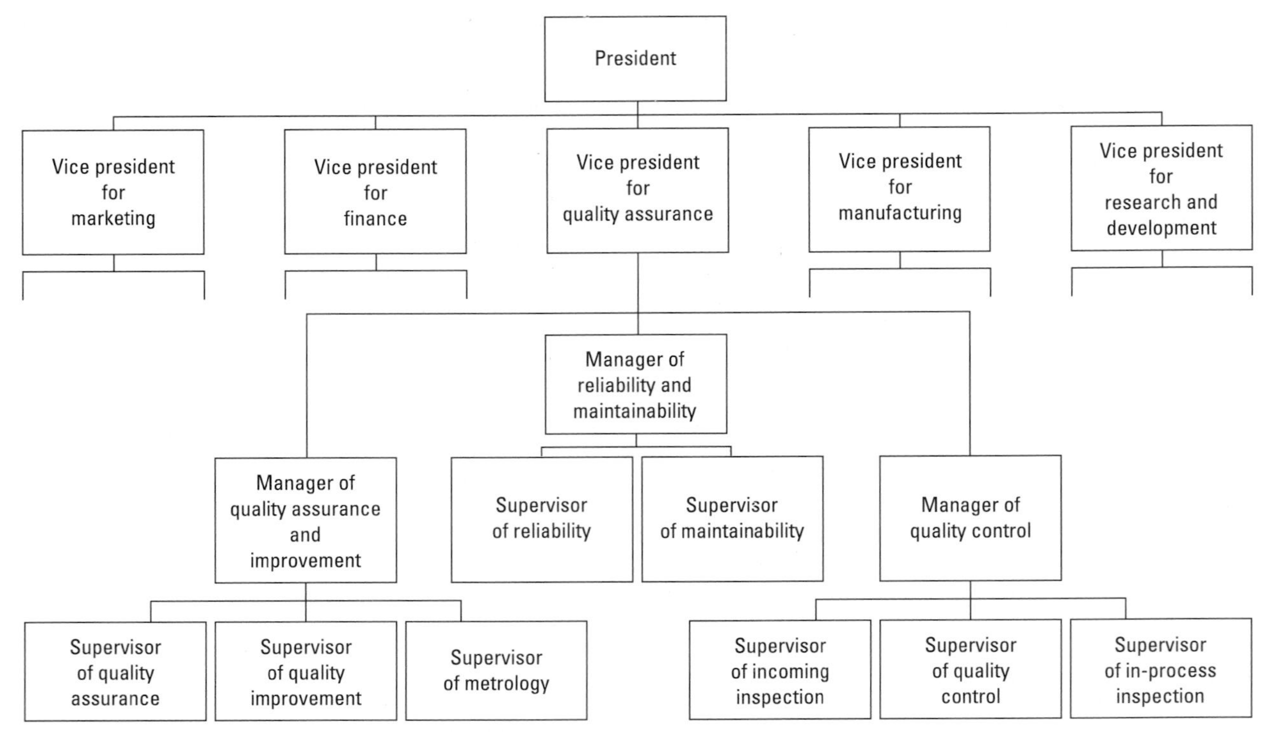

FIGURE 1-9 Quality assurance in a staff organizational structure.

A matrix organizational structure usually exists in situations involving large and complex projects. The duration of these projects is often long. For example, projects may deal with the construction of a space shuttle or a nuclear submarine. Tasks are therefore quite specialized. In such cases, there may be a quality assurance manager for each project. Figure 1-10 shows the truncated organizational structure for this situation. Reporting to the quality assurance manager of each project may be the respective supervisors of quality assurance and quality control for that project. It is possible for a person at a lower level of the matrix structure to report to more than one person.

FIGURE 1-10 Quality assurance in a matrix organizational structure.

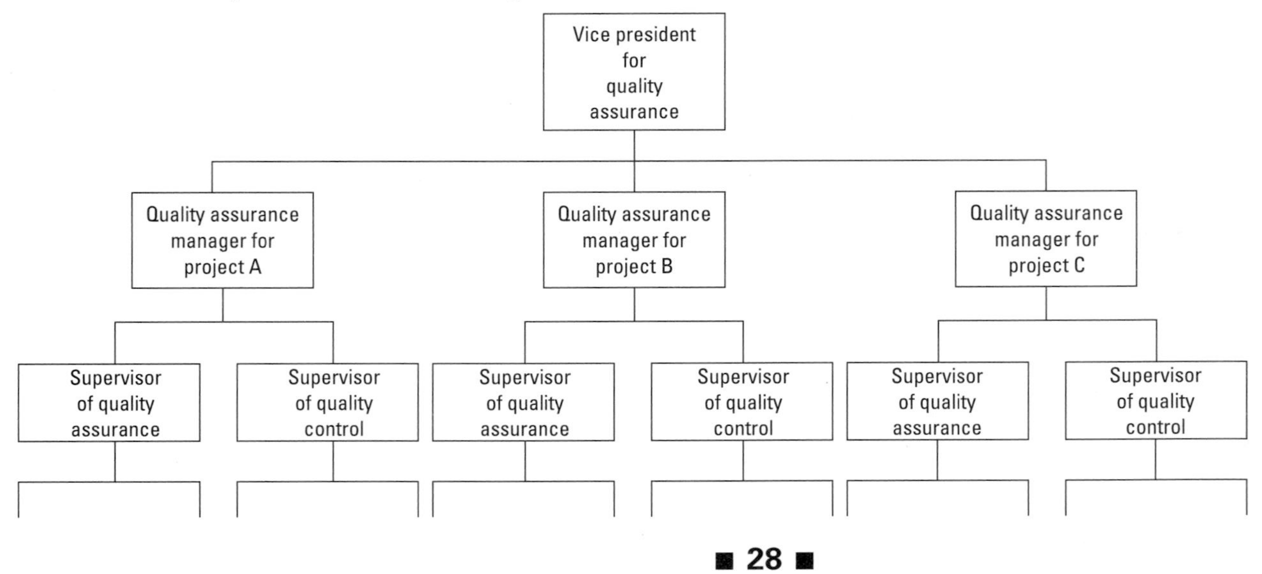

This happens when the person is involved in more than one project. For jobs of a highly specialized nature, a person with the necessary skills may work on multiple projects. Also, as projects are completed, a person may be assigned to other ongoing projects and thus report to different persons.

Staffing for Quality

The success of a companywide quality program depends on all personnel being committed to the quality philosophy. Everyone must develop a sense of ownership in the production of a quality item. Placing personnel with the appropriate training in statistical quality control techniques in key positions facilitates the implementation of a quality program.

To ensure the achievement of a certain level of knowledge regarding quality, certification programs are administered by the American Society for Quality Control. Examples of such certification programs are those for the Certified Quality Engineer (CQE) and the Certified Reliability Engineer (CRE). An organization should staff its managerial positions with candidates certified in the appropriate quality area. Additionally, a company can send its personnel to workshops on quality, and can give them incentives to become certified (either through salary increases, promotion, or acknowledgment through awards). Motivating employees to produce a quality item is critical to the success of the quality program. Employees must be contributing members who are given opportunities to suggest changes that improve quality. Deming's points for management (as discussed in Chapter 2) give specific guidelines for motivating employees.

Directing for Quality

Once quality has been incorporated into the design of the product, the production phase takes measures to ensure that the produced item conforms to the design specifications. The quality of purchased raw materials and components is important in this phase, so vendors must be monitored for quality control. Companies often require vendors to document their use of statistical quality control procedures. Those vendors unable to produce such evidence can lose accounts. A company should strive to develop long-term relationships with its vendors. Such relationships promote a more uniform and acceptable quality level of incoming goods.

Because every company needs to monitor process quality, the procedures for conducting process control, such as those involving control charts, must be in place. The appropriate data forms should be designed. The inspection points where the data is to be collected should be chosen, along with the inspection frequency. If final inspection is to be conducted, such procedures must be finalized. Qualified inspectors for both intermediate and final inspection should be selected. Inspectors must be trained and certified. In the case of sampling inspection, rules for the acceptance or rejection of lots should be developed.

Controlling for Quality

The control aspects occur during the production and postproduction phases. The objective is to determine out-of-control process conditions and product nonconformance as early in the production phase as possible. To achieve this end, procedures must be established to obtain relevant process information on a timely basis. The ability to process information in real time will expedite the identification of process problems. Possible sets of rectifying actions must be identified in case the process is found to be out of control. Such rectifying actions must be taken when appropriate. A feed-forward (as opposed to feed-back) approach should be considered when the effect of the controllable process variables on the product's output characteristics is known. For example, suppose we know the impact on tensile strength of furnace temperature and the manganese content in the product. If we are limited to certain settings of the furnace temperature, we can predetermine and control the manganese content such that the product will have a desired tensile strength.

Another important concern deals with controlling the costs of quality. As mentioned previously, the four main categories of costs are prevention, appraisal, internal failure, and external failure. Actions should be taken to minimize the total cost of quality; this may be attained through sufficient reductions of internal and external failures.

1-15 QUALITY AND PRODUCTIVITY

A misconception that has existed among businesses (and is hopefully in the process of being debunked) is the notion that quality decreases productivity. On the contrary, the relationship between the two is positive: Quality *improves* productivity. Making a product right the first time lowers total costs and improves **productivity.** More time is available to produce defect-free output because items do not have to be reworked and extra items to replace scrap do not have to be produced. In fact, doing it right the first time increases the available capacity of the entire production line. As waste is reduced, valuable resources—people, equipment, material, time, and effort—can be utilized for added production of defect-free goods or services. The competitive position of the company is enhanced in the long run, and an improvement is observed in profits.

Effect on Cost

As discussed previously, quality costs can be grouped into the categories of prevention, appraisal, internal failure, and external failure. Improved productivity may affect each of these costs differently.

Prevention and Appraisal Costs

With initial improvements in productivity, it is possible that prevention and appraisal costs will increase. As adequate process control procedures are installed, they contribute to prevention and appraisal costs. Furthermore, process improvement procedures may also increase costs in these two categories. These are thus called the costs of conformance to quality requirements. With time, a reduction in appraisal costs is usually observed. As process quality improves, it leads to efficient and simplified operations. This may yield further improvements in productivity.

Internal and External Failure Costs

A major impact of improved quality is the reduction in internal and external failure costs. In the long run, decreasing costs in these two categories usually offset the increase in prevention and appraisal costs. The total cost of quality thus decreases. Moreover, as less scrap and rework is produced, more time is available for productive output. The company's profitability increases. As external failures are reduced, customer satisfaction improves. Not only does this emphasis on quality reduce the tangible costs in this category (such as product warranty costs and liability suits), it also significantly affects intangible costs of customer dissatisfaction. Figure 1-11 shows how improved quality leads to reduced costs, improved productivity, and eventually increased profits.

As noted previously, management must focus on long-term profits rather than on short-term profits. A frequently cited reason for not adopting a total quality system is management's emphasis on short-term profits. As is well known, short-term profits can be enhanced by postponing much-needed investment in process improvement equipment and methods, by reducing research and development, or by delaying preventive maintenance. These actions eventually hurt competitiveness and profitability.

Effect on Market

An improvement in quality can lead to increased market shares, improved competitive position, and increased profitability.

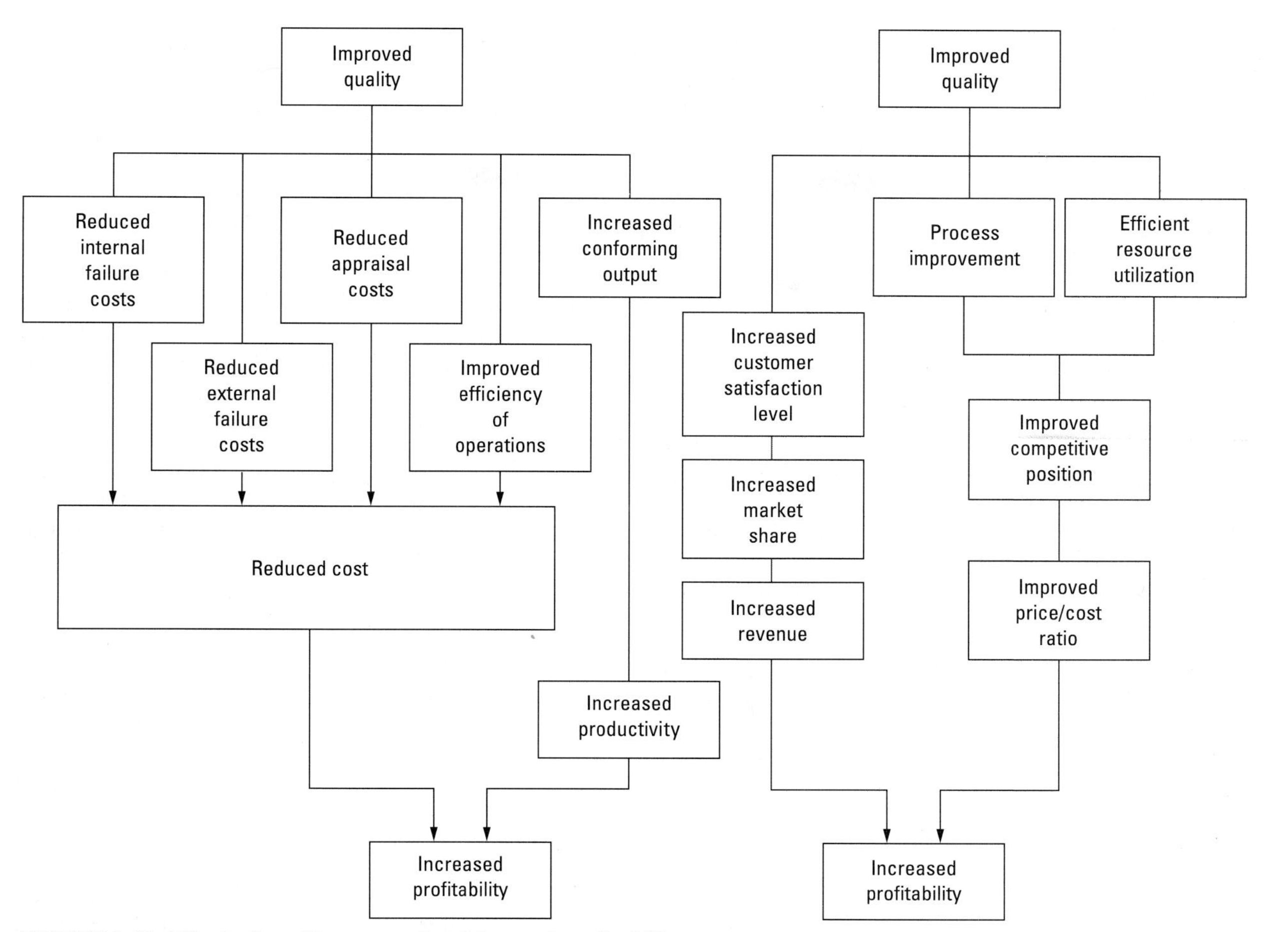

FIGURE 1-11 Effect of quality on productivity and profitability.
(Gabriel A. Pall, *Quality Process Management*, © 1987, pp. 34, 46. Adapted by permission of Prentice Hall, Inc., Upper Saddle River, New Jersey.)

Market Share

With a reduction in external failure costs and improved performance of the product in its functional phase, the company is in a position to raise the satisfaction level of its customers. Many of them return to buy the product again. Satisfied customers spread the word about good quality, which leads to additional customers. Market share goes up as the quality level goes up. Figure 1-11 demonstrates the effect of quality improvement on profitability via market share.

Competitive Position

All organizations want to stay competitive and to improve their market position. Efforts to improve quality are crucial in attaining these goals. Figure 1-11 shows that through process control and improvement and efficient resource utilization (reduced production of scrap and rework), a firm can minimize its costs. So, even if the selling price remains fixed, an improved price/cost ratio is achieved. Alternatively, as quality improves, the firm may be able to charge a higher price for its product, although customer satisfaction and expectations ultimately determines price. In any event, an improved competitive position paves the way for increased profitability.

1-16 TOTAL QUALITY ENVIRONMENTAL MANAGEMENT

In recent years, we have witnessed the emergence of many national and regional standards in the environmental management field. Some companies, of course, have long felt a social responsibility to operate and maintain safe and adequate environmental conditions, regardless of whether outside standards required it. Xerox Corporation, for one, is an example of a large corporation that takes its social obligations toward the environment seriously. The company has undertaken a major effort to reduce pollution, waste, and energy consumption. The quality culture is reflected in Xerox's protection of the environment; their motto is to reuse, remanufacture, and recycle. Company goals are aimed at creating waste-free products in waste-free factories using a "Design for the Environment" program. To support their environmental management program, Xerox uses only recyclable and recycled thermoplastics and metals.

With the concern for protection of the environment that is mandated in regional and national standards, standards need to be developed in environmental management tools and systems. British Standards Institute's BSI 7750 standards on environmental management is one such example; the European Union's (EU) eco-label and Eco-Management and Auditing Scheme (EMAS) are other examples. Both of these rely on consensus standards for their operational effectiveness. Similarly, in the United States, technical environmental standards under the sponsorship of the American Society of Testing and Materials (ASTM) have been published that address the testing and monitoring associated with emission and effluent pollution controls.

The International Organization for Standardization (ISO) based in Geneva, Switzerland, has long taken the lead in providing quality management standards. Its *ISO 9000* standards, discussed in Chapter 3, have become a benchmark for quality management practices. U.S. companies have adopted *ISO 9000* standards and have found this particularly beneficial in doing business with or trading in the European Union. Technical Committee (TC) 207 of ISO is in the process of developing standards in the field of environmental management tools and systems; their document will be called *ISO 14000: An International Environmental Management Standard* and will deal with developing management systems for the day-to-day operations that have an impact on the environment (Clements 1996). It will also address issues such as Environmental Management System Environmental Auditing, Environmental Performance Evaluation Life Cycle Assessment, and Environmental Labeling Environmental Aspects in Product Standards.

Environmental management began as a regulation-based and compliance-driven system. It has subsequently evolved into a voluntary, environmental stewardship process where companies have undertaken a continuous improvement philosophy to set goals that go beyond the protection levels required by regulations. The *ISO 14000* standards promote this philosophy with the objective of developing uniform environmental management standards that do not create unnecessary trade barriers. These standards are therefore not product standards. Also, they do not specify performance or pollutant/effluent levels. Specifically excluded from the standards are test methods for pollutants and setting limit values for pollutants or effluents.

Environmental management systems and environmental auditing span a variety of issues. These include top management commitment to continuous improvement; compliance and pollution prevention; creating and implementing environmental policies; setting appropriate targets and achieving them; integrating environmental considerations in operating procedures; training employees on their environmental obligations; and conducting audits of the environmental management system.

Several benefits will accrue from the adoption of the environmental management system standards. First and foremost is the worldwide focus on environmental management that the standards will help to achieve. At the commercial level, *ISO 14000* will have an impact on creating uniformity in national rules and regulations, labels and methods. They will minimize trade barriers and promote a policy that is consistent.

Not only will the standard help in maintaining regulatory compliance, they will help create a structure for moving beyond compliance. Management commitment and the creation of a system that reflects the goal to maintain a self-imposed higher standard will pave the way for continuous improvement in environmental management.

1-17 PROFILE OF A COMPANY—CRI

Custom Research, Inc. (CRI), a national marketing research firm, pursues individualized service and satisfied customers with an intensive focus on customer satisfaction, a team-oriented workforce, and information technology advances. The firm is a 1996 winner of the *Malcolm Baldrige National Quality Award* in the small business category (U.S. Department of Commerce 1996). Since 1988, when CRI adopted its highly focused customer-as-partner approach, client satisfaction has risen from already high levels, and gains in productivity, sales volume, and profits have outpaced industry averages.

CRI's steering committee is responsible for crafting CRI's goals and strategies and views customer loyalty as the firm's most valuable business asset. With all CRI employees as members of customer-focused teams, a flat organizational structure makes managers immediately accessible to employees, customers, and suppliers. Well-developed systems are in place for understanding customer expectations, soliciting customer feedback, and monitoring each facet of company, team, and individual performance. Together, these systems, which can serve as a model for other professional services firms, set the course for CRI efforts to meet or exceed customer expectations.

About CRI

Founded in 1974 and based in Minneapolis, privately owned CRI conducts survey marketing research for a wide range of firms. The bulk of its projects assist clients with new product development in consumer, medical, and service businesses. Revenues of more than $21 million in 1996 place CRI among the 40 largest firms in the highly fragmented, $4 billion marketing research industry, which is characterized by low entry costs and tough competition. The firm credits its reputation for quality for making it one of only a handful of companies that has remained independent while growing over the past two decades.

Besides its Minneapolis headquarters, the firm has electronically linked offices in San Francisco and Ridgewood, New Jersey, as well as telephone interviewing centers in St. Paul, Minnesota, and Madison, Wisconsin. It employs approximately 100 full-time staff members, most of whom are cross-trained to create the flexibility needed to accommodate the demands and schedules of research projects. Interviewing services assist CRI in doing personal interviewing.

Surprising and Delighting Customers

In recent years, CRI senior management aimed for a new level of consistency and competence in delivering quality services by organizing, systematizing, and measuring quality. The firm distilled requirements for each research project to four essentials: accurate, on time, on budget, and meeting or exceeding client expectations. Before the first survey data are collected, criteria defining these requirements are determined in consultation with clients where CRI managers and project team leaders interview clients—and they do that extensively.

The company was reorganized to make maximum use of customer-focused teams and to merge support departments to reduce cycle time—a growing client priority. All CRI teams have the same goal of "surprising and delighting" their clients. CRI captures the essence of this goal in its Star icon. Quality at CRI is client-driven—the center of the Star—and is integrated into the company's business system as captured by the five key business drivers that are the points of the star: people, processes, requirements, relationships, and results. With extensive staff involvement, the steering committee annually sets corporate goals for the company, which then tie to the goals for each work unit. With the steering committee, account teams review business plans and results for each client quarterly.

Meeting customer-specified requirements depends on efficient execution of well-documented, measurable processes. Most professional service firms believe their services cannot be "standardized." At CRI, while each project is custom-designed, the process for handling it flows through essentially the same steps across all projects. CRI developed a project implementation manual for interviewing and uses it heavily. Internal "project quality recap" reports completed for every study track errors in any step of the project flow. CRI measures the accuracy of results and the quality of personal and telephone interviewing. For example, over the last several years, ratings for interviewers show sustained average quality scores of approximately 95 points out of 100, up from 83 points in 1990.

Clients have ample opportunity to advise and critique CRI. At the end of each project, clients are surveyed to solicit an overall satisfaction rating based on their expectations. Each month, the results of the client feedback are summarized and distributed to all staff members. Internally, end-of-project evaluations are also conducted for CRI support teams and key suppliers. Personal interviewing services, for example, are evaluated on performance and are expected to contribute ideas for improving the quality of their services.

People Make the Difference

CRI uses a "high-tech/high-touch" approach to satisfying customers. On the high-touch side, CRI uses its flat organizational structure and relatively small size to assure that information flows freely within the company. Just as importantly, they view continuous improvements as part of their jobs. Staff members are surveyed annually, giving CRI senior managers specific feedback, including data on their own performance as viewed by CRI-ers.

A variety of recognition programs; bonuses based on achievement of corporate, team, and individual goals; and a peer-review system for evaluating personal performance serve to reinforce worker commitment to continuous improvement.

CRI has a companywide education plan, which is used to align individual training with business and quality goals. Each employee has a development plan that sets annual and long-term goals for improvement and helps identify training needs. In 1996, the average CRI employee received over 134 hours of training. All new people receive companywide and job-specific training that addresses quality and service issues. CRI bases companywide training requirements on client feedback, performance reviews, CRI's education plan, CRI development plans, on-the-job reviews, interviewer monitoring, and employee surveys.

A Technology-Driven Approach

The high-tech component of CRI's business is reflected by its openness to technological opportunities to improve its performance or to devise new services that respond to customer needs. "Managing work through technology-driven processes" is one of CRI's key business drivers. CRI led, for example, in the use of computers to assist in telephone interviewing, data collection, and analysis. Software enables CRI to use technology to integrate all stages of a project: Produce a questionnaire for computer-assisted interviewing; control the sampling and autodialing for interviewing; edit and then tabulate the answers from the questionnaires; display them in tabular format; and generate report-ready table for the final report and presentation. CRI views its major software supplier as a key partner. The long-standing relationship extends to annual planning sessions during which CRI shares its goals and the two firms determine how the software maker can contribute to meeting the goals. These and other quality-promoting actions—including an unconditional satisfaction guarantee—aim to build client confidence and loyalty, which, in turn, generate a variety of business benefits.

Quality and Business Performance Achievements

- Since 1988, feedback from clients on each of its projects shows steadily improved overall project performance. CRI is now "meeting or exceeding" clients' expectations on 97 percent of its projects. Seventy percent of CRI's clients say the company exceeds expectations. CRI is rated by 92 percent of its clients as better than competition on the key dimension of "overall level of service."
- Managing work through technology-driven processes is one of CRI's key business drivers. CRI uses a standardized nine-step process to deliver customized proposals and projects to clients. Within this process, CRI's integrated software system links database questionnaires to the coding, tabulation, and reporting functions so that no re-keying is necessary after survey data are received from the field. As a result, cycle time for data tabulation has dropped from two weeks to one day.
- Trends for on-time delivery of final reports and data tables to clients have been favorable since 1993, with 1995 results showing 99 percent of final reports and 96 percent of data tables being delivered on time.
- Revenue per full-time employee has risen 70 percent since 1988 when CRI reorganized into cross-functional, goal-directed teams. Since then, it has been substantially higher than CRI's key competitors on this productivity measure.
- CRI reduced its overall client base from 138 in 1988 to 67 in 1995 as part of the company's strategy to better serve its biggest clients and build partnerships with them. Since then, the number of larger clients has increased from 25 to 34, and revenue has continued to grow.
- Ninety-four percent of CRI's employees agree that, "All things considered, this is a good place to work." This is significantly above the norm of 76 percent nationally for business service companies. An important part of its employee development is a focus on training, which currently runs at over 120 hours per employee per year.
- CRI celebrates company successes and creates an atmosphere that encourages recognition. This includes monthly "Good News Meetings" and companywide trips to celebrate achieving major business goals.
- CRI has played a key role within its industry in formulating the Council of Marketing & Opinion Research (CMOR). CMOR speaks as one voice for the industry on key issues, such as increasing the public's cooperation in marketing research studies.

1-18 SUMMARY

This chapter has examined the detailed framework on which the concept of the total quality system is based. It has introduced some of the basic terminology and provided an overview of the design, production, and implementation phase of the quality concept. It traces the evolution of quality control and presents specifics on the benefits of quality control, who is responsible for it, and how it is to be adopted. The subject of quality costs is explored, and the tradeoffs among the cost categories that take place with the successful usage of a total quality system are discussed. A critical consideration in the whole scheme is the management of the quality function. Subfunctions within this area are outlined, along with the role of management. An important outcome of improvement in quality is an increase in productivity and profitability. The relationship between quality and productivity has been demonstrated in this chapter.

CASE STUDY

Motorola's Quest for Quality*

The Total Customer Satisfaction (TCS) worldwide competition showcases the quality achievements of Motorola teams. All employees are given an equal opportunity to participate. Like many companies, Motorola uses teams to solve problems. In fact, almost half of Motorola's employees are on teams. But Motorola takes it one step further and gives teams the opportunity to compete with one another and share firsthand what they've accomplished, allowing them to see how their achievements impact the organization through their TCS competition. Winning team members from all over the world are treated like royalty for a few days and are given the opportunity to make a presentation to top executives of the company.

THE NEED FOR TCS

Why develop such a competition? The Chairman of the Board says the first few years of Motorola's quality journey "were carved by the idea of the Malcolm Baldridge National Quality Award." But after winning the award in 1988, "the company needed something to carry the momentum." This led the company to develop the TCS team competition. Teams already existed within Motorola, and the idea of a competition was met with enthusiasm. Since its inception eight years ago, Motorola estimates this quality program has resulted in savings of $2.4 billion a year. A savings that is essential for the company to remain competitive when its products have a price learning curve of 15% to 35% a year. In addition to the dollar savings, TCS has helped develop a company of empowered workers. "I'm not sure if maybe the whole empowerment aspects of what we did with the team process is not more important than the individual savings that we've generated," comments the Director of Corporate Quality for business systems. "Nothing has empowered the work force faster than the team process—it makes the difference between a good company and a bad company." That opinion is echoed by the Chairman of the Board who says, "The numbers are impressive, but the numbers are not what counts."

The TCS competition is based on the following objectives:

- Renew emphasis on the participative process at all levels of the organization, worldwide.
- Recognize and reward outstanding performance at the team level.
- Reaffirm the environment for continuous improvement.
- Demonstrate the power of focused team effort.
- Communicate the best team achievements throughout Motorola.

HOW IT WORKS

The competition starts with preliminary contests held for each of Motorola's business units. As many as 5,000 teams take part initially, incorporating roughly 65,000 of Motorola's 142,000-plus employees. The number of teams has grown from approximately 1,500 seven years ago and has increased every year. Depending on the size of each regional competition, one to five teams are selected to move forward to the worldwide finals.

This year's one-day competition featured 24 teams from countries that included the United States, Ireland, the Philippines, Israel, Taiwan, China, Malaysia, and Japan. Teams took command of a large stage with four video screens and microphones at both ends. Each team had 12 minutes to present its accomplishments to a panel of judges that included the company's top 15 executives; teams that went over time lost points (Table 1-2 shows the scoring form). The presentations were well rehearsed and proceeded like clockwork, with many teams having committed their entire presentations to memory—very impressive considering all of the presentations were given in English, a language some of the team members didn't even speak.

The TCS teams generally consist of 10 to 12 members, all of whom participate in the presentation. Teams are awarded points in the following seven categories:

1. **Project selection.** The project must be tied to Motorola's key initiatives and should use specific customer input. Projects should last from three to 12 months.

*Adapted from: L. A. Klaus (1997), "Motorola Brings Fairy Tales to Life," *Quality Progress*, 30(6): 25–28.

TABLE 1-2 1996 TCS Team Competition Scoring Form

Organization:

Team Name: Date:

Key initiative Six-sigma quality:________ Profit improvement:______________ Product, manufacturing, and environment leadership:__________

Total cycle time:________ Participative management:________

Category	*Criteria/Recommended Distribution*		*Score 0 to 10*	*Weight*	*Weighted Score*
Project selection (10 points)	• Criteria and a methodology for selection evident. Project clearly defined.	30%			
	• Aggressive goals with linkage to key initiatives established	30%		1.0	
	• Customer identified. Customer requirements and metrics defined.	40%			
Teamwork (10 points)	• Team participation that demonstrates commitment to the project	20%			
	• Appropriate team membership evident in the improvement process	40%		1.0	
	• Participative practices reinforced	40%			
Analysis techniques (20 points)	• Thorough and appropriate analysis techniques used and understood	50%			
	• Benchmarking of best practices evident	30%		2.0	
	• Innovative use of fundamental tools and/or progression to more advanced tools. Team growth evident.	20%			
Remedies (20 points)	• Alternative solutions seriously explored	30%			
	• Remedies consistent with the analysis	20%		2.0	
	• Implementation plans thorough and well defined	20%			
	• Innovation in the remedies or implementation evident	30%			
Results (20 points)	• Verified improvements measured favorably against the difficulty of achievement	40%			
	• Ancillary effects identified and characterized	20%		2.0	
	• Customer satisfaction results evident	40%			
Institutionalization (15 points)	• Improvements sustainable and permanent	40%			
	• Solutions adopted by and from other groups	40%		1.5	
	• Team's growth in the problem-solving process evident	20%			
Presentation (5 points)	• Clear and concise	60%			
	• Improvement process followed	40%		0.5	
				Weighted score	__________
				Deductions (–1 point for each minute over 12)	__________
				Total team score	__________

2. **Teamwork.** The team must handle the project from selection through implementation. Participation of customers and/or suppliers is encouraged and all team members are expected to contribute to all phases of the project.

3. **Analysis.** Analysis techniques used should support appropriate analytical processes for the project, lead to a root cause, identify alternative solutions, and reflect innovative use of analytical tools.

4. **Remedies.** The team must defend its choice of remedies from the alternatives, and remedies should be consistent with the analysis. Creative and innovative solutions are especially noted.

5. **Results.** Results should be compared to the original goals and requirements. The degree of achievement of these goals is considered by the judges.

6. **Institutionalization.** Teams must demonstrate that improvement is maintainable over time. They are

encouraged to adapt solutions from other teams and spread their success throughout the company. Teams should emerge as leaders in their own right.

7. **Presentation.** Presentations must be clear and concise, with overhead graphs and charts that are clear and easy to read. Listeners should be able to easily follow the team's thinking through the entire process.

IMPRESSIVE RESULTS

The accomplishments of these teams are truly impressive. A team from Motorola's Automotive Energy and Controls group in Sequin, Texas, achieved a savings of $1.8 million in 1996 by reducing polyimide delamination for electronic circuits—an 85% improvement in six months.

Another cross-functional team from the company's General Systems Sector in Hong Kong set out to make the best cellular phone on the market in China and increase production capacity by 50% in just eight weeks. It also corrected a design problem that prevented users from ending their phone calls when closing the lower flap on the cellular phone.

Motorola's Land Mobile Product Sector in Schaumberg, Illinois, created a cartoon character named Eugene and a site on the World Wide Web to help improve its responsiveness to the Motorola service station community. The team's work resulted in 86% growth in new-account setups, 99% improvement in cycle-time reduction, and 90% improvement in customer satisfaction.

THE REAL REWARD

Even though team members are there to compete, the world-wide final is more than a competition—it's a celebration. It's a way for Motorola to thank its employees and vice versa. It's no coincidence that the 1996 worldwide competition was held at the Phoenician in Scottsdale, Arizona, a five-star resort. To the presenters, the real prize was just being there and being a part of something so grand.

At Motorola, all the teams are considered winners. "We do say, and mean it, that everybody wins because you're here," stated the Chief Executive Officer (CEO). Following the competition, an awards banquet was held to honor all of the participants. Company executives were visible throughout, mingling in conversation with team members and other guests. After dinner, each team was called up on stage to be recognized and photographed with the President and CEO. These photos were just a few of some 2,000 pictures taken throughout the event that will be compiled in a TCS yearbook for all participants.

The excitement and enthusiasm of these Motorola employees was evident. "This is probably the grandest display of our efforts to be global and the way we manage and think about our business," said the CEO. He continued by saying that "This event reflects all of the important aspects of the corporation."

The winner of this year's customer satisfaction competition was a team from a manufacturing plant in Boynton Beach, Florida, with a history of noteworthy quality improvements. The team had members representing Motorola's Messaging, Information, and Media Sector, and its goal was to develop and implement a low-cost, reliable packaging system that demonstrates environmental leadership and corrects problems identified by the team. Team members found $1.2 million in hidden packaging costs and identified three root causes: lack of packaging standards, stock outages, and inefficient reuse of materials. After tackling each problem individually, the team developed a standardized packaging tray that could be used to hold both finished pagers and incoming housing. It also created a central database to track packaging requirements. The result of their efforts was a 29-cent-per-unit total cost reduction, expected to save $6.1 million in 1997.

AN OPEN INVITATION

How do Motorola employees become a part of the contest? The TCS competition is open to all Motorola functional or cross-functional teams, 98% of which are self-forming. Since all team members are required to participate in all phases of the project, teams with representatives located around the world rely heavily on e-mail, telephone, and other communications technology. Motorola also realizes the challenges of working with employees from different countries and offers cultural diversity classes to help employees prepare for these differences.

For every team that enters the competition, there are numerous others that have also made significant accomplishments, perhaps on a smaller scale. Roughly 40% of Motorola's teams go on to present. Many of the teams choose not to compete, but participate in what Motorola calls "showcase days" at their facilities, where teams set up booths and display their accomplishments to facility managers and co-workers.

To help them learn how to solve quality problems as a team, Motorola employees are trained in quality techniques and teamwork. Additionally, many teams, particularly those in Asia, have big sisters or big brothers who act as sponsors for newly formed teams. These sisters and brothers are more experienced workers who help direct the teams and offer expertise. The company also has TCS process manuals that describe quality tools, and one of Motorola's business units even developed a CD-ROM training tool that creates graphs and visual aids to use in the presentations. Finally, Motorola University offers quality training in areas such as quality processes and teamwork.

NEVER GOOD ENOUGH

In keeping with Motorola's philosophy of continuous improvement, the competition has changed over the years. For instance, in past years, all teams that made it to the worldwide finals were presented either a gold or silver award, but the silver award winners went away feeling like losers, even though they were really winners. So in 1996, after recognizing each team for its work, the company gave away one diamond award to the overall winner.

To pinpoint areas for improvement, team members are usually surveyed at the worldwide competition for suggestions. Some of the regional competitions have been shifted to different countries to allow more employees to experience other cultures. While the events have been refined so that they are good for the employees, they have been designed to retain the travel and the excitement.

SHARING THE WEALTH

Motorola encourages other companies to learn from its success. This year, Motorola extended the TCS competition to its suppliers and for the first time held a formal supplier contest in conjunction with the Motorola competition. Fifty-one supplier teams competed in three regional competitions, and the three winners participated in the worldwide supplier competition, which was held the day prior to the TCS competition. The winning team from Varitronix presented at the Motorola competition as a showcase team. These supplier teams were eager to learn from Motorola; many of them had heard about quality processes, but didn't know how to implement them until Motorola stepped forward.

Representatives from other countries, educators, foreign government representatives, and customers were also invited to attend the competition. Sun Microsystems started a similar competition a few years ago based on TCS. Additionally, to help others learn about the process, Motorola offers quality briefings to the public through Motorola University that address total customer satisfaction. Motorola University Consulting and Training Services offers quality briefings to the public that explain the six-sigma story, total cycle time reduction, and total customer satisfaction teams.

Equally impressive as the competition is the company's ability to motivate its employees. It has much to do with Motorola's culture. The company stresses the importance of trust before implementing something similar to TCS. Teams and empowerment will not work without trust. The employees have to trust management and they have to trust each other. Without trust, it just won't work. A team from Dublin, Ireland was motivated by the opportunity to show others in Motorola what they had accomplished. And, the chance to win a trip didn't hurt. For TCS team members, it is an opportunity to get away, meet new people, and learn, but also to have fun. The day following the contest is usually set aside for recreation, which this year included hiking, rafting, mountain biking, and golf.

Whether employees are executives or factory workers, Motorola works to show them that they are valued. The TCS competition only reinforces these feelings. The experience emphasizes even more the value of each individual in the company. Finalists say that they treasure every one of their TCS memories. It's worth remembering time and again—even forever.

QUESTIONS FOR DISCUSSION

1. What is the most important asset of a company? How does Motorola ensure maintaining this asset?
2. Describe the role of cross-functional teams in the process of quality improvement. What are some actions taken by Motorola to promote such teams?
3. Discuss the advantages to be gained by companies conducting TCS–type competitions.
4. Discuss the importance of organizational culture and its adoption at all levels in striving for quality improvement.
5. Discuss how Motorola motivates its employees to strive for continuous improvement.
6. The concept of employee-empowerment is an important part of process control and improvement. How is this accomplished at Motorola?
7. What type of an organizational structure would best promote the quality activities at Motorola?

Key Terms

- acceptance sampling plans
- attributes
- cause-and-effect diagram
- causes
  - common or chance
  - special or assignable
- customer service
- defect
- fishbone diagram
- inspection
- Ishikawa diagram
- liability
  - product
  - strict
- management of quality
  - planning for quality
  - organizing for quality
  - staffing for quality
  - directing for quality
  - controlling for quality
- manufacturing
- manufacturing engineering
- marketing
- nonconformity
- nonconforming unit
- off-line quality control
- productivity
- product design
- product planning
- purchasing
- quality
  - quality of design
  - quality of conformance
  - quality of performance
  - responsibility for quality
- quality assurance
- quality characteristic
- quality circles
- quality control
  - benefits of quality control
  - responsibility for quality control
- quality cost measurement
  - labor base index
  - cost base index
  - sales base index
  - unit base index
- quality costs
  - prevention costs
  - appraisal costs
  - internal failure costs
  - external failure costs
- quality improvement
  - commitment stage
  - consolidation stage
  - maturity stage
  - teams
- reliability
- sampling
- specification
  - specification limits
- standard
- statistical process control
  - on-line statistical process control
- systems approach
- total quality environmental management
- total quality system
- variables
- zero defects program

Exercises

1. Define quality, and explain its role in the modern business environment.
2. What are the three major aspects of quality? Explain the relationship between them.
3. Distinguish the philosophical difference of the underlying concepts of statistical process control and acceptance sampling plans.
4. Explain the role of the quality assurance function.
5. What is a quality circle? Describe an application where it could be used.
6. What are the benefits of quality control?
7. Discuss the responsibilities of the different departments of an organization as far as the quality function is concerned.
8. Define a quality system.
9. Discuss the different elements of a total quality system. Give an example of the implementation of such a system with respect to a selected company.
10. Explain the difference between quality control and quality improvement.
11. Describe the stages of quality improvement. Also, discuss the costs incurred in each stage.
12. What are the major categories of quality costs? Explain each of them, and give examples.
13. Explain the various indices for measuring quality costs. Discuss the relative advantages of each.
14. Discuss the role of top management in the planning function associated with the management of quality.
15. What are the various forms of organizational structures? Discuss how the quality function is accomplished through such structures.
16. Select a company of your choice. Discuss the organizational structure that exists to achieve the quality commitment.
17. Explain how quality affects productivity. Discuss the implications on cost.
18. Select a company for which you have access to profit and cost data. Draw a graph of total quality costs over the past five years, and indicate the

amount in each of the four cost categories. Compare this data with the profit for the same period. What inferences can you draw?

19. Discuss the role of teamwork in arriving at feasible solutions through an interdisciplinary approach.
20. Discuss the role of management in reducing total quality costs. As management becomes involved, which costs actually rise and which fall? Explain in the context of an organization.
21. Loss of customer goodwill due to unsatisfying product or service is often cited as a quality cost. What category does this fall in and what are the difficulties in measuring it?
22. What type of an organizational structure would be suitable for promoting information flow between various functions?
23. Select a company and identify its organizational structure. Analyze the strengths and weaknesses of such a structure.
24. Is it feasible to increase productivity and reduce quality costs at the same time? Explain.
25. Describe the role of the standards *ISO 14000* in promoting good environmental management practices.

References

American Society for Quality Control (1971). *Quality Costs—What and How,* 2nd ed. Milwaukee, Wis.: ASQC.

American Society for Quality Control (1979). *ANSI/ASQC Standard Z-1.5-1979, Generic Guidelines for Quality Systems.* Milwaukee, Wis.: ASQC.

American Society for Quality Control, Statistics Division (1983). *Glossary and Tables for Statistical Quality Control,* 2nd ed. Milwaukee, Wis.: ASQC.

American Society for Quality Control (1987). *ANSI/ASQC Standard A3-1987, Quality Systems Terminology.* Milwaukee, Wis.: ASQC.

Banks, J. (1989). *Principles of Quality Control.* New York: John Wiley.

Besterfield, D. H. (1990). *Quality Control,* 3rd ed. Upper Saddle River, N.J.: Prentice Hall.

Clements, R. B. (1996). *Complete Guide to ISO 14000.* Upper Saddle River, N.J.: Prentice Hall.

Crosby, P. B. (1979). *Quality Is Free.* New York: McGraw-Hill.

Evans, J. R., and W. M. Lindsay (1989). *The Management and Control of Quality.* St. Paul, Minn.: West.

Feigenbaum, A. V. (1983). *Total Quality Control.* New York: McGraw-Hill.

Garvin, D. A. (1984) "What Does Product Quality Really Mean?" *Sloan Management Review,* 26(1): 25–43.

Juran, J. M., ed. (1974). *Quality Control Handbook,* 3rd ed. New York: McGraw-Hill.

Klaus, L. A. (1997). "Motorola Brings Fairy Tales to Life." *Quality Progress,* 30(6): 25–28.

Pall, G. A. (1987). *Quality Process Management.* Upper Saddle River, N.J.: Prentice Hall.

Schilling, E. G. (1984). "The Role of Statistics in the Management of Quality." *Quality Progress,* 17(8): 33.

Taguchi, G. (1986). *Introduction to Quality Engineering: Designing Quality into Products and Processes.* Asian Productivity Organization (available in North America, U.K., and Western Europe from the American Supplier Institute, Inc., and UNIPUB/Kraus International Publications, White Plains, New York).

Taguchi, G., and Yuin Wu (1979). *Introduction to Off-Line Quality Control.* Central Japan Quality Control Association, Nagoya.

U.S. Department of Commerce, National Bureau of Standards (1983). NBS Handbook 130, *Model State Laws and Regulations,* Gaithersberg, Md.

U.S. Department of Commerce (1996). *Malcolm Baldrige National Quality Award—Profiles of Winners,* National Institute of Standards and Technology, Gaithersberg, Md.

Wadsworth, H. M., K. S. Stephens, and A. B. Godfrey (1986). *Modern Methods for Quality Control and Improvement.* New York: John Wiley.

Walsh, L., R. Wursten, and R. J. Kimber, eds. (1986). *Quality Management Handbook.* New York: Marcel Dekker.

CHAPTER

2

Some Philosophies and Their Impact on Quality

Chapter Outline

2-1 INTRODUCTION

Several people have made significant contributions in the field of quality control. In this chapter we look at the philosophies of three people—W. Edwards Deming, Philip B. Crosby, and Joseph M. Juran. Pioneers in the field of quality control, they are largely responsible for the global adoption and integration of quality assurance and control in industry. Through their teachings, articles, books, and consultation, they have convinced management around the world of the significance of this approach.

Management commitment is key to a successful program in quality, and convincing them of the need to use these programs on a companywide basis did not occur overnight. In fact, it took U.S. industry about 30 years. Thirty years in which U.S. market share in key industries shrank alarmingly. However, due in no small part to the quality programs, industries in the United States have turned around and are once again world leaders in productivity, quality, and profitability.

A change has taken place in corporate culture. The idea of an ongoing quality control and improvement program is now widely accepted.

In this chapter we examine Deming's philosophy in depth. His **14 points for management** are fundamental to the implementation of any quality program. These points, which constitute a "road map," should be thoroughly understood by all who undertake the implementation of such programs. We also discuss Crosby's and Juran's philosophies and then compare the three philosophies. The goal is the same in all three—creating and adopting a world-class quality business culture. Although the paths they describe are slightly different, companies should look closely at each approach before embarking on a quality program.

2-2 W. EDWARDS DEMING AND HIS CONTRIBUTION

Historical Background

W. Edwards Deming is credited with the impressive turnaround in Japanese industry after World War II. Deming's broad background in physics, mathematics, and statistics landed him in the Bureau of the Census and thereafter the U.S. War Department in 1939.

The Allies had a problem: The sheer logistics of the war effort was overwhelming. Such a massive endeavor had never before been attempted, and coordination was key to its success. What was needed was nothing less than a way to absolutely guarantee high quality and keep to schedules—people's lives were on the line. Deming supplied it; his revolutionary technique helped win the war. He called it "quality control."

After the war, Deming tried to convince U.S. industry of the advantages of this approach, but he got nowhere. However, in 1950, the Union of Japanese Scientists and Engineers (JUSE) invited Deming to address their leading industrialists. Japan had a reputation for poor quality. Their domestic economy was shattered and JUSE was looking at ways to jump-start the Japanese economy. Deming introduced his techniques, and many in the audience were skeptical initially. However, they took Deming's advice to heart and set about totally revamping their industrial culture. Through their total philosophical commitment to Deming's methods, Japanese industry moved to the forefront.

What we see today are the results of using Deming's philosophy over the last four decades. Improvement in quality did not happen overnight, but Deming knew with certainty that it would happen. The Japanese continue to use Deming's ideas, and they constantly seek ways to improve their processes and products.

Deming was awarded the prestigious Second Order Medal of the Sacred Treasure by Emperor Hirohito for his contributions to the Japanese economy. He was a hero, a household name, in Japan and almost totally unknown in the United States. In 1951, JUSE instituted the coveted Deming Award, an annual award given to the individual or firm contributing most to the advancement of industrial quality. Recipients of this award include companies like Toyota, Nissan, Nippon Steel, and Hitachi. The first U.S. company to win the Deming Award was Texas Instruments in 1985.

It is ironic that the biggest name in Japanese quality control is American. At the end of World War II, complacent U.S. companies shunned Deming's advice. During that period, demand for U.S. goods was high, and quality was acceptable. Whatever was produced was devoured by the consumer. There was little competition in our domestic market, and in the foreign markets, there was less—U.S. products reigned supreme.

Given this setting, major companies thought it would be foolish to change their way of doing things. They failed to realize that this lack of competition was the result of a world devastated by war. U.S. products were not inherently superior—they were the *only* products because we were the only ones left standing. Our industrial capacity was intact. This situation would not last long, and U.S. industrial leaders failed to realize this.

The plight of U.S. industry in the seventies and eighties is now, of course, history. Today, U.S. industry is listening, and Deming's message is being heard. The big automakers such as Ford Motor Company and General Motors have adopted Deming's teachings. Other companies are following suit. The larger companies are demanding the same stringent quality from their vendors as they demand from themselves. It's a global economy out there, and to many the quality philosophy will be their means to survival—there will be no other alternative.

Philosophy

Deming's philosophy emphasizes the role of management. Of the problems that industry faces, Deming said that 85% can be solved only by management. These involve changing the system of operation and are not influenced by the workers. In Deming's world, workers' responsibility lies in communicating to management the information they possess regarding the system, for both must work in harmony. The Deming's ideal management style is holistic: The organization is viewed as an integrated entity. The idea is to plan for the long run and provide a course of action for the short run. Too many U.S. companies in the past (and in the present) focused on short-term gains.

Deming believed in the adoption of a total quality program and emphasized the neverending nature of quality control in the quality improvement process. Such a program achieves the desired goals of improved quality, customer satisfaction, higher productivity, and lower total costs in the long run. A commonly held myth—that an improvement in quality and increase in productivity cannot go hand in hand—was smashed by Deming's approach. He demonstrated that an improvement in quality inevitably leads to increased capacity and greater productivity.

As experience demonstrates, these desirable changes occur over an extended period of time. Thus, Deming's approach is not a "quick fix" but rather a plan of action to achieve long-term goals. He stressed the need for firms to develop a corporate culture where short-term goals such as quarterly profits are abandoned.

Deming's approach demands nothing less than a cultural transformation—it must become a way of life. The principles may be adapted and refined based on the experience of a particular organization, but they still call for total commitment.

At the heart of this philosophy is the need for management and workers to speak a common language. This is the language of statistics—statistical process control. The real benefits of quality programs will accrue only when everyone involved understands their statistical underpinning. Therefore, Deming's fundamental ideas require the understanding and use of statistical tools and a change in management attitude. The 14 points that we will look at shortly identify a framework for action. This framework must be installed for the quality program to be successful. Management must commit to these points in thought, word, and deed if the program is to work.

Deming advocated certain key components that are essential for the journey toward continuous improvement. The following four components comprise the basis for what Deming called the **System of Profound Knowledge:**

1. *Knowledge of the system and the theory of optimization.* Management needs to understand that optimization of the *total system* is the objective, not necessarily the optimization of individual subsystems. In fact, optimizing subsystems can lead to a sub-

optimal total system. The total system consists of all constituents—customers, employees, suppliers, shareholders, the community, and the environment. A company's long-term objective is to create a win-win situation for all of its constituents.

2. *Knowledge of the theory of variation.* All processes exhibit variability, the causes of which are of two types: **special causes** and **common causes.** Special causes of variation are external to the system. It is the responsibility of operating personnel and engineering to eliminate such causes. Common causes, on the other hand, are due to the inherent design and structure of the system. They define the system. It is the responsibility of management to reduce the common causes. A system that exists in an environment of common causes only is said to be stable and in control. Once a system is considered to be in control, its capability can be assessed and predictions on its output made.

3. *Exposure to the theory of knowledge.* Information, by itself, is not knowledge. Knowledge is evidenced by the ability to make predictions. Such predictions are based on an underlying theory. The underlying theory is supported or invalidated when the observed outcome is compared to the predicted value. Thus, experience and intuition are not of value to management unless they can be interpreted and explained in the context of a theory. This is one reason why Deming stressed a data analysis–oriented approach to problem solving where data is collected to ascertain results. The results then suggest what remedial actions should be taken.

4. *Knowledge of psychology.* Managing people well requires a knowledge of psychology because it helps us understand the behavior and interactions of people, and also the interactions of people with their work environment. Also required is a knowledge of what motivates people. People are motivated by a combination of intrinsic and extrinsic factors. Job satisfaction and the motivation to excel are intrinsic. Reward and recognition are extrinsic. Management needs to create the right mix of these factors to motivate employees.

2-3 EXTENDED PROCESS

The **extended process** envisioned by Deming expands the traditional organizational boundaries to include suppliers, customers, investors, employees, the community, and the environment. Figure 2-1 shows an extended process. An organization consists of people, machines, materials, methods, and money. The extended process adds a key

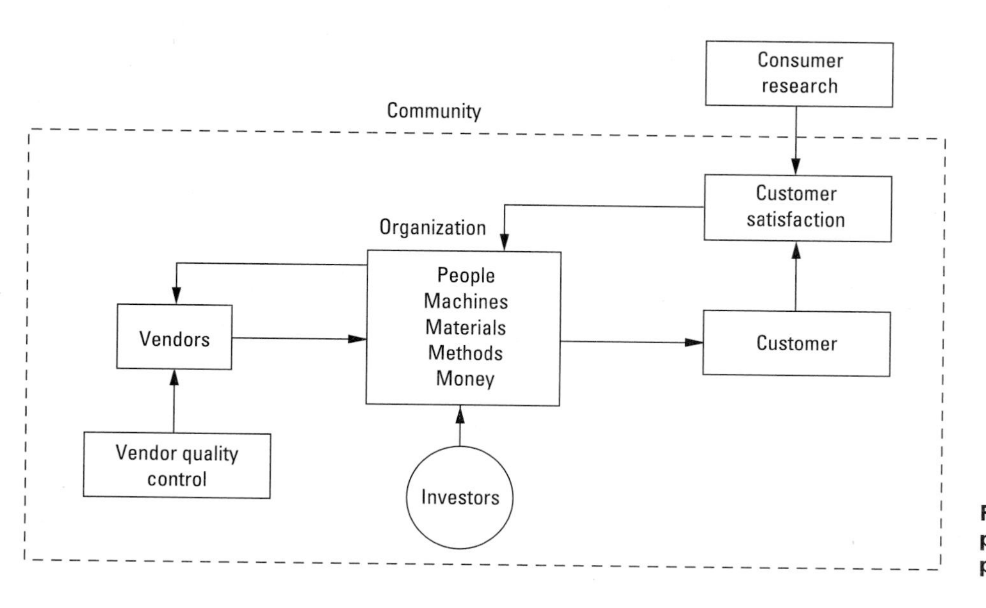

FIGURE 2-1 An extended process in Deming's philosophy.

entity—the consumer. An organization is in business to satisfy the consumer. This should be its primary goal. Goals such as providing the investors with an acceptable rate of return are secondary. Achieving the primary goal—customer satisfaction—automatically causes secondary goals to be realized. This primary goal is especially relevant to a service organization; here the customer is more obviously central to the success of the organization.

The community in which the organization operates is also part of this extended process. This community includes consumers, employees, and anyone else who is influenced by the operations of the company, directly or indirectly. An accepting and supportive community makes it easier for the company to achieve a total quality program. Community support ensures that there is one less obstacle in the resistance to the changes proposed by Deming.

Vendors are another component of the extended process. Because the quality of raw materials, parts, and components influences the quality of the product, efforts must be made to ensure that vendors supply quality products. In Deming's approach, a long-term relationship between the vendor and the organization is encouraged, to their mutual benefit.

As an example, consider Armstrong World Industries. This company, which is profiled in Section 2-8, demonstrates how the extended process works. Its suppliers, distributors, and carriers are full partners in its quality improvement process. The driving theme of all operations is increased customer satisfaction; thus, the customer is the centerpiece of the extended process as continuous feedback from the customer is obtained from each of Armstrong World's market categories. In its decision making, the company embraces the notion of the extended process by incorporating activities that add value to its customers, employees, vendors, and investors.

2-4 DEMING'S 14 POINTS FOR MANAGEMENT

The focus of Deming's philosophy is management. Since a major proportion of problems can be solved by management, Deming noted that management cannot "pass the buck." Only a minority of problems can be attributed to suppliers or workers, so in Deming's view, what must change is the fundamental style of management and the corporate culture.

In Deming's ideal organization, workers, management, vendors, and investors are a team. However, without management commitment, the adoption and implementation of a total quality system will not succeed. It is management who creates the culture of workers' "ownership" and their investment in the improvement process. It is management who creates the **corporate culture** that enables workers to feel comfortable enough to recommend changes. It is management who develops the long-term relationship with vendors. And finally, it is management who convinces investors of the long-term benefits of a quality improvement program. A corporate culture of trust can only be accomplished with the blessings of management.

Deming's 14 points for management provide the necessary sense of direction. The adoption of these points will sustain productivity and competitiveness of the company in the long run. Books by Deming (1982, 1986), Scherkenbach (1986), Gitlow and Gitlow (1987), Walton (1988, 1990), and Fellers (1992) provide in-depth explanations and interpretations of each of the 14 points.

Deming's Point 1

> *Create and publish to all employees a statement of the aims and purposes of the company or other organization. The management must demonstrate constantly their commitment to this statement.**

---

*Deming's 14 points (January 1990 revision) reprinted from *Out of Crisis* by W. Edwards Deming by permission of MIT and The W. Edwards Deming Institute. Published by MIT Center for Advanced Engineering Study, Cambridge, MA 02139. Copyright 1986 by The W. Edwards Deming Institute. Reprinted from *Quality Productivity and Competitive Position* by W. Edwards Deming by permission of MIT and The W. Edwards Deming Institute. Published by MIT Center for Advanced Engineering Study, Cambridge, MA 02139. Copyright 1982 by The W. Edwards Deming Institute.

This principle stresses the need to create long-term strategic plans that will steer the company in the right direction. Mission statements should be developed and clearly expressed so that everyone in the organization understands them. This includes not only everyone in the organization but also vendors, investors, and the community at large. Mission statements address such issues as the continued improvement of quality and productivity, competitive position, stable employment, and reasonable return for the investors. It is not the intent of such statements to spell out the finances required; however, the means to achieve the goals should exist. To develop strategic plans, management must encourage input from all levels. If the process through which the plan is developed is set up properly, companywide agreement with, and commitment to, the strategic plan will be a natural outcome.

Product Improvement Cycle

Committing to a specific rate of return on investment should not be a *strategic* goal. In this philosophy, the customers—not the investors—are the driving force in the creation of strategic goals. The old approach of designing, producing, and marketing a product to customers without determining their needs is no longer valid. Instead, the new approach is a four-step cycle that is customer-driven. Figure 2-2 shows this cycle, which includes designing a customer needs–based product, making and testing it, selling it, determining its in-service performance with market research, and using this information to start the cycle again. This approach integrates the phases of quality of design, conformance, and performance (discussed in Chapter 1).

Determining customer needs in clearly understood terms is key to improving product quality. For instance, noting that the customer prefers a better refrigerator that is less noisy and can produce ice quickly is not sufficient. What noise level (stated in decibels) is acceptable? How fast (in minutes) does the customer expect the ice to be produced? Only when specific attributes are quantified can the product be made better.

One reason that U.S. industries lost their market share in the eighties was that companies were too busy looking at each other. They ignored the importance of satisfying customer needs. By contrast, Japanese companies were looking at the needs of customers. They improved the processes and products, and that led to satisfied customers who came back for more. Merely meeting the competition is not a long-term strategy. The company must focus on the customer, not the competition. In the case of a stable process, improvements in the system can be created by management alone.

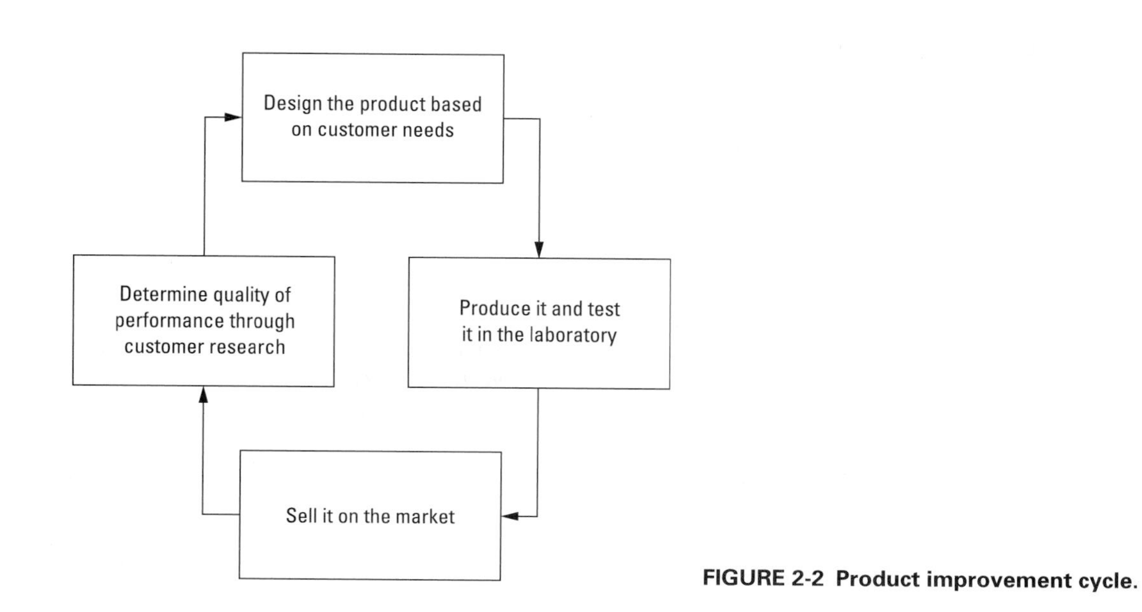

FIGURE 2-2 Product improvement cycle.

Constancy and Consistency of Purpose

Foresight is critical. Management must maintain a **constancy of purpose.** This implies setting a course (for example, all departments within the organization will pursue a common objective of quality improvement) and keeping to it. Too often, a focus on short-term results such as weekly production reports or quarterly profits deters management from concentrating on the overall direction of the company. Actions that create a profit now can have a negative impact on profits 10 years from now.

Management must be innovative. They must allocate resources for long-term planning, including consumer research and employee training and education that addresses not only work performance but also the new philosophies. They must ensure that there are resources available to cover the new costs of determining quality of performance and for new methods of production or changes in equipment. The priorities are research and education and a constant improvement in the product or service.

Organizations need to reevaluate their time frames. They should strive to create and maintain a competitive position for years to come by taking appropriate actions today. Of course, blame for emphasizing the short term rests with top management. Managers are often fired for not maximizing short-term profits. This has to change; the new corporate culture must focus on long-term survival. Actions in support of this new philosophy will provide employment security for all and create a stable environment conducive to improved productivity.

In addition to constancy of purpose, there should be a **consistency of purpose.** This means that the company should not digress from long-term objectives. For example, are all units of the company working toward the company goal to improve quality synergistically? Or are they working for a departmental subgoal, which may be to increase production? Management should also accept the fact that variability exists and will continue to exist in any operation. What they must try to do is determine ways in which this variation can be reduced. And this is a neverending process. Figure 2-3 depicts the concepts of constancy and consistency of purpose. As Deming put it, "Doing your best is not good enough. You have to know what to do. Then do your best."

Deming's Point 2

Learn the new philosophy, top management and everybody.

The new attitude must be adopted by everyone. Quality consciousness must be everything to everyone. Previously acceptable levels of defects should be abandoned; the idea that improvement is a neverending process must be embraced wholeheartedly.

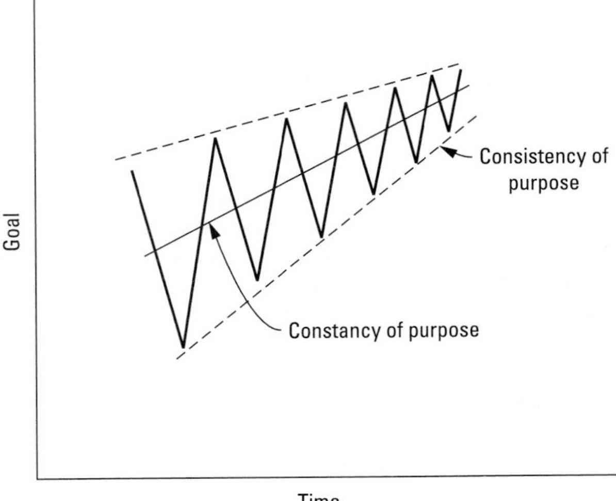

FIGURE 2-3 Constancy and consistency of purpose.

Human beings are resistant to change. Managers who have been successful under the old system where certain levels of defects were acceptable may find it difficult to accept the new philosophy. Overcoming this resistance is a formidable task, and it is one that only management can accomplish. The idea is not only to continually reduce defects but also to address the needs of the customer.

Declaring any level of defect to be acceptable promotes the belief that defects *are* acceptable. Say a contract specifies that a defective rate of 4 units in 1000 is acceptable; this *ensures* that 0.4% will be defective. This philosophy must be abandoned and a no-defects philosophy adopted in its place.

Deming's Point 3

Understand the purpose of inspection, for improvement of processes and reduction of cost.

Quality has to be *designed* into the product; it cannot be *inspected* into it. Creating a "design for manufacturability" is imperative because producing the desired level of quality must be feasible. Inspection merely separates the acceptable from the unacceptable. It does not address the root cause of the problem—that is, what is causing the production of nonconformities and how to eliminate them. The emphasis is on defect prevention, not on defect detection.

The production of unacceptable items does not come without cost. Certain items may be reworked, but others will be scrapped. Both are expensive. The product's unit price increases, and the organization's competitiveness decreases. Market share and available capacity are inevitably affected.

Drawbacks of Mass Inspection

Mass inspection does not prevent defects. In fact, depending on mass inspection to ensure quality *guarantees* that defects will continue. Even 100% inspection will not eliminate all the defectives if more than one person is responsible for inspection. When several points of inspection are involved, it is only human to assume that others will find what you have missed. This is *inherent* to the mass inspection system. Inspector fatigue is another factor in 100% inspection. Thus, defects can only be prevented by changing the *process.*

Mass inspection implies that workers cannot be trusted; it does nothing to improve the process. Reliance on it implies that defects are expected and thus acceptable. Deming believed that "routine 100% inspection is the same thing as planning for defects, acknowledgment that the process cannot make the product correctly, or that specifications make no sense in the first place" (Deming 1982). The mentality that allows the production of defects is no longer tolerable. Instead, feedback from the customer and from the process itself should be used to monitor and control defects.

Although action based on feedback should be done on a real-time basis, caution is the operative concept regarding the time frame for the action. Using individual measurements to adjust the process rather than statistical signals will cause even more variability. The result is overcontrol, and overcontrol does not bring stability to the process. Suppose we are producing a lathe-turned unit. To meet specifications, the diameter of the shaft must be 50 ± 2 mm. An operator produces a shaft with a 51-mm diameter and lowers the setting on the machine to reduce the diameter of the next unit. If the operator continues to make similar adjustments (that is, lowering and raising the machine setting with each new unit), the end result will be increased variability in the diameter. On the other hand, if adjustments are based on a statistical control chart that states when to leave the process alone and when to take corrective action, the effect will be an improved process. Details on statistical control charts will be discussed in later chapters.

Deming's Recommendation

If inspection must be performed, Deming advocated a plan that minimizes the total cost of incoming materials and thus the final product. His plan is an inspect-all-or-none rule. Its basis is statistical evidence of quality, and it is applied to a stable process. This is the *kp rule* (see Chapter 10).

Deming's Point 4

End the practice of awarding business on the basis of price tag alone.

Many companies, as well as state and federal governments, award contracts to the lowest bidder as long as they satisfy certain specifications. This practice should cease. Companies should also review the bidders' approaches to quality control. What quality assurance procedures do the bidders use? What methods do they use to improve quality? What is the attitude of management toward quality? Answers to these questions should be used, along with price, to select a vendor, because low bids do not always guarantee quality.

Unless the quality aspect is considered, the effective price per unit that a company pays its vendors may be understated and, in some cases, unknown. Knowing the fraction of nonconforming products and the stability of the process provides better estimates of the price per unit.

Suppose three vendors A, B, and C submit bids of $15, $16, and $17 per unit, respectively. If we award the contract on the basis of price alone, vendor A will get the job. Now let's consider the existing quality levels of the vendors.

Vendor B has just started using statistical process-control techniques and has a rather stable defect rate of 8%. Vendor B is constantly working on methods of process improvement. The effective price we would pay vendor B is $\$16/(1 - 0.08) = \$17.39$ per unit (assuming that defectives cannot be returned for credit).

Vendor C has been using total quality management for some time and has a defect rate of 2%. Therefore, the effective price we would pay vendor C is $\$17/(1 - 0.02) = \$17.35$ per unit.

Vendor A has no formal documentation on the stability of its process. It does not use statistical procedures to determine when the process is out of control. The outgoing product goes through sampling inspection to determine which ones should be shipped to the company. In this case, vendor A has no information on whether the process is stable, what its capability is, or how to improve the process. The effective price we would pay vendor A for the acceptable items is unknown. Thus, using price as the only basis for selection controls neither quality nor cost.

A flagrant example where the lowest-bidder approach works to the detriment of quality involves urban municipal transit authorities. These agencies are forced to select the lowest bidder in order to comply with the policy set by the Urban Transit Authority of the United States. The poor state of affairs caused by complying with this policy is visible in many areas around the country.

Principles of Vendor Selection

Management must change the process through which **vendor selection** is conducted. The gap between vendor and buyer must be closed; they must work as a team to choose methods and materials that improve customer satisfaction.

In selecting a vendor, the total cost (which includes the purchase cost plus the cost to put the material into production) should be taken into account. A company adhering to Deming's philosophy buys not only a vendor's products but also its process. Purchasing agents play an important role in the extended process. Knowing whether a product satisfies certain specifications is not enough. They must understand the precise problems encountered with the purchased material as it moves through the extended process of manufacture, assembly, and eventual shipment to the consumer. The buyers must be familiar with statistical methods so as to assess the quality of the vendor's plant. They must be able to determine the degree of customer satisfaction with the products, tell what features are not liked, and convey all related information to the vendor. Such information enables the vendor to improve its product.

Another important principle involves reducing the number of suppliers. The goal is to move toward single-supplier items. Companies in the United States have had multiple vendors for several reasons, including a fear of price increases, a vendor's inability to meet projected increases in demand, and a vendor's lack of timeliness in meeting delivery schedules.

There are several disadvantages to this policy. First, it promotes a feeling of mistrust between buyer and vendor and thereby creates a short-term, price-dependent relationship between them. Furthermore, vendors have no motivation to change their process to meet the buyer's specifications. Price, not quality, is the driving factor because another vendor may be selected if their price is lower. A long-term commitment cannot exist in such a situation.

Other disadvantages involve cost. Increased paperwork leads to increased order preparation costs. Travel costs of the vendor to purchaser sites increase. Volume discounts do not kick in because order sizes are smaller when there is more than one vendor. Setup costs go up because the buyer's process changes when the incoming supplier changes. Machine settings may have to be adjusted along with tooling. In continuous process industries, such as chemical companies producing sulfuric acid, differences in raw materials may require changes in the composition mix. Multiple set-up periods mean idle production and therefore reduced capacity. Also, training the people who work with vendors costs more with multiple vendors.

One major disadvantage is the increased variability in incoming quality, even if individual vendor's processes are stable. Figure 2-4 explains this concept. Suppose we purchase from three vendors, A, B, and C, each quite stable and having a small dispersion as far as the quality characteristic of interest is concerned (say, density of a red color pigment in producing a dye). However, the combined incoming product of three good suppliers may turn out to be mediocre. This happens because of the intervendor variability.

Many benefits are gained by moving to single-vendor items. The disadvantages mentioned in the preceding paragraphs can be eliminated when a long-term relationship with a quality-conscious vendor is developed. A long-term vendor can afford to change its process to meet the needs of the buyer because the vendor does not fear losing the contract. Such a relationship also permits open contract negotiations, which can also reduce costs.

Deming's Point 5

Improve constantly and forever the system of production and service.

In Deming's philosophy, companies move from defect detection to defect prevention and continue with process improvement to meet and exceed customer requirements on a neverending basis. Defect prevention and process improvement are carried out by the use of statistical methods. Statistical training is therefore a necessity for everyone and should be implemented on a gradual basis.

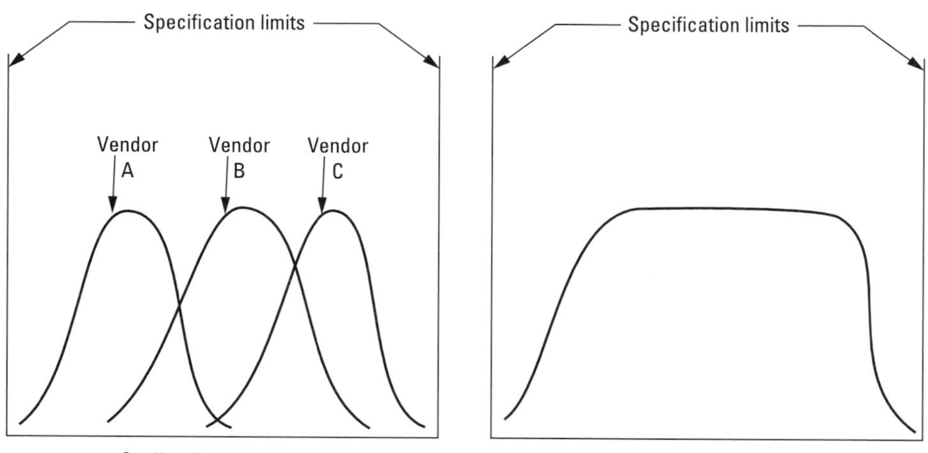

FIGURE 2-4 Mediocre incoming quality due to multiple vendors.

Deming Cycle

The continuous cycle of process improvement is based on a scientific method originally called the *Shewhart cycle* after its originator Walter A. Shewhart. He also developed control charts. In the 1950s, the Japanese renamed it the **Deming cycle.** It consists of four basic stages—*plan, do, check,* and *act* (the PDCA cycle). The Deming cycle is shown in Figure 2-5.

Plan Stage In this stage (depicted in Figure 2-6), opportunities for improvement are recognized and operationally defined. A framework is developed that addresses the effect of controllable process variables on process performance. Since customer satisfaction is the focal point, the degree of difference between customer needs satisfaction (as obtained through market survey and consumer research) and process performance (obtained as feedback information) is analyzed. The goal is to reduce this difference. Possible relationships between the variables in the process and their effect on outcome are hypothesized.

Suppose a company that makes paint finds that one major concern for customers is drying time; the preferred time is 1 minute. Feedback from the process says that the actual drying time is 1.5 minutes. Hence, the opportunity for improvement, in operational terms, is to reduce the drying time by 0.5 minute.

The next task is to determine how to reduce drying time by 0.5 minute. The paint components, process parameter settings, and interaction between them are examined to determine their precise effect on drying time. Quality of design and quality of conformance studies are undertaken. Company chemists hypothesize that reducing a certain ingredient by 5% in the initial mixing process will reduce the drying time by 0.5 minute. This hypothesis is then investigated in the following stages.

Do Stage The theory and course of action developed in the plan stage is put into action in the do stage. Trial runs are conducted in a laboratory or prototype setting. Feedback is obtained from the customer and from the process. At this stage, our paint company will test the proposed plan on a small scale. They reduce the ingredient by 5% and obtain the product.

Check Stage Now the results are analyzed. Is the difference between customer needs and process performance reduced by the proposed action? Are there potential drawbacks relating to other quality characteristics that are important to the customer?

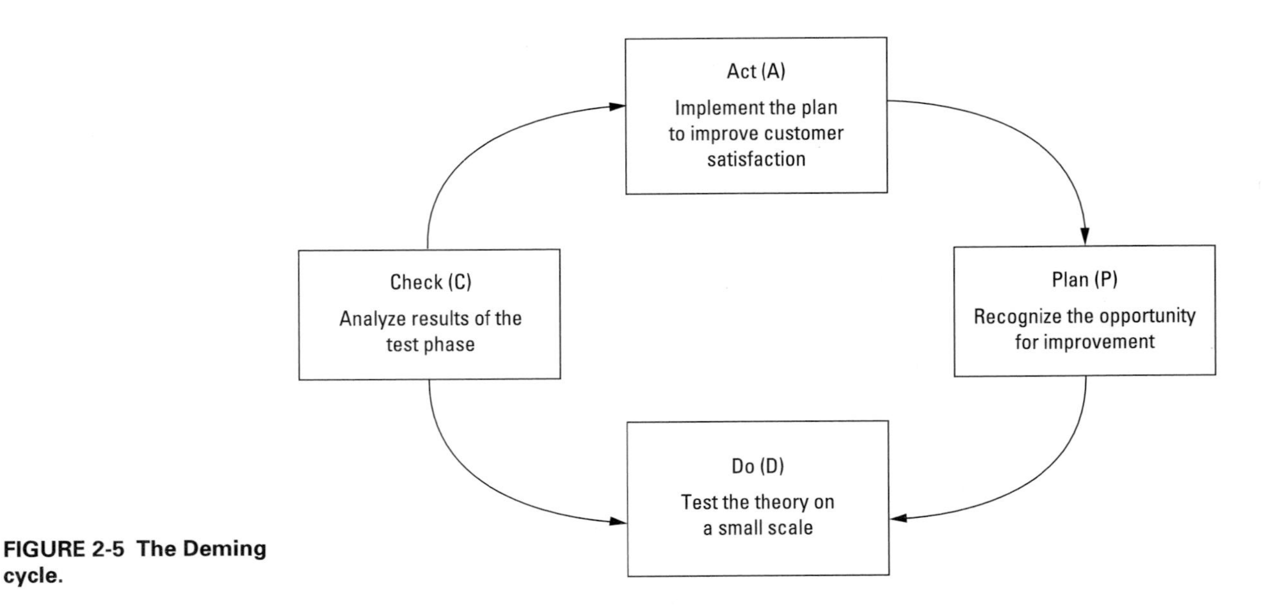

FIGURE 2-5 The Deming cycle.

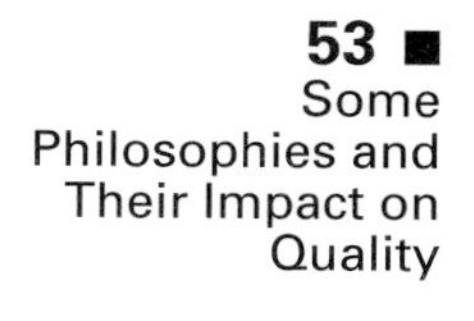

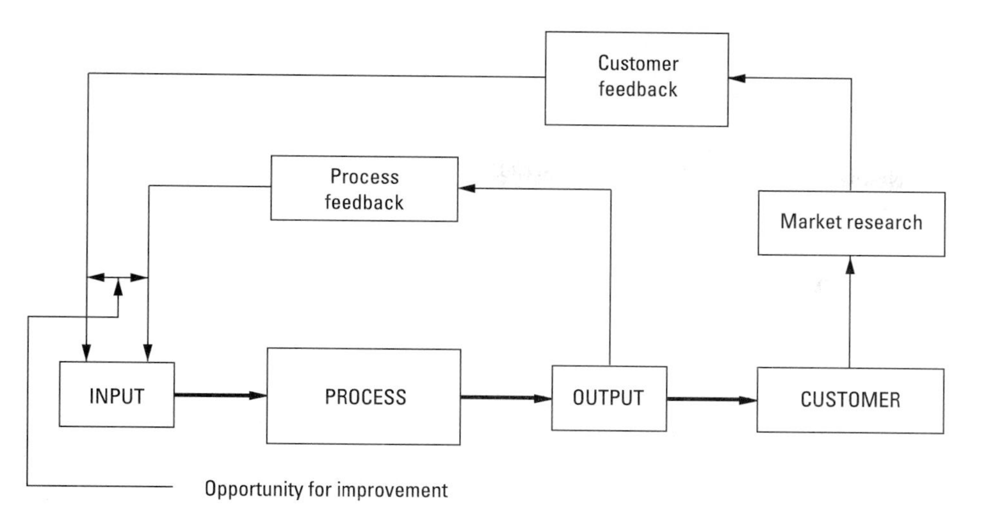

FIGURE 2-6 Plan stage of Deming cycle.

Statistical methods will be used to find these answers. As our paint company attempts to reduce drying time by 0.5 minute, samples are taken from the modified output. The mean drying time and the variability associated with it are determined; the analysis yields a mean drying time of 1.3 minutes with a standard deviation of 0.2 minute. Prior to the modification, the mean was 1.5 minutes with a standard deviation of 0.3 minute. The results thus show a positive improvement in the product.

Act Stage In the act stage, a decision is made regarding implementation. If the results of the analysis conducted in the check stage are positive, the proposed plan is adopted. Customer and process feedback will again be obtained after full-scale implementation. Such information will provide a true measure of the plan's success. If the results of the check stage show no significant improvement, alternative plans must be developed, and the cycle continues.

In our paint example, the proposed change in paint mix reduced the drying time, and the decision to change the mix is then implemented on a full-scale. Samples produced by this new process are now taken. Is the mean drying time still 1.3 minutes with standard deviation of 0.2 minute, as in the check stage? Or has full-scale implementation caused these statistics to change? If so, what are they? What can be done to further reduce the mean drying time and the standard deviation of drying time?

Customers' needs are not constant. They change with time, competition, societal outlook, and other factors. In the paint example, do customers still desire a drying time of 1 minute? Do they have other needs higher in priority now compared to the drying time? Therefore, the proposed plan is continuously checked to see whether it is keeping abreast of these needs. Doing so may require further changes; that is, the cycle continues, beginning once again with the plan stage.

Variability Reduction and Loss Function

Deming's philosophy calls for abandoning the idea that everything is fine if specifications are met. The idea behind this outmoded attitude is that there is no loss associated with producing items that are off-target but within specifications. Of course, just the opposite is true, which is the reason for striving for continual process improvement. Reducing process variability is an ongoing objective that minimizes loss.

Following the course set by Deming, Genichi Taguchi of Japan formalized certain **loss functions** in 1960. He based his approach on the belief that economic loss accrues with any deviation from the target value. Achieving the target value wins high praise from the customer and yields no loss. Small deviations yield small losses. However, the losses increase in a nonlinear relationship (say, quadratic) with larger deviations from the target.

Figure 2-7 demonstrates the new loss function along with the old viewpoint. The new loss function ties into the idea that companies must strive for continual variability reduction; only then will losses be reduced. Any deviation from the customer target will not yield the fullest possible customer satisfaction. Losses may arise because of such problems as lost opportunities, warranty costs, customer complaint costs, and other tangible and intangible costs. There is even a loss associated with the customers not praising the product even though they are not unhappy with it. It is important to get people to praise the product or service because this affects public perception of the product and hence of the company.

Deming's Point 6

Institute training.

Employee **training** is integral to proper individual performance in the extended process setting. If employees function in accordance with the goals of the company, an improvement in quality and productivity results. This in turn reduces costs and increases profits.

Employees are the fundamental asset of every company. When employees are hired, they should be carefully instructed in the company's goals in clear-cut operational terms. Merely stating that the company supports a total quality program is not sufficient. Instead, employees must know and buy into the company's long-term goals. Understanding these goals is essential to performing adequately. Employees' individual goals may not always be compatible with those of the company. For example, an employee's desire to produce 50 items per day may not be consistent with the company's goal of defect-free production. Instruction enables the employee to understand what his or her responsibilities are for meeting customers' needs.

Training must be presented in unambiguous operational terms. Employees must know exactly what is to be done and its importance in the entire process. Such an approach will not only improve their understanding of the job but will also create an atmosphere in which workers take pride in their work, feel more secure, have better morale, and improve their productivity. Even the employee who performs only one operation of the many that a product goes through must understand the needs of the customer and the role of the supplier in the extended process.

Training is ongoing. It should start when employees are hired and continue throughout their time with the company. It may be done in segments, and procedures may vary from company to company. Training may be done through such means as

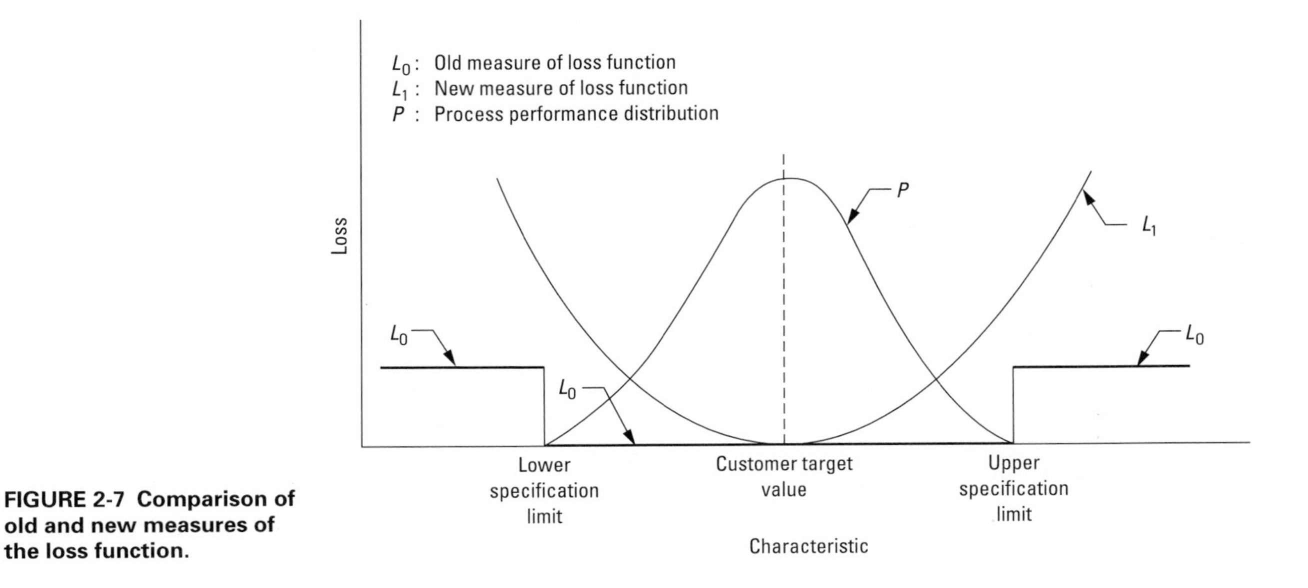

FIGURE 2-7 Comparison of old and new measures of the loss function.

classroom instruction, handing out of instructional materials, discussion of experiential situations, or exposure to quality improvement procedures. It is imperative to remember that human beings are the most important asset of an organization.

Statistical concepts and techniques play a central role in the Deming program. Consequently, employees must be trained in several statistical tools; these include flow diagrams, histograms, control charts, cause and effect diagrams, pareto diagrams, scatter diagrams, and design of experiments. We will examine these tools in detail later. Training programs should be evaluated on a statistical basis to determine whether they promote significant improvement in employee output. Furthermore, statistical methods should also be used to determine when to end the training program, since no statistically significant improvements arise from overtraining. If, after training is complete, employees are still unable to meet the job requirements, they may have to be reassigned to more appropriate jobs.

Certain attitudes reduce the benefits that accrue from training. These can be companywide or individually held. For instance, claiming that training is only useful for others, or that it applies only to those in manufacturing, is detrimental. Also, the idea that experience determines reliability is not appropriate because employees may not make the best decisions if they are not in an optimal work environment. Careful continuous training achieves better solutions than the quick-fix, make-do approach.

Deming's Point 7

Teach and institute leadership.

Supervisors serve as vital links between management and the workers. Since there is usually no direct contact between these two groups, supervisors have the difficult job of maintaining communication channels. Thus, they must understand both the problems of the workers and top management's goals. Communicating management's commitment to quality improvement to the workers is a key function of supervisors. To be effective leaders, the supervisors must no longer think punitively—they must think in terms of helping workers do a better job. Shifting to this positive attitude creates an atmosphere of self-respect and pride for all concerned.

Supervisors need to be trained in statistical methods; they are positioned to provide crucial leadership and instruction in these areas. The Deming philosophy dictates that supervisors create a supportive atmosphere in which all workers benefit. This improves employee morale, promotes teamwork by establishing communication links, and helps achieve the overall goal of quality improvement. Using modern methods of leadership helps the workers understand their place in the extended process and creates a culture of trust and support.

Supervisors are in the best position to identify common causes inherent to the system, causes for which the workers should not be blamed. It is management's responsibility to minimize the effects of common causes. Special causes, such as poor quality of an incoming raw material, improper tooling, and poor operational definitions should be eliminated first. Identification of these special causes can be accomplished through the use of control charts, which will be discussed in later chapters. Supervisors must first be trained in the Deming philosophy before they can serve as role models or leaders. Supervisors often end up managing things (equipment, for instance) and not people. Such an approach overlooks the fundamental asset of an organization—people.

Deming's Point 8

Drive out fear. Create trust. Create a climate for innovation.

Functioning in an environment of fear is counterproductive, because employee actions are dictated by behavior patterns that will please supervisors rather than meet the long-term goals of the organization. The economic loss associated with fear in organizations is immense. Employees are hesitant to ask questions about their job, the methods involved in production, the process conditions and influence of process parameters, the operational definition of what is acceptable, and other such important issues. The wrong signal is given when a supervisor or manager gives the impression that asking these questions is a waste of time.

A fear-filled organization is wasteful. Consider an employee who produces a quota of 50 parts per day—without regard to whether they are all acceptable—just to satisfy the immediate supervisor. Many of these parts will have to be scrapped, leading to wasted resources and a less than optimal use of capacity. Fear can cause physical or physiological disorders, and poor morale and productivity can only follow. A lack of job security is one of the main causes of fear.

What could be more important to the worker than to know exactly what his or her job entails? How can employees perform effectively if they cannot get clarification about the process? If they think that their suggestions to management regarding process improvement are a waste of time, how can they identify with the company and feel like a part of the team? Therefore, creating an environment of trust is a key task of management. When this trust embraces the entire extended process—when workers, vendors, investors, and the community are included—only then can an organization strive for true innovation.

As management starts to implement the 14 points, removing or reducing fear is one of the first tasks to tackle, because an environment of fear will impede implementation of the other points. The process of eliminating fear starts at the top. The philosophy of managing by fear is totally unacceptable; it destroys trust, and it fails to remove barriers that exist between different levels of the organization.

Deming's Point 9

Optimize toward the aims and purposes of the company the efforts of teams, groups, staff areas.

Organizational Barriers

Organizational barriers (Figure 2-8) impede the flow of information. Internal barriers exist within the organization; these include barriers between organizational levels (for instance, between the supervisor and workers) and those between departments (perhaps between engineering and production, or between product design and marketing). The presence of such barriers impedes the flow of information, prevents each entity in the extended process from perceiving organizational goals, and fosters the pursuit of individual or departmental goals that are not necessarily consistent with the organizational goals.

External barriers include those between the vendors and the company, the company and the customer, the company and the community, and the company and its investors. The detrimental effects of any of these barriers is obvious. If the company does not have a clear and concise understanding of the needs of its customer, how can it come up with acceptable product designs? The very survival of the company may be in question if it does not incorporate the sentiments of the community in which it exists. The relationships of the company to its customers, the vendors, and community must be harmonious.

Poor communication is often a culprit in barrier creation. Perhaps top management fails to model open and effective communication. A fear-ridden atmosphere builds barriers. Another reason barriers exist is the lack of cross-functional teams. Interdisciplinary teams promote communication between departments and functional areas. They also result in innovative solutions to problems. Employees must feel that they are part of the same team trying to achieve the overall mission of the company.

Breaking down barriers takes time; it requires changing attitudes and attitudes do not change overnight. However, they can and do change when everyone involved is convinced of the advantages of doing so and of the importance of a team effort in achieving change. At Ford Motor Company, for instance, this concept is found at every level of their design approval process. This process incorporates input from all related units such as design, sales and marketing, advance product planning, vehicle engineering, body and assembly purchasing, body and assembly manufacturing, product plan-

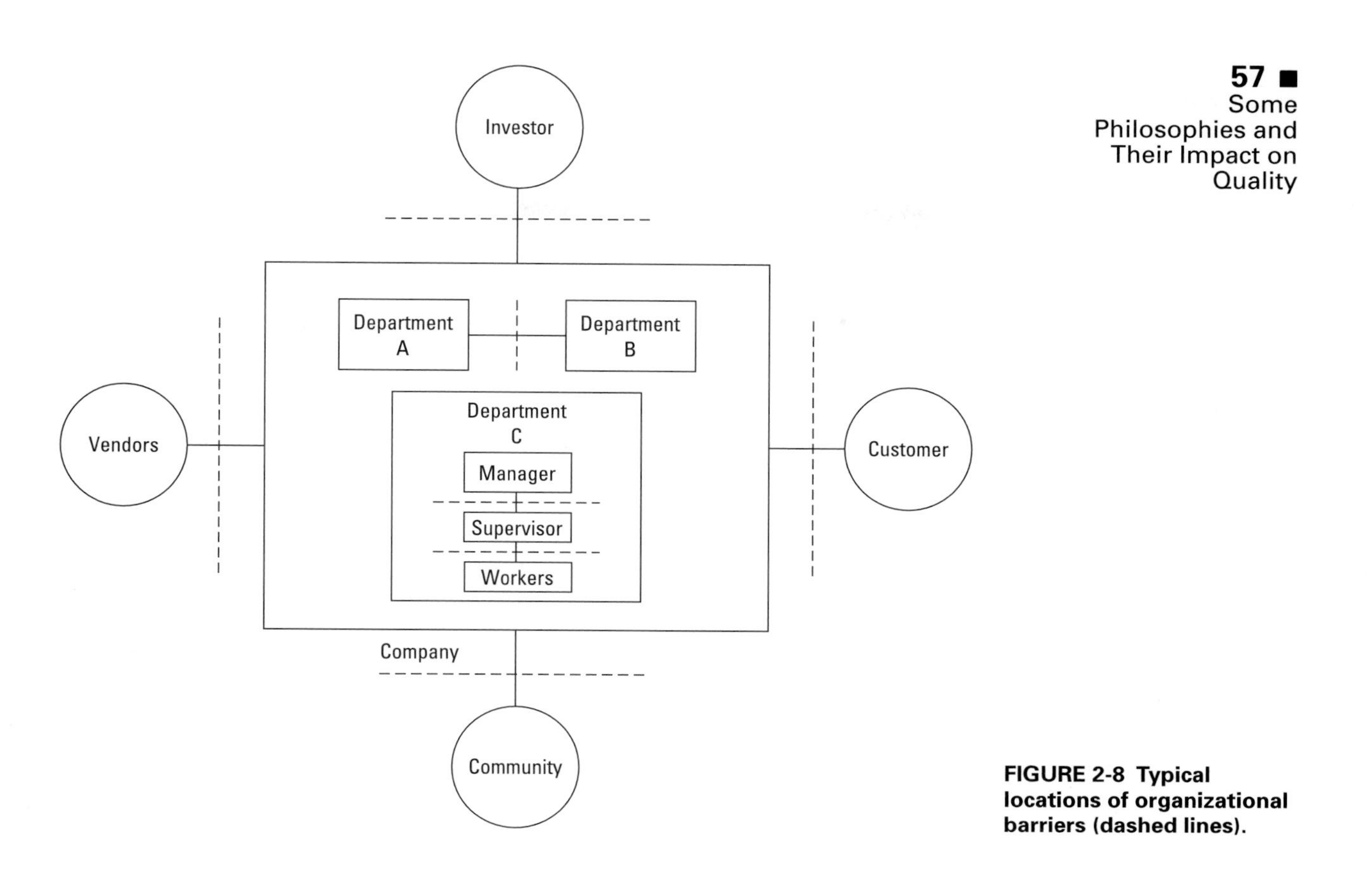

FIGURE 2-8 Typical locations of organizational barriers (dashed lines).

ning, and others. Management emphasizes open lines of communication at all levels and among different departments. The reward system used to facilitate this process is based on teamwork rather than an individual person's production.

Deming's Point 10

Eliminate exhortations for the workforce.

Numerical goals such as a 10% improvement in productivity set arbitrarily by management have a demoralizing effect. Rather than serving to motivate, such standards have the opposite effect on morale and productivity.

Consider, for example, a company producing the housing for a certain bearing. Based on customer needs, the specifications for the housing diameter are 20 ± 0.2 mm. The company tracks the proportion of nonconforming housings daily. A plot of the proportion of nonconforming housings over time is shown in Figure 2-9. The figure shows that the proportion of nonconforming items hovers around 6%. Suppose management now sets a goal of 4% for the maximum proportion of unacceptable housings. What are the implications of this? First, what is the rationale behind the goal of 4%? Second, is management specifying ways to achieve the goal? If the answers to these questions are unsatisfactory, employees can only experience frustration when this goal is presented.

If the process is stable, the employees have no means of achieving the goal unless management changes the process or the product. Management has to come up with a feasible course of action so the desired goal can be achieved. Failing to do so will lower morale and productivity. By taking such an action, management is basically shirking its responsibility. Hence, goals should be set by management in a participative style, and procedures for accomplishment should be given.

Management should demonstrate its commitment to the neverending process of quality improvement. Rather than providing slogans, they should describe precisely

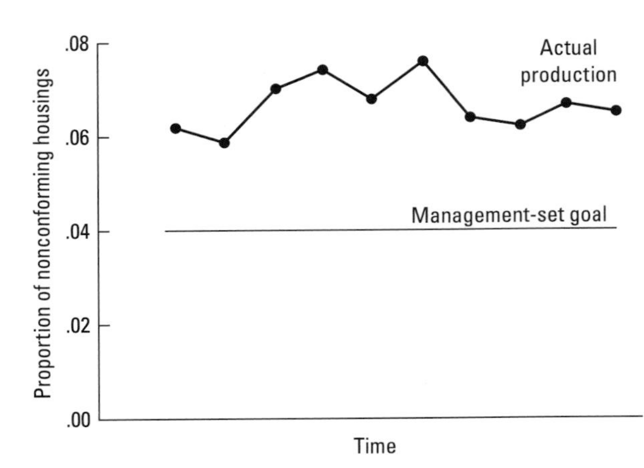

FIGURE 2-9 Arbitrarily established numerical goals.

what they are doing to implement this long-term goal to the employees. Only then will employees sense management's commitment, which will motivate the employees to make every effort to achieve it. Intuitively set goals should be avoided. Instead, goals should be based on information from every level. Bottom-up planning is recommended, assuming that the process is in a state of statistical control. Being in a state of statistical control allows the employees to make projections as to what might be expected from the process. This information can then be used by management to set goals and allocate resources.

Deming's Point 11

(a) Eliminate numerical quotas for production. Instead, learn and institute methods for improvement. (b) Eliminate M.B.O. [management by objectives]. Instead, learn the capabilities of processes, and how to improve them.

Work standards are typically established by someone other than those who perform the particular job in question. They are based on quantity without regard to quality. According to Deming, setting such work standards guarantees inefficiency and increases costs. The numerical quota defined by a work standard may derail implementing the Deming improvement cycle because people naturally strive to meet the quota rather than to produce acceptable goods. As such, numerical quotas actually promote the production of nonconforming items.

Another drawback of work standards is that they give no information about the procedure that might be used to meet the quota. Is the numerical value a feasible one? Usually, when determining work standards, an allowance is made for the production of nonconforming items, but this often simply ensures that a certain proportion of defectives will be produced. The company thus moves further from the desirable goal of continuous improvement. Quotas provide no game plan for implementing a quality system.

A third drawback of the quota system is that it fails to distinguish between special causes and common causes when improvements in the process are sought. Consequently, workers may be penalized for not meeting the quota when it is really not their fault. As discussed previously, common causes can be eliminated only by management. Thus, improvements in process output cannot occur unless a conscientious effort is made by management. If the quota is set too high, very few workers will meet the objectives. This will lead to the production of more defective units by workers, because they will try to meet the numerical goal without regard for quality. Furthermore, workers will experience a loss of pride in their work, and worker morale and motivation will drop significantly.

Work standards are typically established through union negotiation and have nothing to do with the capability of the process. Changes in process capability are not pursued, so the standards do not reflect the potential of the current system. Workers who surpass a standard that has been set too high may be producing several defectives, and they may know it. They realize they are being rewarded for producing nonconforming items—which is totally against Deming's philosophy. On the other hand, if quotas are set too low, productivity will be reduced. A worker may meet the quota with ease, but once the quota is achieved, he or she may have no motivation to exceed it; if management finds out that several people's output meets or exceeds the quota, the quota will likely be increased. This imposes an additional burden on the employee to meet the new quota, without the aid of improved methods or procedures.

The work standard system is never more than a short-term solution, if it is that. On the other hand, using control charts to analyze and monitor processes over time offers a proper focus on long-term goals. Statistical methods are preferable over arbitrary work standards, because they help an organization stay competitive.

Deming's Point 12

Remove barriers that rob people of pride of workmanship.

A **total quality system** can exist only when all employees synergistically produce output that conforms to the goals of the company. Quality is achieved in all components of the extended process when the employees are satisfied and motivated, when they understand their role in the context of the organization's goals, and when they take pride in their work. It is management's duty to eliminate barriers that prevent these conditions from occurring. A direct effect of pride in workmanship is increased motivation and a greater ability for employees to see themselves as part of the same team—the team that makes good things happen.

Factors That Cause a Loss of Pride

Several factors diminish worker pride. First, management may not treat employees with dignity. Perhaps they are insensitive to workers' problems (personal, work, or community). Happy employees are productive, and vice versa. Happy employees don't need continuous monitoring to determine whether their output is acceptable.

Second, management may not be communicating the company's mission to all levels. How can employees help achieve the company's mission if they do not understand what that mission is?

Third, management may assign blame to the employees for failing to meet company goals when the real fault lies with management. If problems in product output are caused by the system (such as poor-quality raw materials, inadequate methods, or inappropriate equipment), then employees are not at fault and should not be penalized (even though the best employee might be able to produce a quality product under these circumstances). Assigning blame demoralizes employees and affects quality. As Deming noted, the problem is usually the system, not the people.

Focusing on short-term goals compounds these problems. Consider daily production reports. Different departments dutifully generate pertinent data, but the focus is wrong and they know it: Top management is "micromanaging" and not attending to long-term goals. The pressure to constantly increase quantity on a daily basis does not promote the notion of quality. How many times have we heard of a department manager having to explain why production dropped today by say, 50 units, compared to yesterday? Such a drop may not even be statistically significant. Inferences should be based on sound statistical principles.

Performance Classification Systems

Inadequate performance evaluation systems rob employees of their pride in workmanship. Many industries categorize their employees as excellent, good, average, fair, or unacceptable, and they base pay raises on these categorizations. These systems fail because there are often no clear-cut differences between categories, which lead inevitably to inconsistencies in performance evaluation. A person may be identified as "good" by one manager and "average" by another. This is not acceptable.

A major drawback is that management does not have a statistical basis for saying that there are significant differences between the output of someone in the "good" category and someone in the "average" category. For instance, if a difference in output between two workers were statistically insignificant (that is, due to chance), then it would be unfair to place the two individuals in two different categories. Figure 2-10 shows such a classification system composed of five categories. The categories numbered 1 through 5 (unacceptable through excellent) have variabilities that are not due to a fundamental difference in the output of the individuals. In fact, the employees may not even have a chance to improve their output because of system deficiencies. Thus, two employees may be rated by their supervisors as 3 (average) and 4 (good), respectively, with the employee rated as 4 to be considered superior in performance to the other. However, both of these employees may be part of the same distribution, implying that there is no statistically significant difference between them. With this particular system, whatever aggregate measure of evaluation is being used to lump employees into categories, there are no statistically significant differences between the values for categories 1, 2, 3, 4, or 5. The only two groups that may be considered part of another system, whose output therefore varies significantly from those in categories 1–5, are categories 0 and 6. Those placed in group 0 may be classified as performing poorly (so low that they are considered part of a different system). Likewise, those placed in group 6 may be classified as performing exceptionally well.

Categorizing employees according to "good," "average," and so on is demoralizing. What is worse is that it may be statistically unsound to place them in these categories in the first place. Because the differences between categories is not statistically significant, there is no scientifically sound basis for moving workers from one category to another, and this too is frustrating to employees trying to improve the quality of their work. Note that the above discussion is based on the assumption that the numerical measure of performance being used to categorize individuals into groups is normally distributed as in Figure 2-10. Only those significant deviations from the mean (say, three or more standard deviations on either side) may be considered extraordinary, that is, not belonging to the same system that produces ratings of 1–5. Under these circumstances, category 0 (whose cutoff point is three standard deviations below the mean) and category 6 (whose cutoff point is three standard deviations above the mean) would be considered different from categories 1–5 because they are likely from a different distribution.

Thus, statistically speaking, the only categories should be the three groups corresponding to the original categories 0, 1–5, and 6. In Deming's approach, teamwork is considered extremely important. Consideration should be given to this point when

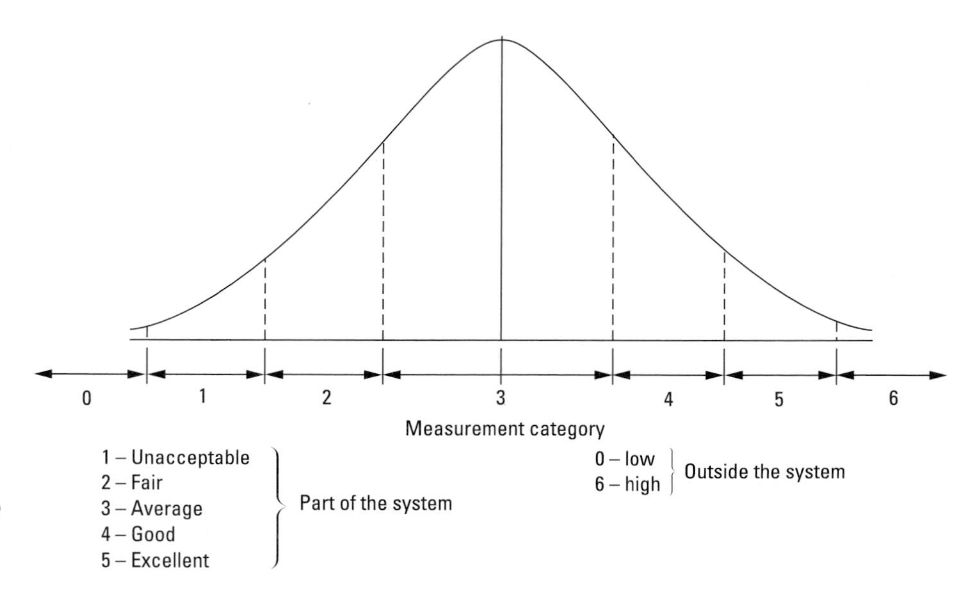

FIGURE 2-10 Improper classification system where categories are part of the same system.

conducting performance appraisal. The procedure should be designed so that promoting teamwork is a necessary criterion for placing someone in the highest category. Such individuals should be rewarded through merit raises. For those placed in categories 1–5, monies set aside for investment in the system should be evenly distributed.

Deming's Point 13

Encourage education and self-improvement for everyone.

Deming's philosophy is based on long-term, continuous process improvement. To meet this goal, the organization's most important resource, its people, have to be motivated and adequately trained. This is the only way the company can survive in today's highly competitive business environment. Management must commit resources for education and retraining. This represents a sizable investment in both time and money. However, an educated workforce, with its eye on future needs, can safeguard improvements in quality and **productivity** and help the company maintain a competitive position.

Point 6 (page 54) deals with training that enables new employees to do well. Point 13, on the other hand, addresses the need for ongoing and continual education and self-improvement for the entire organization. A workforce that is continually undergoing education and retraining will be able to keep up with cutting-edge technology. Today's workplace is changing constantly and knowing that they can and are adapting to it gives employees a sense of security. A company that invests in its employees' growth and well-being will have a highly motivated workforce—a win–win situation for everyone. Education is an investment in the future.

Advantages of Education and Retraining

Investments in the workforce, such as educating and retraining employees at all levels, serve a multifold purpose. First, employees (who believe that if the company is willing to incur such expenditures for their benefit, it must be interested in their well-being) are likely to be highly motivated employees. Such a belief fosters the desire to excel at work.

Second, ongoing education programs enable the company to inculcate its employees in the goals of the organization. When everyone is familiar with the company's goals, work can proceed in a synergistic manner. The content and duration of the educational process may vary among employees. Top management may have to be instructed periodically on the missions of the company and its philosophy. All employees may undergo instruction on basic statistical methods and their relevance to quality improvement. Technical personnel may need further training in advanced statistical techniques and applications. As processes and methods change, retraining will be necessary for certain workers.

Third, education and retraining keeps the employees up-to-date on the latest techniques and promotes teamwork, which is important in meeting the company's goals. As employees learn more about the product and process, they realize the significance of tasks different from their own.

Fourth, as employees grow with the company and their job responsibilities change, retraining provides a mechanism to ensure adequate performance in their new jobs. Such progressive action by the company makes it easier to adapt to the changing needs of the consumer.

Fifth, providing education and retraining to employees reduces the number of people who "job-hop." The average tenure of managers and technical personnel in many U.S. companies is quite low. By investing in their employees through seminars, workshops, and managerial and technical training, the company gains employee loyalty, and loyal employees stay with their employers.

Education should always include instruction on Deming's 14 points. Also, education and retraining applies to people in nonmanufacturing or service areas as well. Thus, employees in accounting, purchasing, marketing, sales, or maintenance need this instruction to participate in the neverending process of improvement. Industry should work with academia to aid in this process of education and retraining. Management, of course, should be aware that the process of education and retraining is a lengthy one.

Deming's Point 14

Take action to accomplish the transformation.

Point 14 involves accepting the Deming philosophy and committing to seeing its implementation in the extended process. A structure must be created and maintained for the dissemination of the concepts associated with the first 13 points. Responsibility for creating this structure lies with top management. Besides being totally committed to the idea of quality, management must be visionary and knowledgeable as to its potential impacts, and they must be in it for the long run.

For the implementation to succeed, people must be trained in statistics at all levels of the organization. A quality company requires statistical assistance at every level. Figure 2-11 shows how integral statistical training is to the successful implementation of Deming's philosophy. Numerous people trained in statistical methods are needed to maintain this organizational structure, the underlying theme of which is the integration of statistical principles in a holistic manner. That is, the first 13 points must become a way of life.

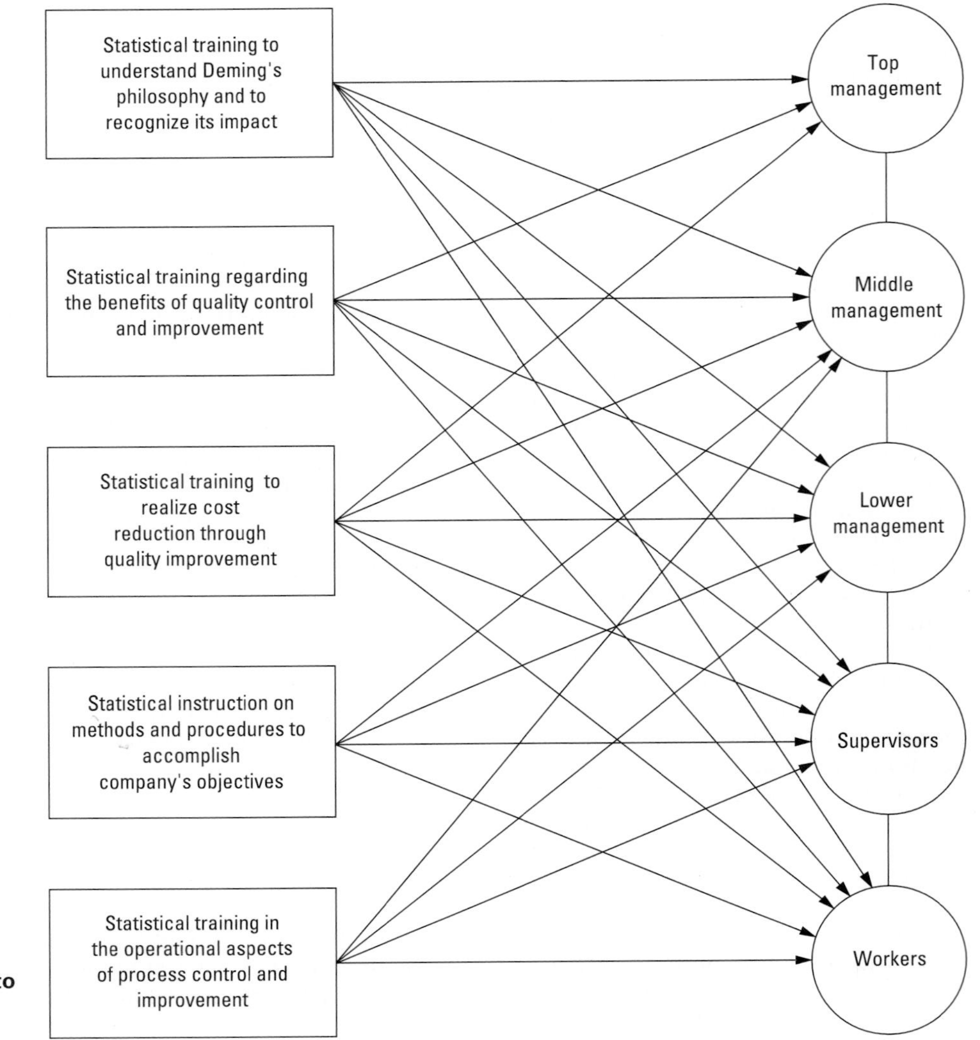

FIGURE 2-11 An organizational structure to promote Deming's philosophy.

Successful companies allow free flow of information among their different levels. An information flow from workers to supervisors to management and eventually to top management facilitates determining which areas need improvement. As plans are developed and then implemented, vendors and customers are brought into the process because they are part of the extended process.

Deming's Deadly Diseases

Deming's 14 points for management provide a road map for continuous quality improvement. In implementing these points, certain practices of management, which Deming labels as deadly diseases or sins, must be eliminated. Most of **Deming's deadly diseases** involve a lack of understanding of variation. Others address management's failure to understand who (they and only they) can correct common causes in the system. The five deadly diseases, which Deming's 14 points seek to stamp out, are as follows:

- Management by visible figures only

This deadly disease is also known as "management by the numbers." Visible figures are such items as monthly production, amount shipped, inventory on hand, and quarterly profit. Emphasizing short-term visible figures doesn't always present an accurate picture of the financial state of an organization. Visible figures can be easily manipulated to show attractive quarterly profits when, in fact, the company's competitive position is in jeopardy. How many managers would deny that it is easy to show a large amount of production shipped at the end of the month and to hide the numbers of defectives returned? When quantity is the only figure on which the manager is judged, this is no small impetus to manipulate numbers and report large values, although such figures represent total output only, not the proportion of conformance. Cutting back research, training, and education budgets is amazingly tempting when quarterly profits are disappointing. While these actions will all show desirable short-term profits, they can seriously sabotage the company's long-term goals of remaining competitive and of improving constantly and forever. As Deming stated many a time, "he who runs his company on visible figures alone will in time have neither company nor figures."

Properly selected statistical data, unbiased and as objective as possible, is *the* quality assessment tool of the quality company. Some examples include data on customer satisfaction; employee satisfaction with, and perception of the company; and employee morale and motivation. Often, however, the most pertinent data is unknown or difficult to measure. How does a company measure loss of market share due to customer dissatisfaction? How does a company measure loss of goodwill because of the unmet needs of its customers who decide to switch? How does it measure the damage caused by an unmotivated workforce? How does it measure the losses that accrue when management fails to create an atmosphere of innovation? These losses are difficult, if not impossible, to measure; we could call them the invisible figures. They are invisible and real, and they must be addressed in the transformation process.

Some of Deming's 14 points for management seek to eliminate the deadly visible figures only disease. Consider Point 5, which deals with constant, neverending cycles of improvement. This suggests that a system should first be brought to a state of statistical control by eliminating the special or assignable causes. Subsequently, cross-disciplinary teams could be used to solve the common causes. While variability will always be present in any system, it is important that common causes not be treated as special causes. Point 8 stresses eliminating fear and creating a climate of trust and innovation. Implementing this point promotes teamwork because teams obviously cannot function in fear-driven organizations. Then and only then can workers drop the "every person for himself or herself" attitude and work for the common good of the company. Once unrealistic numerical goals, which cannot be accomplished under the existing system, are eliminated, employee fear dissolves, and effective quality improvement teams can get to work.

- Lack of constancy of purpose

This disease prevents organizations from making the transformation to the long-term approach. Top management must rid itself of myopia. Management must return again and again to the company's long-term goals. Management cannot mouth a long-term vision but enforce a short-term approach to operational decisions. The slogan of the legendary founder of Wal-Mart, Sam Walton, applies here: "Management must walk their talk." Only then will employees be persuaded to pursue the quality work ethic.

Constancy of purpose ensures survival. A healthy company, doing everything it can to stay in for the long haul, provides a sense of security to its employees. No company can guarantee lifetime employment; however, when employees see that the company has a long-term vision that it lives by, and they see daily demonstrations of this vision, they also see stability.

Mission statements play an important role here. Unless the mission statement is propagated through all levels of the company and understood and adapted by all, it is a useless document. But, when employees understand the company's mission statement and know that management and other employees believe in it *and* use it, it becomes the embodiment of the corporate **quality culture.** The wording key: The emphasis must be on the *process* and not on the end result. For example, a mission statement that talks about maximizing profits and return on investment, does little to guide employees. Profits and such are an end result, not the means. The mission statement is the means, the vision, the road map writ large. Managing the process will automatically take care of the end result. Mission statements must be simple and provide vision and direction to *everyone.*

Effective mission statements are formulated only after considerable input from all constituencies, but top management has the ultimate say. Mission statements are usually nonquantitative. Here is an excerpt from General Electric Company's mission: *"Progress is our most important product."*

The mission is implemented "bottom-up." New recruits should be hired to further the mission, and current employees must demonstrate their support for the mission. The wise company also makes its customers aware of its mission.

- Performance appraisal by the numbers

In such systems, the focus is on the short term and on the outcome rather than on the process. One tremendous disadvantage of these systems is that they promote rivalry and internal competition. Employees are forced to go up against each other and about 50% must be ranked as below average whether their work is satisfactory or not. In our society, being ranked "below average" is a strong message—strong enough to cause depression and bitterness. It demoralizes and devastates. Whatever spirit of cooperation that may have existed among employees previously begins to erode.

The company must endeavor to stay competitive, even as it completes the transformation process. Such is the message of Points 1 and 5. Numerical performance appraisals make this difficult because they encourage counterproductive behavior among employees. Rivalry within weakens the battle on rivalry without—that is, rivalry in the marketplace. For the company to stay in business, prosper, and create jobs, a concerted effort must be made by everyone to work as a team. Individual-centered performance, whose sole objective is to demonstrate one employee's superiority over another, sabotages this effort. Numerical performance appraisal thus has a detrimental effect on teamwork.

Synergism—effective teamwork—is attained only when employees share the common goal of improving companywide performance. It is possible for individual employees to dispatch their duties "by the book," yet fail to improve the competitive position of the company. Point 9 focuses on breaking down organizational barriers and promoting teamwork. Cross-functional teams, which focus on the process, are a means of continuous improvement. Obviously, successfully implementing such teams is difficult when employees are evaluated on individual performance.

- A short-term orientation

Many companies choose less-than-optimal solutions because they focus on near-term goals and objectives. Decisions dictated by quarterly dividends or short-term profits can diminish a company's chances of long-term survival. This is in direct contradiction to Points 1 and 2.

A reason often given for adopting a short-term approach is the need to satisfy shareholders. Companies are fearful of hostile takeovers and believe that demonstrating short-term profits can discourage this activity. Top-down pressure to create short-term profits causes many undesirable actions: monthly or quarterly production may be manipulated; training programs may be cut; and other investments on employees may be curtailed. Although such actions may defer costs and show increased short-term profits, they can have a devastating effect on the company's long-term survival. Moreover, the real impact of such schemes is not always tangible. The connection between a disheartened workforce and cutting back on quality control doesn't always manifest itself in the visible numbers.

Long-term partnerships can have a sizable impact on cost-savings. IBM used to contract with about 200 carriers to move parts to different locations around the country. Thousands of invoices had to be processed annually, and stability among carriers was nonexistent as they struggled to stay in business in spite of fluctuating orders. By reducing the number of carriers over tenfold, IBM achieved significant savings in order processing costs. While the initial goal may have been to reduce invoicing headaches, in reality, relationships with a few selected vendors have positively affected IBM's competitive position. The concept of the extended process becomes a reality as the vendor's technical problems are also that of the company's, and vice versa.

Linking executive salaries with annual bottom-line profit figures also promotes a short-term focus. In many companies, shareholders approve this policy, and when this prevails, senior managers sometimes postpone much-needed expenditures to demonstrate short-term gains. Solutions to this problem include basing compensation and bonus packages on 5-year moving averages of corporate earnings.

- Mobility of management

A major cause of organizational instability is the short tenure of management. When the average tenure of a mid-level manager is 4–5 years, continuity is difficult, if not impossible. Getting acquainted with, and accustomed to, a quality culture requires major paradigm shifts, and this takes time. Frequent changes in management sabotage constancy of purpose. An atmosphere of trust is difficult to maintain when the "ground shifts" with each new change in management. To reduce mobility, top management can promote from within, institute job enrichment programs, and practice job rotation.

Companies must demonstrate to managers their concern for their advancement. Recall that Point 13 deals with providing education and self-improvement for everyone. By investing in management training and seminars that develop skills for expanded responsibilities, the company also gains loyalty from its managers. Attention to salaries, relative to industry norms, is important. When managers sense that the company is doing all it can to maintain salaries at competitive levels, they are often motivated to stay. Managers know that there are priorities, and they will stay if they are convinced that commitment to employees is the top one.

2-5 PHILIP B. CROSBY'S PHILOSOPHY

Philip B. Crosby founded Philip Crosby Associates in 1979. Prior to that, he was a corporate vice president for ITT, where he was responsible for worldwide quality operations. Crosby has a particularly wide-ranging understanding of the various operations in industry because he started as a line inspector and worked his way up. Such first-hand experience has provided him with a keen awareness of what quality is, what the obstacles to quality are, and what can be done to overcome them. Crosby has trained and consulted with many people in manufacturing and service industries, and he has written many books (Crosby 1979, 1984, 1989) on quality management.

The Crosby approach begins with an evaluation of the existing quality system. His quality management grid (Crosby 1979) identifies and pinpoints operations that have potential for improvement. Figure 2-12 is an example of a **quality management maturity grid.** The grid is divided into five stages of maturity, and six measurement categories aid in the evaluation process.

Four Absolutes of Quality Management

In order to understand the meaning of quality, Crosby (1979) has identified four absolutes of quality management.

FIGURE 2-12 Crosby's quality management grid.
(Philip B. Crosby. *Quality Is Free*, copyright 1979, McGraw-Hill, Inc. Reprinted with permission.)

	Stages of Maturity				
Measurement Categories	*Stage I: Uncertainty*	*Stage II: Awakening*	*Stage III: Enlightenment*	*Stage IV: Wisdom*	*Stage V: Certainty*
Management understanding and attitude	No comprehension of quality as a management tool. Tend to blame quality department for "quality problems."	Recognizing that quality management may be of value but not willing to provide money or time to make it all happen.	While going through quality improvement program learn more about quality management. Recognize their personal role in continuing emphasis.	Participating. Understand absolutes of quality management. Recognize their personal role in continuing emphasis.	Consider quality management as essential part of company system.
Quality organization status	Quality is hidden in manufacturing or engineering departments. Inspection probably not part of organization. Emphasis on appraisal and sorting.	A stronger quality leader is appointed but main emphasis is still on appraisal and moving the product. Still part of manufacturing or other.	Quality department reports to top management, all appraisal is incorporated, and manager has role in management of company.	Quality manager is an officer of company, effective status reporting and preventive action. Involved with consumer affairs and special assignments.	Quality manager on board of directors. Prevention is main concern. Quality is a thought leader.
Problem handling	Problems are fought as they occur; no resolution; inadequate definition; lots of yelling and accusations.	Teams are set up to attack major problems. Long-range solutions are not solicited.	Corrective action communication established. Problems are faced openly and resolved in an orderly way.	Problems are identified early in their development. All functions are open to suggestion and improvement.	Except in the most unusual cases, problems are prevented.
Cost of quality as a percentage of sales	Reported: unknown Actual: 20%	Reported: 3% Actual: 18%	Reported: 8% Actual: 12%	Reported: 6.5% Actual: 8%	Reported: 2.5% Actual: 2.5%
Quality improvement actions	No organized activities. No understanding of such activities.	Trying obvious "motivational" short-range efforts.	Implementation of the 14-step program with thorough understanding and establishment of each step.	Continuing the 14-step program.	Quality improvement is a normal and continued activity.
Summary of company quality posture	"We don't know why we have problems with quality."	"Is it absolutely necessary to always have problems with quality?"	"Through management commitment and quality improvement we are identifying and resolving our problems."	"Defect prevention is a routine part of our operation."	"We know why we do not have problems with quality."

- Definition of quality: Quality means conformance to requirements.
- System for achievement of quality: The rational approach is prevention of defects.
- Performance standard: The only performance standard is zero defects.
- Measurement: The performance measurement is the cost of quality. In fact, Crosby emphasizes the costs of unquality such as scrap, rework, service, inventory, inspection, and tests.

14-Step Plan for Quality Improvement

Crosby has a **14-step plan,** discussed briefly here to help businesses implement a **quality improvement program.*** The reader who is interested in a detailed examination of these steps should consult the suggested references.

1. *Management commitment.* For quality improvement to take place, commitment must start at the top. The emphasis on defect prevention has to be communicated, and a quality policy that states the individual performance requirements needed to match customer requirements must be developed.

2. *Quality improvement team.* Representatives from each department or division form the quality improvement team. These individuals serve as spokespersons for each group they represent. They are responsible for ensuring that suggested operations are brought to action. This team brings all the necessary tools together.

3. *Quality measurement.* Measurement is necessary to determine the status of quality for each activity. It identifies the areas where corrective action is needed and where quality improvement efforts should be directed. The results of measurement, which are placed in highly visible charts, establish the foundation for the quality improvement program. These principles apply to service operations as well, such as counting the number of billing or payroll errors in the finance department, the number of drafting errors in engineering, the number of contract or order description errors in marketing, and the number of orders shipped late.

4. *Cost of quality evaluation.* The cost of quality (or rather unquality) indicates where corrective action and quality improvement will result in savings for the company. A study to determine these costs should be conducted through the comptroller's office, with the categories that comprise quality costs precisely defined. This study establishes a measure of management's performance.

5. *Quality awareness.* The results of the cost of nonquality should be shared with all employees, including service and administrative people. Getting everybody involved with quality facilitates a quality attitude.

6. *Corrective action.* Open communication and active discussion of problems creates feasible solutions. Furthermore, such discussion also exposes other problems not identified previously and thus determines procedures to eliminate them. Attempts to resolve problems should be made as they arise. For those problems without immediately identifiable remedies, discussion is postponed to subsequent meetings. The whole process creates a stimulating environment of problem identification and correction.

7. *Ad hoc committee for the zero defects program.* The concept of **zero defects** must be communicated clearly to all employees; everyone must understand that the achievement of such a goal is the company's objective. This committee gives credibility to the quality program and demonstrates the commitment of top management.

8. *Supervisor training.* All levels of management must be made aware of the steps of the quality improvement program. Also, they must be trained so they can explain the program to the employees. This ensures the propagation of the quality concepts from the chief executive officers to the hourly worker.

---

*Philip B. Crosby. *Quality Is Free,* copyright 1979, McGraw-Hill, Inc. Reprinted with permission.

9. *Zero defects (ZD) day.* The philosophy of zero defects should be established companywide and should originate on one day. This ensures a uniform understanding of the concept for everyone. Management has the responsibility of explaining the program to the employees, and they should describe the day as signifying a "new attitude." Management must foster this type of quality culture in the organization.

10. *Goal setting.* Employees, in conjunction with their supervisors, should set specific measurable goals. These could be 30-, 60-, or 90-day goals. This process creates a favorable attitude for people to ultimately achieve their own goals.

11. *Error cause removal.* The employees are asked to identify reasons that prevent them from meeting the zero defects goal—not to make suggestions but to list the problems. It is the task of the appropriate functional group to come up with procedures for removing these problems. Reporting problems should be done quickly. An environment of mutual trust is necessary so that both groups work together to eliminate the problems.

12. *Recognition.* Award programs should be based on recognition rather than money and should identify those employees who have either met or exceeded their goals or have excelled in other ways. Such programs will encourage the participation of everyone in the quality program.

13. *Quality councils.* Chairpersons, team leaders, and professionals associated with the quality program should meet on a regular basis to keep everyone up to date on progress. These meetings create new ideas for further improvement of quality.

14. *Do it over again.* The whole process of quality improvement is continuous. It repeats again and again as the quality philosophy becomes ingrained.

2-6 JOSEPH M. JURAN'S PHILOSOPHY

Joseph M. Juran founded the Juran Institute, which offers consulting and management training in quality. Juran has worked as an engineer, labor arbitrator, and corporate director in the private sector, and as a government administrator and a university professor in the public sector. He has authored many books on the subjects of quality planning, control, management, and improvement (Juran and Gryna 1980; Juran 1986, 1988a, 1988b, and 1989).

Like Deming, he visited Japan in the early 1950s to conduct training courses in quality management. He eventually repeated these seminars in over 40 countries, on all continents. In the 1980s, Juran met the explosive demand for his services with offerings through the Juran Institute. His books and videotapes have been translated into many languages, and he has trained thousands of managers and specialists.

Juran believes that management has to adopt a unified approach to quality. Quality is defined as "fitness for use." The focus here is on the needs of the customer.

Certain nonuniformities deter the development of a unified process. One is the existence of multiple functions—such as marketing, product design and development, manufacture, and procurement—where each function believes itself to be unique and special. Second, the presence of hierarchical levels in the organizational structure creates groups of people who have different responsibilities. These groups vary in their background and may have different levels of exposure to the concepts of quality management. Third, a variety of product lines that differ in their markets and production processes can cause a lack of unity.

Juran proposes a universal way of thinking about quality, which he calls the **quality trilogy:** quality planning, quality control, and quality improvement. This concept fits all functions, levels of management, and product lines.

Quality Trilogy Process

The quality trilogy process starts with quality planning at various levels of the organization, each of which has a distinct goal. At the upper management level, planning is termed *strategic quality management.* Broad quality goals are established. A structured approach is selected in which management chooses a plan of action and allocates resources to achieve the goals. Planning at the middle management level is termed *operational quality management.* Departmental goals consistent with the strategic goals are established. At the workforce level, planning involves a clear assignment to each worker. Each worker is made aware of how his or her individual goal contributes to departmental goals.

After the planning phase, quality control takes over. Here, the goal is to run the process effectively such that the plans are enacted. If there are deficiencies in the planning process, the process may operate at a high level of chronic waste. Quality control will try to prevent the waste from getting worse. If unusual symptoms are sporadically detected, quality control will attempt to identify the cause behind this abnormal variation. Upon identifying the cause, remedial action will be taken to bring the process back to control. Figure 2-13 shows the three phases of planning, control, and improvement and the cost trends associated with poor quality. The objectives of the control phase are to eliminate the causes associated with the sporadic spikes and to bring the process output within the zone of quality control.

The next phase of the trilogy process is **quality improvement,** which deals with the continuous improvement of the product and the process. This phase is also called the **quality breakthrough sequence.** Such improvements usually require an action on the part of upper and middle management. They deal with actions such as creating a new design, changing methods or procedures of manufacturing, and investing in new equipment. As can be seen from Figure 2-13, quality improvement will usually cause a reduction in the cost of poor quality. The chronic waste drops to a lower level due to quality improvement. Repeating the whole cycle lets the company strive for further improvements. Readers should consult the listed references (Juran 1986, 1988a, 1988b, 1989) for an elaborate treatment of the details of each phase.

Quality Planning

1. *Identify the customer—both external and internal.* Juran has a concept similar to Deming's extended process. Juran's includes vendors and customers. He stresses the importance of identifying the customer. In cases where the output from one department flows to another, the customer is considered internal.

2. *Determine customer needs.* Long-term survival of the company is contingent upon meeting the needs of the customer. Conducting analysis and research, surveying clients and nonclients, and studying customer needs are a few examples of activities in this category.

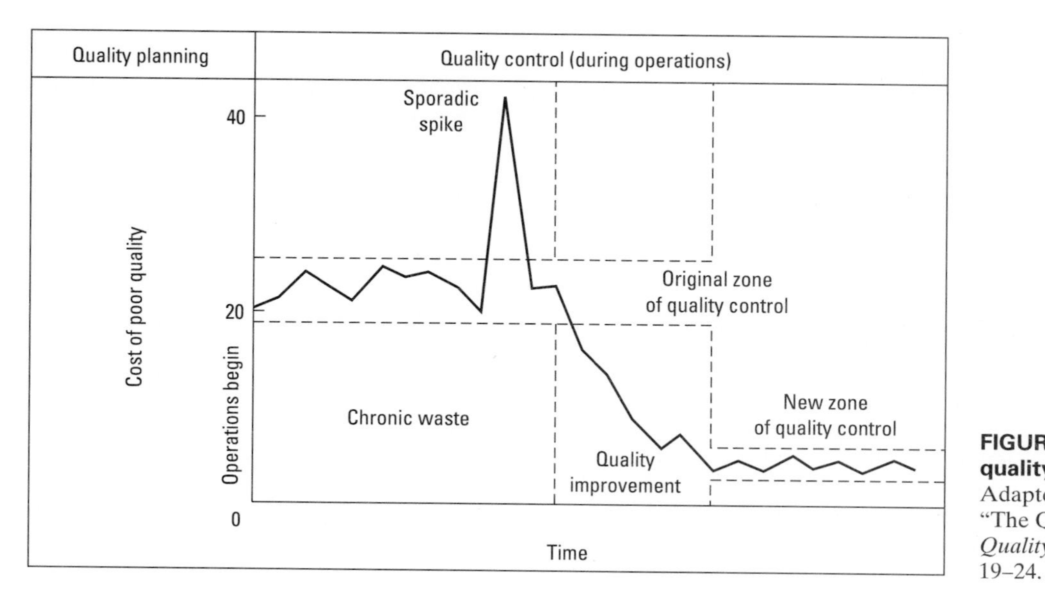

FIGURE 2-13 Juran's quality trilogy.
Adapted from J. M. Juran, "The Quality Trilogy," *Quality Progress* (Aug. 1986): 19–24.

3. *Develop product features that respond to customer needs.* With customer satisfaction as the utmost objective, the product or service should be designed to meet the customer requirements. As customer needs change, the product should be redesigned to conform to these changes.

4. *Establish quality goals that meet the needs of customers and suppliers alike, and do so at a minimum combined cost.* This point embraces the concept of the extended process involving the vendors and customers as well as the organization. Pursuit of individual or departmental goals should be avoided. The total cost from an organizational point of view should be minimized, and corresponding goals should be determined.

5. *Develop a process that can produce the needed product features.* A product is designed based on a knowledge of customer needs. This step deals with the manufacturing process of that product. Methods must be developed, and adequate equipment must be available to make the product match its design specifications.

6. *Prove* **process capability.** The task here is to establish whether the given process is adequate for making a product that will conform to the design specifications. This task may require analyzing output from a stable process and determining its level of operation. If the operating level meets the desirable goals, the process is labeled as capable.

Quality Control

1. *Choose control subjects.* Product characteristics that are to be controlled in order to make the product conform to the design requirements should be chosen. For instance, a wheel's control characteristics may be the hub diameter and the outside diameter. Selection is done by prioritizing the important characteristics that influence the operation or appearance of the product and hence impact the customer.

2. *Choose units of measurement.* Based on the quality characteristics that have been selected for control, appropriate units of measure should be chosen. For example, if the hub diameter is being controlled, the unit of measurement might be millimeters.

3. *Establish measurement.* Procedures for taking measurements must be defined. These procedures should take into account the equipment to be used, the way in which it is to be used, who should take the measurements, how the measurements should be taken (that is, what sampling plan, if any, is to be used), and other related issues. Care must be taken to ensure that all measuring instruments are properly calibrated and that persons taking measurements are adequately trained.

4. *Establish standards of performance.* These standards should be based on customer requirements. For instance, a standard of performance for the hub diameter could be 20 ± 0.2 mm. A hub with a diameter in this range would be compatible in final assembly and would also contribute to making a product that will satisfy the customer.

5. *Measure actual performance.* This phase of quality control is concerned with the measurement of the actual process output. Measurements are taken on the previously selected control subjects (or quality characteristics). Such measurements will provide information on the operational level of the process.

6. *Interpret the difference (actual versus standard).* This involves comparing the performance of the process with the established standard. If the process is stable and capable, then any differences between the actual and the standard may not be significant.

7. *Take action on the difference.* In the event that a discrepancy is found between the actual output of the process and the established standard, remedial action needs to be taken. In Figure 2-13, the sporadic spike shows a significant difference between what is observed and the standard. It is management's responsibility to suggest a remedial course of action.

Quality Improvement

1. *Prove the need for improvement.* Juran's breakthrough sequence tackles the chronic problems that exist because of a change in the current process; this task requires management involvement. First, however, management has to be convinced of the need for this improvement. Problems such as rework and scrap could be converted to dollar figures to draw management's attention. It would also help to look at problems as cost savings opportunities.

2. *Identify specific projects for improvement.* Because of the limited availability of resources, not all problems can be addressed simultaneously. Therefore, problems should be prioritized. A Pareto analysis (discussed in Chapters 3 and 5) is often used to identify vital problems. Juran's quality improvement process works on a project-by-project basis. A problem area is identified as a project, and a concerted effort is made to eliminate the problem.

3. *Organize to guide the projects.* The organizational structure must be clearly established so projects can be run smoothly. As mentioned previously, authority and responsibility are assigned at all levels of management to facilitate this. Top management deals with strategic responsibilities, and lower management deals with the operational aspects of the actions. Furthermore, the structure should establish precise responsibilities for the following levels: guidance of overall improvement program, guidance for each individual project, and diagnosis and analysis for each project.

4. *Organize for diagnosis—for discovery of causes.* Juran defines a **diagnostic arm** as a person or group of persons brought together to determine the causes (*not* remedies) of the problem. The organization needs to enlist the right people and to ensure that the required tools and resources are available. This is accomplished through a **steering arm.** This investigation may require technical skills that managers may not possess. In such instances, use of professional specialists is appropriate.

5. *Find the causes.* This is often the most difficult step in the whole process. It involves data gathering and analysis to determine the cause of a problem. The symptoms surrounding the defects are studied, and the investigator then hypothesizes causes for the symptoms. Finally, an analysis is conducted to establish the validity of the hypotheses.

6. *Provide remedies.* The diagnostic step identifies the cause-and-effect relationship. Here, remedial actions are developed to alleviate the chronic problems. Remedies may deal with problems that are controllable by management or those that are controllable by operators. Changes in methods or equipment should be considered by management and may require substantial financial investment. Frequently, the return on investment is analyzed. Remedies here may involve a change in the standards. If tighter specifications have no real impact on the performance of the product, can the standards be relaxed? Errors controllable by the operators may be inadvertent, due to lack of skill or knowledge, or willful.

7. *Prove that the remedies are effective under operating conditions.* This is the real test of the effectiveness of the remedies proposed in the previous step. Can the suggested actions be implemented, and do they have the beneficial effect that has been hypothesized? The breakthrough process requires overcoming resistance to change. Changes may be technological or social in nature. The proposed procedure may require new equipment, and operators may have to be trained. Management commitment is vital to the effective implementation of the changes. By the same token, social changes, which deal with human habits, beliefs, and traditions, require patience, understanding, and the participation of everyone involved.

8. *Provide control mechanisms to hold the gains.* Once the remedial actions have been implemented and gains have been realized, there must be a control system to sustain this new level of achievement. In other words, if the proportion of nonconforming items has been reduced to 2%, we must make sure that the process does not revert to

the former nonconformance rate. A control mechanism is necessary; for example, audits may be performed in certain departments. Such control provides a basis for further process improvement as the whole cycle is repeated.

2-7 THE THREE PHILOSOPHIES COMPARED

We have now briefly examined the quality philosophies of three experts—Deming, Crosby, and Juran. All three philosophies have the same goal of developing an integrated total quality system with a continual drive for improvement. Although there are many similarities in these approaches, some differences do exist. A good discussion of these three philosophies may be found in the article by Lowe and Mazzeo (1986).*

Definition of Quality

Let's consider how each expert defines **quality.** Deming's definition deals with a predictable uniformity of the product. His emphasis on the use of statistical process control charts is reflected in this definition. Deming's concern about the quality of the product is reflected in the quality of the process, which is the focal point of his philosophy. Thus, his definition of quality does not emphasize the customer as much as do Crosby's and Juran's. Crosby defines quality as conformance to requirements. Here, requirements are based on customer needs. Crosby's performance standard of zero defects implies that the set requirements should be met every time. Juran's definition of quality—fitness of a product for its intended use—seems to incorporate the customer the most. His definition explicitly relates to meeting the needs of the customer.

Management Commitment

All three philosophies stress the importance of top management commitment. Deming's first and second points (creating a constancy of purpose toward improvement and adopting the new philosophy) define the tasks of management. In fact, his 14 points are all aimed at management, implying that management's undivided attention is necessary to create a total quality system. Point 1 in Crosby's 14-step process deals with management commitment. He stresses the importance of management communicating its understanding and commitment. Crosby's philosophy is focused on the creation of a "quality culture," which can be attained through management commitment. Juran's quality planning, control, and improvement process seeks management support at all levels. He believes in quality improvement on a project basis. The project approach gets managers involved and assigns responsibilities to each. Thus, in all three philosophies, the support of top management is crucial.

Strategic Approach to a Quality System

Deming's strategy for top management involves their pursuing the first 13 points and creating a structure to continually promote the 13 points in a neverending cycle of improvement (that is, Point 14). Crosby's approach to quality improvement is sequenced. His second step calls for the creation of quality improvement teams. Under Juran's philosophy, a quality council guides the quality improvement process. Furthermore, his quality breakthrough sequence involves the creation of problem-solving steering arms and diagnostic arms. The steering arm establishes the direction of the problem-solving effort and organizes priorities and resources. The diagnostic arm analyzes problems and tracks down their root causes.

---

*Ted A. Lowe and Joseph M. Mazzeo, "Three Preachers, One Religion," *Quality* (Sept. 1986): 22–25. Adapted with permission from *Quality,* a publication of Hitchcock Publishing, a Capital Cities/ABC Inc., company.

Measurement of Quality

All three philosophies view quality as a measurable entity, although in varying degrees. Often, top management has to be convinced of the effects of good quality in dollars and cents. Once they see it as a cost-reducing measure, offering the potential for a profit increase, it becomes easier to obtain their support. A fundamental aim of the quality strategy is to reduce and eliminate scrap and rework, which will reduce the cost of quality. A measurable framework for doing so is necessary. The total cost of quality may be divided into subcategories of prevention, appraisal, internal failure, and external failure. One of the difficulties faced in this setting is the determination of the cost of nonquality, such as customer nonsatisfaction. Notice that it is difficult to come up with dollar values for such concerns as customer dissatisfaction—which is one of Deming's concerns in deriving a dollar value for the total cost of quality. Crosby believes that quality is free; it is "unquality" that costs.

Neverending Process of Improvement

These philosophies share a belief in the neverending process of improvement. Deming's 14 steps repeat over and over again to continuously improve quality. Deming's PDCA cycle (plan-do-check-act) sustains this neverending process, as does Juran's breakthrough sequence. Crosby also recommends continuing the cycle of quality planning, control, and improvement.

Education and Training

Fundamental to quality improvement is the availability of an adequate supply of people who are educated in the philosophy and technical aspects of quality. Deming specifically referred to this in his sixth point, which talks about training all employees, and in his thirteenth point, which describes the need for retraining to keep pace with the changing needs of the customer. Deming's focus is on education in statistical techniques. Education is certainly one of Crosby's concerns as well; his eighth step deals with quality education. However, he emphasizes developing a quality culture within the organization so that the right climate exists. Juran's steps do not explicitly call for education and training. However, they may be implicit, because people must be knowledgeable to diagnose defects and determine remedies.

Eliminating the Causes of Problems

In Deming's approach, **special causes** refer to problems that arise because something unusual has occurred, and **common causes** refer to problems that are inherent to the system. Examples of special causes are problems due to poor quality from an unqualified vendor or use of an improper tool. With common causes, the system itself is the problem. Examples of common causes are inherent machine variability or worker capability. These problems are controllable only by management. Both Deming and Juran have claimed that about 85% of problems have common causes. Hence, only action on the part of management can eliminate them; that is, it is up to management to provide the necessary authority and tools to the workers so the common causes can be removed.

The heart of Deming's philosophy are the statistical techniques that identify special causes and common causes—especially statistical process control and control charts. Variations *outside* the control limits are attributed to special causes. These variations are worker-controllable, and the workers are responsible for eliminating these causes. On the other hand, variations *within* the control limits are viewed as the result of common causes. These variations require management action.

Juran's approach is similar to Deming's. In his view, special causes create **sporadic problems** and common causes create **chronic problems.** Juran provides detailed guidelines for identifying sporadic problems. For example, he categorizes operator error as being inadvertent, willful, or due to inadequate training or improper technique.

He also specifies how the performance standard of zero defects that Crosby promotes can be achieved. Crosby, of course, suggests a course of action for error cause removal in his eleventh step, whereby employees identify reasons for nonconformance.

Goal Setting

Deming was careful to point out that arbitrarily established numerical goals should be avoided. He asserted that such goals impede, rather than hasten, the implementation of a total quality system. Short-term goals based mainly on productivity levels without regard to quality are unacceptable. By emphasizing the neverending quality improvement process, Deming saw no need for short-term goals. On the other hand, both Crosby and Juran call for setting goals. Crosby's tenth point deals with goal setting; employees (with guidance from their supervisors) are asked to set measurable goals for even short-term periods such as 30, 60, or 90 days. Juran recommends an annual quality improvement program with specified goals. He believes that such goals help measure the success of the quality projects undertaken in a given year. The goals should be set according to the requirements of the customer. Juran's approach resembles the framework of management by objectives, where performance is measured by achievement of stipulated numerical goals.

Structural Plan

Deming's 14-point plan emphasizes using statistical tools at all levels. Essentially a bottom-up approach, the process is first brought into a state of statistical control (using control charts) and then improved. Eliminating special causes to bring the process under control takes place at the lower levels of the organizational structure. As these causes are removed and the process assumes a state of statistical control, further improvements require the attention of upper-level management.

Crosby, on the other hand, takes a top-down approach. He suggests changing the management culture as one of the first steps in his plan. Once the new culture is ingrained, a plan for managing the transition is created.

Finally, Juran emphasizes quality improvement through a project-by-project approach. His concept is most applicable to middle management.

Because each company has its own culture, companies should look at all three approaches and select the one (or combination) that is most suited to its own setting.

2-8 PROFILE OF A COMPANY—ARMSTRONG WORLD INDUSTRIES

Building Products Operations

Quality is ingrained at Armstrong World Industries' Building Products Operations (BPO), a winner of the 1995 Malcolm Baldrige National Quality Award in the manufacturing category. Upholding a commitment to quality made more than a century ago by its corporate founder, the manufacturer of acoustical ceilings and wall panels is aligning every facet of its business with the exhaustively researched product and service requirements of its customers.

All quality-focused changes initiated by the Pennsylvania-based company—from redesigning jobs and operations at manufacturing plants to reorganizing its sales force—are driven by thoroughly evaluated expectations of increases in customer value. In turn, BPO's Strategic Management Process completes the chain of cause and effect by determining how customer-focused goals impact market and financial performance and shareholder value. Across eight market categories in 1994, at least 97% of its customers gave BPO an overall rating of good or better. As it pursues increasingly ambitious levels of customer satisfaction, BPO is also reducing operating costs. Scrap rates, for example, have been cut by 38% since 1991. Manufacturing output per employee has jumped 39% over the same span, exceeding company goals.

About BPO

Based at Armstrong's corporate headquarters in Lancaster, Pennsylvania, BPO employs about 2,400 people, 85% of whom work at the operation's seven manufacturing plants in six states. The world's largest cork company, founded in the late 1800s, Armstrong now has six core businesses, which make and market hundreds of products for both home and commercial interiors and industry. The world's largest manufacturer of acoustical ceilings, BPO accounted for nearly one-fourth of Armstrong's sales in 1994. Major commercial market customers include wholesale distributors and large subcontractors, as well as architects and others who use Armstrong products in building projects. In the smaller, but rapidly growing residential market, BPO ceilings are marketed through wholesalers, lumberyards, home centers, and corporate retail accounts.

Strategic Management Process

In the early 1980s, with no signs of a business crisis on the horizon, BPO's executives chose not to leave well enough alone. First, they launched plans to increase profitability and market share. Next, they embarked on a quality initiative focused on improving internal processes and operations. But quality plans and business plans did not converge until 1990, when the highest-ranking person at each BPO organization was charged with leading quality improvement efforts. BPO's Strategic Management Process established customer satisfaction and value as both the targets and reference points of all improvement efforts.

Since this crucial integration, organizational change has become a nearly constant feature. The pace of change is quickening as managers and workers become more skilled practitioners of quality methods and set evermore ambitious business improvement goals. Within BPO, overall responsibility rests with its 10-member Quality Leadership Team (QLT) composed of senior executives and headed by the BPO President, who also serves on Armstrong's Corporate Business Excellence Team. The QLT emphasizes shared leadership and thus fully shares its responsibility for identifying and realizing improvement opportunities within the entire organization.

In each of the past 5 years, over half of the BPO workforce has participated in the more than 250 improvement teams that operate at any given time. The objectives of teams range from correcting specific operational problems at one plant to improving key business processes that enhance the entire organization. At each plant, the Quality Improvement Team, led by the facility's top manager, monitors the progress of all team efforts and reports on the results to the QLT. All Quality Improvement Teams are required to develop specific action plans and set goals that will have a measurable impact on one or more of BPO's five "key business drivers": customer satisfaction, sales growth, operating profit, asset management, and high-performance organization (human resources capabilities).

Change is purposeful, guided by information and evaluations that point the way to improvements that will make a major difference in customer value, employee value, and shareholder value. Over the past few years, BPO has made substantial investments to optimize its information-gathering analytical capabilities. It has also stepped up its benchmarking studies, conducting 89 in 1994, or more than twice the number performed during the previous year. The principal return on these efforts, according to the company, has been an everimproving understanding of the dynamics of BPO's markets, competitors' performance, and its own business results.

The QLT performs fact-based assessments on how well it stacks up against its competitors in each of BPO's eight market segments. Then, the team defines BPO's full potential in each segment. Drawing on this and other information, such as the results of customer surveys, the QLT sets goals and devises action plans so that BPO will grow to reach its full potential. Along with organizationwide goals, each functional unit develops and deploys action plans to every BPO employee. Relevant BPO goals and supporting process objectives are incorporated into the various incentive plans that now cover more than 93% of hourly and salaried workers.

Committed to promoting a high-performance work environment, BPO is immersed in work and job redesign efforts intended to provide workers with the skills, tools, information, flexibility, and authority needed to respond quickly and effectively to customer requirements. For example, pricing of ceiling jobs, once performed by headquarters, is now performed by sales people in the field, eliminating two approval steps and cutting turnaround time in half. The field sales force can also approve insurance claims of up to $5,000, accounting for about 90% of all claims.

At all seven manufacturing plants, employees are organized into natural work teams or business unit teams whose individual members can perform a variety of jobs. As of 1995, six plants pay workers for mastering new skills and knowledge. Six plants also offer gainsharing, which links measures of safety, customer satisfaction, process effectiveness, and other aspects of performance areas to a portion of each employee's compensation.

BPO has also made its suppliers, distributors, and carriers full partners in its quality improvement process. In 1985, the company established a supplier quality management process that has entailed assessing the quality systems of 135 suppliers. Overall, notices of nonconformance sent to suppliers have been declining, falling 32% from 1992 to 1994.

EXAMPLES OF BPO'S QUALITY IMPROVEMENT CHARACTERISTICS.

- In its manufacturing process, BPO uses a significant amount of waste materials from other industries, including scrapped newsprint and mineral wool, a by-product of steel production. In addition, most of the scrap generated during manufacturing is reused in the process.
- Each year for the past 5 years, BPO has had more than 300 improvement teams operating at any given time. "Best practices" generated by these teams are shared among plants through conference calls, computer networks, and through "Functional Excellence" conferences.
- Output per manufacturing employee and annual sales per manufacturing employee are critical measures of performance for BPO. As a result of employee involvement, recognition, and gainsharing and eliminating "nonvalue-added" activity, output per manufacturing employee has improved by 39% since 1991, and annual sales per manufacturing employee have risen by 40%.
- BPO continuously monitors its carriers' on-time delivery performance, a key customer satisfier, and provides them with monthly "report cards." Since 1992, on-time delivery has improved from 93% to 97.3%, while BPO carriers reduced their arrival time window from four hours to 30 minutes.
- Safety is integrated into BPO's improvement process. In 1994, BPO employees worked 3 million hours without a lost-time injury, a company and industry record.

2-9 SUMMARY

This chapter discusses the quality philosophies of Deming, Crosby, and Juran, with the emphasis on Deming's 14 points for management. Many companies are adopting the Deming approach to quality and productivity. The quality philosophies of Crosby and Juran provide the reader with a broad framework of the various approaches that exist for management. All three approaches have the same goal, with slightly different paths.

Whereas Deming's approach emphasizes the importance of using statistical techniques as a basis for quality control and improvement, Crosby's focuses on creating a new corporate culture that deals with the attitude of all employees toward quality. Juran advocates quality improvement through problem-solving techniques on a project-by-project basis. He emphasizes the need to correctly diagnose the root causes of a problem based on the observed symptoms. Once these causes have been identified, Juran focuses on finding remedies. Management, upon understanding all three philosophies, should select one or a combination of these approaches to best fit their own environment. These philosophies of quality have had a global impact.

CASE STUDY

Continuous Improvement at Texas Instruments*

Texas Instruments (TI) Defense Systems & Electronics (DS&E) has always considered quality an essential element of its business, and winning the Malcolm Baldrige National Quality Award in 1992 was the culmination of a quality journey that began in the early 1980s. During that time, DS&E had many elements of a total quality culture, but it needed a unifying framework. DS&E chose the Baldrige Award criteria to serve as that framework because it saw the criteria as a way to integrate initiatives and increase momentum toward total quality. Today, total quality is part of a comprehensive DS&E business strategy geared toward satisfying the customer. Providing customer satisfaction through total quality keeps DS&E accelerating toward its objective of business excellence.

BALDRIGE AWARD PROVIDED LESSONS FOR THE FUTURE

Since DS&E wanted to customize its strategy to fit the organization's goals, objectives, culture, and values, no single program or theory was a perfect fit. Early in its quality journey, DS&E used ideas from W. Edwards Deming, Joseph M. Juran, Philip Crosby, and others to define a quality philosophy, and it defined its strategic approach to total quality in three key areas: customer satisfaction, continuous improvement, and people involvement.

To assess the effectiveness of its defined strategy and its business, DS&E chose the Baldrige Award's criteria. The criteria represent world-class performance in all aspects of a business. They complemented the elements of DS&E's total quality strategy and were important to evolving the strategy to where it is today.

Winning the Baldrige Award was a milestone—a step along the way in attaining business excellence. As a result of lessons learned during the award process, DS&E leadership has deployed communications that make group-level priorities and initiatives clearer to everyone in the organization, which makes it easier for employees to align their work with the organization's goals. Evaluating business processes against the criteria has increased DS&E's focus on common goals and information sharing. The walls dividing functional organizations within DS&E have come down, and teamwork has improved. The information gained during assessments also helps DS&E refine key performance metrics.

By continuing to regularly assess its business against worldclass quality criteria and by sharing proven practices, DS&E increases its competitive advantage. Participating in the Baldrige Award process energized improvement efforts. That energy resulted from the team motivation that occurs when pursuing a common goal and the trend has continued.

DS&E has reduced the number of in-process defects to one-tenth of what they were at the time it won the Baldrige Award. Production processes that took four weeks several years ago now take only one week, and their cost is now 20% to 30% less in many manufacturing areas compared to several years ago. In January 1997, Texas Instruments announced the sale of Defense Systems & Electronics to Raytheon Company. DS&E's transition to Raytheon TI Systems is expected to be final by the end of second quarter 1997.

BUSINESS EXCELLENCE CONTINUES COMPANYWIDE

In 1994, TI, as a corporation, launched the Texas Instruments Business Excellence Standard (TI-BEST). TI-BEST is a four-step assessment and improvement process that grew out of DS&E's Baldrige Award experience. It was a natural evolution for TI to take what was learned through DS&E's experience and develop an internal process for assessment and improvement. Using the TI-BEST process, TI businesses worldwide assess themselves using world-class quality criteria appropriate for the region or country in which they do business. Some of the criteria used include the Baldrige Award, the European Foundation for Quality Management, and the Singapore National Quality Award. The four steps of TI-BEST are:

1. Define business excellence for the company
2. Assess progress
3. Identify improvement opportunities
4. Establish and deploy an action plan

All too often, companies focus only on assessment. TI has found, however, that defining the future state of the business (step 1), periodically assessing progress (step 2), performing a gap analysis and identifying improvement needs (step 3), and developing and deploying a plan to close business gaps

*Adapted from A.B. Rich (1997), "Continuous Improvement: The Key to Future Success," *Quality Progress* 30(6): 33–36.

(step 4) are all necessary for continuous improvement. TI-BEST provides a systematic approach to quality improvement efforts and benefits the organization by:

- Providing a framework that ties efforts together
- Providing a vehicle for identifying best practices
- Providing a structure for sharing knowledge and learning the methods and techniques others have used to make improvements
- Allowing employees to speak the same language regarding quality, which increases communication and organizational alignment toward common goals
- Fostering teamwork across the company
- Improving the ability to measure improvements by documenting processes and results
- Providing a process to accelerate improvement across the organization
- Involving every employee in continuous improvement toward world-class benchmarks

DS&E sees value in all aspects of an assessment and improvement process, including the written assessment that is reviewed by Baldrige Award-trained examiners and the site visit that is performed to evaluate the written assessment. Many of DS&E's groupwide improvement efforts—including enhanced customer surveys, a focus on education, and vigorous benchmarking—have come directly from examiner feedback.

CHANGING BUSINESS PRIORITIES CONTINUE TO BE A QUALITY FOCUS

Market trends directly affect the priority and deployment of DS&E business strategies. In the past, DS&E focused on leading-edge technology that would provide customers with a performance advantage. Today, defense customers are using contractors who can provide the desired performance with affordable, low-cost solutions. Two of DS&E's current initiatives are:

Drive for cost leadership. DS&E knows that cost is a major differentiator in winning new business. DS&E's drive for cost leadership (DCL) initiative, which is the motivating force behind many of its activities, is focused on lowering costs and has a continued emphasis on defect and cycle-time reduction. DCL is aimed at involving every DS&E employee in reducing cost and showing individuals how they can affect the bottom line.

Acquisition reform. In 1993, the U.S. Department of Defense began a major reform to streamline its process for acquiring materials and services for the armed forces. DS&E was the first defense contractor to sign a block change agreement with the Defense Department, which allowed for a common factory process for more than 700 contracts. At the same time, working with the Joint Group on Acquisition Pollution Prevention (JG-APP), DS&E signed a block change agreement to eliminate high volatile organic compounds from paint and primer process specifications and substitute DS&E process specifications for alternative coatings. DS&E was the first defense contractor to complete a prototype pollution prevention and single process initiative (SPI) project. Both block change agreements allow DS&E to offer these changes to its suppliers or to accept suppliers' alternate recommendations. DS&E's most recent block change, which was also accomplished jointly with JG-APP, provides DS&E with a performance-based paint system specification that replaces four different military specifications.

DS&E is committed to maintaining a leadership position in acquisition reform; it participates on joint government and industry councils and shares processes with other Defense Department contractors and subcontractors. In March 1997, it received an SPI award from the Defense Department in the supplier mentoring category for implementing acquisition reform initiatives.

MANAGING SUPPLIERS FOR COMPETITIVE ADVANTAGE

Since DS&E's competitive advantage is directly affected by how well its suppliers perform, it stresses continuous improvement in all aspects of supply process management. As part of its supply strategy, DS&E identifies and targets suppliers in key areas and works to build strong, long-term partnerships. DS&E's supply management strategy is focused in three primary areas: parts, suppliers, and improvement.

- **Parts strategies.** These strategies include a parts selection process and a recommended parts list that has a focus on parts standardization and reuse.
- **Supplier strategies.** These strategies include supplier alliance relationships, recommended manufacturers and distributors, quality process improvement, small disadvantaged and women-owned business development, and the early supplier integration (ESI) process.

  With the ESI process, suppliers are integrated early in the design concept phase of the program. Early supplier involvement maximizes parts standardization and increases the use of proven practices, which reduces DS&E's cost of ownership in the production phases of the program. When implemented consistently over time, ESI has reduced product costs by as much as 70% and reduced cycle time by as much as 50%.

- **Improvement strategies.** These strategies include a supplier optimization process, a lead-time reduction initiative, and a supplier productivity initiative. As part of the supplier productivity initiative, DS&E has a supplier training and development program to enhance suppliers' ability to meet product, quality, and delivery-time requirements. Courses in the training program include design to cost, continuous flow manufacturing, design for manufacturing and assembly, six sigma, and statistical process control.

IMPROVING PRODUCTION CYCLE TIME

In 1993, DS&E adapted a process called continuous flow manufacturing (CFM) that examines and improves production cycle time in fabrication and assembly areas. Production operations at DS&E and at their suppliers have applied CFM to achieve typical cycle-time reductions of 50% and as high as 88%. DS&E uses cycle-time metrics to identify processes that need improvement and to determine critical processes. CFM tools help production areas avoid suboptimization and improve overall performance by identifying and managing the constraint. Although the focus is on cycle time, DS&E achieves improvements across all operational performance metrics. Cycle-time and CFM specialists are strategic resources aimed at full deployment of these initiatives.

DEVELOPING PRODUCTS WITH CONCURRENT ENGINEERING

DS&E's approach to concurrent engineering is its integrated product development process (IPDP). The IPDP documents the process tasks for the entire product life cycle and identifies and describes best practices for a typical or generic program. The process emphasizes design reuse and parts standardization. Use of the IPDP ensures strong customer involvement, early supplier involvement, use of leading metrics, and ongoing evaluation and feedback. DS&E found, through a 1995 evaluation of IPDP use among its programs, that programs with better use of IPDP had a higher overall financial health.

REDUCING DEFECTS WITH SIX SIGMA

In 1991, as a result of feedback received in the Baldrige Award assessments, DS&E initiated a six sigma program based on Motorola's six sigma quality program. DS&E first applied six sigma to its manufacturing processes in an effort to reduce defects. Since 1991, DS&E has seen an improvement rate of approximately 30% each year. Currently, DS&E is focusing six sigma on improving its design process by applying statistical techniques so it can anticipate and avoid defects prior to production.

DS&E has developed a Six Sigma Technology Center that is responsible for the development and deployment of statistical tools and techniques for the organization. The Six Sigma Technology Center facilitates a network of more than 100 six sigma black belts. Black belts receive training in change management and application of statistics to ensure that programs throughout DS&E meet their six sigma goals.

MAKING IMPROVEMENTS THROUGH BENCHMARKING

DS&E uses benchmarking to measure its products, services, and processes against those of its toughest competitors and other best-in-class companies. In 1990, Baldrige Award examiners indicated that DS&E had a business gap in the area of benchmarking. DS&E took the feedback seriously and went to work developing a benchmarking process. In 1992, Baldrige Award examiners noted that benchmarking was a definite strength for the organization.

Since winning the Baldrige Award, DS&E has continued to build its benchmarking program and has had significant improvements. In 1994, DS&E established a benchmarking core team to deploy the process throughout the organization and to focus on benchmarking that advances DS&E toward its strategic objectives.

In 1996, a major DS&E goal was to improve the efficiency of its asset management process. At that time, DS&E asset management involved more than 150,000 tagged assets (excluding computers), and 50,000 to 60,000 of those assets required periodic maintenance. After implementing recommendations from an in-depth benchmarking study on asset management, DS&E saw a 30% reduction in the cost to support and maintain equipment and an 84% reduction in calibration and maintenance cycle time.

Another benchmarking success story is related to a 1996 study to develop a repair depot for high-tech electronics. The customer recognized the benefits of benchmarking, and its contract specified that DS&E perform a benchmarking study of other depots, both government and commercial. The customer approved DS&E's recommendations from the study, and the result was a sale for DS&E in excess of $500,000.

SPREADING THE GOOD NEWS

Best practice sharing focuses on collecting strengths within TI and disseminating those strengths to others in the company who have business gaps. TI's best practice sharing team, through the Office of Best Practices, provides techniques and tools to communicate best practices. The office supports a corporatewide, intranet-accessible, best practice knowledge base that gives employees access to proven best practices across TI. The latest best practice techniques, trends, and policies are communicated through a corporatewide facilitator network. Facilitators are trained to collect best practices and load them into the knowledge base and to use the knowledge base to find solutions. Because different organizations have significantly different maturity levels, TI defines a best practice as "a practice that is best for you." This allows several maturity levels of the same practice to be loaded into the knowledge base. By adapting and adopting best practices from both internal and external sources, TI avoids reinventing the wheel. The increased quantity and speed of TI's learning cycles frees resources that can be used to create innovative products and processes.

ACHIEVING WORLD-CLASS STATUS WITH A PROCESS FOCUS

One of the most significant lessons DS&E learned from its quality journey is that a process focus is essential to achieving the levels of performance consistently demonstrated by worldclass companies. DS&E's top-level business processes are designed to respond to customer needs, and its process focus has helped increase competitive advantage for its customers and for the organization. A process focus allows DS&E to provide customers with higher quality products and services with less expense. Strategic business processes help DS&E understand what work is of value from the customer's perspective and eliminate work that does not provide customer value. A process focus allows DS&E to find and eliminate root causes of waste, defects, delays, and barriers that drive up cost and affect DS&E's ability to respond to customer needs and market changes.

DEVELOPING A WINNING TEAM

DS&E's people strategy is an integral part of its overall business strategy. Therefore, the organization continually revises its people strategy to ensure alignment with key business objectives. The objectives of DS&E's people strategy is to develop a winning team that has the characteristics of innovation, risk taking, agility, efficiency, and disciplined collaboration. The strategy centers on six processes:

- Creating an attractive work environment
- Designing optimal work systems
- Optimizing human resource levels
- Sharing information
- Developing capability
- Aligning reward and recognition systems

Current initiatives related to DS&E's people strategy include developing a long-term people strategy and implementing a people involvement metric.

MEETING BUSINESS GOALS WITH ITS QUALITY STRATEGY

The objective of DS&E's quality strategy is to develop, document, and deploy a quality system that enables the organization to meet current and future business objectives. DS&E's quality strategy is centered on the following processes:

- Developing and defining the quality system
- Effectively partnering with customers and suppliers
- Providing total quality tools
- Assuring effectiveness of the measuring system
- Creating a quality center for excellence (CFE)
- Developing and implementing an environmental, safety, and health strategy

DS&E's quality strategy will continue to evolve and will always support the development of a proactive quality system. DS&E's quality strategy is aimed at achieving sustained competitive advantage, enhanced financial performance, and operational excellence.

As part of its quality strategy, DS&E has developed a quality CFE. The CFE's aim is to develop and grow skill-based competencies in employees, which will enable better execution of business processes. The center has created competency models, developed standard tools, and will continue to make improvements to those models and tools. The CFE also provides a professional development road map and career and technical mentoring. Other CFE services include developing training and training plans, facilitating effective staffing to the business, and providing leadership development and succession planning. The CFE is an important part of DS&E's continuing quality journey.

MEETING COMMON BUSINESS OBJECTIVES IS DS&E'S FOCUS

Prior to DS&E's serious use of the Baldrige Award criteria, its business units (missile systems, electronic systems, advanced products, and advanced technologies and components) seemed as though they were different companies. Today, DS&E's cohesive organizational structure, focused business strategy, and aligned initiatives are focused on meeting common business objectives. Providing customer satisfaction through total quality will continue to be DS&E's method for achieving business excellence.

DS&E's definition of business excellence will continue to evolve as market and customer requirements change. This inevitable change makes continuous improvement the key to DS&E's future success. By using a periodic assessment and improvement process and by focusing on the customer, DS&E will effectively manage that change and sustain continuous improvement.

QUESTIONS FOR DISCUSSION

1. Discuss the elements that comprise TI's strategic approach to quality.
2. Describe the assessment and improvement process developed by TI and known as TI-BEST. How may companies benefit from using this process?
3. How does the DS&E unit change its strategic plans based on market needs? Discuss some specific examples.
4. What measures are taken by DS&E to conform to environmental guidelines?
5. In the context of Deming's extended process, describe how TI incorporates its suppliers in the journey to quality improvement.
6. Discuss how concurrent engineering and benchmarking have helped TI in its quest for continuous improvement.
7. Describe the themes around which DS&E's quality strategy is based. How does it stay proactive?

Key Terms

- causes
  - common
  - special
- chronic problems
- consistency of purpose
- constancy of purpose
- corporate culture
- Crosby's 14-step plan for quality improvement
- Crosby's philosophy
- Deming cycle
- Deming's deadly diseases
- Deming's 14 points for management
- diagnostic arm
- extended process
- Juran's philosophy
- leadership
- loss function
- organizational barriers
- performance classification
- process capability
- product improvement cycle
- productivity
- quality
- quality breakthrough sequence
- quality culture
- quality improvement
- quality management maturity grid
- quality trilogy
- sporadic problems
- steering arm
- system of profound knowledge
- total quality system
- training
- vendor selection
- work standards
- zero defects

Exercises

1. What is the fundamental theme in Deming's philosophy for quality improvement?
2. Briefly state and describe Deming's 14 points for management.
3. Explain the notion of the extended process.
4. Explain the Deming cycle and its role in quality improvement.
5. What is the difference between constancy and consistency of purpose?
6. What are the reasons for mass inspection not being a viable alternative for quality improvement?
7. Describe some characteristics for selecting vendors. Consider a company of your choice. Describe their selection process for vendors, and make recommendations.
8. Explain why there is a loss associated with product meeting specifications but deviating from the target value.

9. Describe the role of managers and supervisors in Deming's approach.
10. Explain the organizational barriers that prevent a company from adopting the total quality philosophy.
11. According to Deming, explain the drawbacks of setting up numerical goals.
12. What are the drawbacks of some traditional performance appraisal systems, and how may they be modified?
13. What is the theme behind Crosby's quality management philosophy?
14. Describe and explain the "four absolutes" of quality management prescribed by Crosby.
15. Explain the role and function of the quality management maturity grid. Select a company, and locate their position on the grid.
16. State and explain Crosby's 14-step plan for quality improvement.
17. Explain the meaning of the term "quality culture." What is the importance of management's role in this context?
18. Describe Juran's quality trilogy program and its significance.
19. What is the difference between quality control and quality improvement? Discuss the role of management in each of these settings.
20. Discuss the importance of the diagnostic process from symptom to cause in Juran's approach.
21. Compare and contrast the philosophies of Deming, Crosby, and Juran.
22. Select a company. Analyze and develop a quality philosophy for this company. How does your recommendation differ from the existing approach?
23. Discuss the five deadly diseases in the context of Deming's philosophy of management.
24. Of Deming's 14 points for management, discuss which ones address the deadly disease of short-term management focus.
25. Discuss the remedial measures that management may pursue to overcome the drawbacks of a numerical performance appraisal system.
26. What actions would you suggest to management to create a long-term constancy of purpose? What are the possible consequences of such actions in the short term?
27. Removing organizational barriers and creating a system that supports free flow of information are imperatives for management. Describe how this can be accomplished. What are some specific action plans?
28. What are the drawbacks of the bottom-line management approach?
29. Discuss the dilemma management faces when they sacrifice short-term profits for long-run stability. What is the recommended approach?
30. Discuss specific actions that senior management may take to promote longer employee tenure.
31. What are some organizational cultural issues that management must address as they strive to create long-run stability and growth?
32. How does the extended process concept aid in meeting the goal of continuous quality improvement?

References

Crosby, Philip B. (1979). *Quality Is Free.* New York: McGraw-Hill.

____ (1984). *Quality Without Team—The Act of Hassle-Free Management.* New York: McGraw-Hill.

____ (1989). *Let's Talk Quality.* New York: McGraw-Hill.

Deming, W. Edwards (1982). *Quality, Productivity, and Competitive Position.* Cambridge, Mass.: Center for Advanced Engineering Study, Massachusetts Institute of Technology.

____ (1986). *Out of the Crisis.* Cambridge, Mass.: Center for Advanced Engineering Study, Massachusetts Institute of Technology.

Fellers, G. (1992). *The Deming Vision: SPC/TQM for Administrators.* Milwaukee, Wis.: ASQC Quality Press.

Gitlow, Howard S., and Shelly J. Gitlow (1987). *The Deming Guide to Quality and Competitive Position.* Englewood Cliffs, N.J.: Prentice-Hall.

Juran, J. M. (1988a). *Juran on Planning for Quality.* New York: The Free Press.

____ (1988b). *Juran's Quality Control Handbook.* New York: McGraw-Hill.

____ (1989). *Juran on Leadership for Quality—An Executive Handbook.* New York: The Free Press.

____ (Aug. 1986). "The Quality Trilogy," *Quality Progress*: 19–24.

Juran, J. M., and Frank M. Gryna, Jr. (1980). *Quality Planning and Analysis—From Product Development through Use.* New York: McGraw-Hill.

Lowe, Ted A., and Joseph M. Mazzeo. "Three Preachers, One Religion," *Quality* (Sept. 1986): 22–25.

Rich, A.B. (1997). "Continuous Improvement: The Key to Future Sucess," *Quality Progress,* 30(6): 33–36.

Scherkenbach, William W. (1986). *The Deming Route to Quality and Productivity.* Washington, D.C.: Ceep Press.

Walton, M. (1988). *The Deming Management Method.* New York: G. P. Putnam.

____ (1990). *Deming Management at Work.* New York: G. P. Putnam.

CHAPTER

3

Quality Management: Practices, Tools, and Standards

Chapter Outline

3-1 INTRODUCTION

The road to a quality organization is paved with the commitment of management. If management is not totally behind this effort, the road will be filled with potholes, and the effort will drag to a halt. A keen sense of involvement is a prerequisite for this journey because, like any journey of import, the company will sometimes find itself in uncharted territory. Company policies must be carefully formulated according to principles of a quality program. Major shifts in paradigms may occur. Resources must, of course, be allocated to accomplish the objectives but this by itself is not sufficient. Personal support and motivation are the key ingredients to reaching the final destination.

In this chapter we look at some of the quality management practices that enable a company to achieve its goals. These practices start at the top, where top management creates the road map, and they continue with middle and line management, who help employees follow the map. With an ever-watchful eye on the satisfaction of the customer, the entire workforce embarks on an intensive study of product design and process design. Company policies on vendor selection are discussed. *Everything* is examined through the lens of quality improvement. Key to this companywide investigation of itself is a variety of statistical tools, which we will discuss here and in later chapters.

We will look at the standards set out by the International Standards Organization (ISO); in particular ISO 9000–9004 standards are presented. Organizations seek to be certified by these standards to demonstrate the existence of a quality management process in their company. We will look at some prime examples of companies that foster quality and the award that gives them recognition for their efforts—the highly prestigious Malcolm Baldrige National Quality Award. Companies who win this award become the benchmarks in their industries.

3-2 MANAGEMENT COMMITMENT

A company's strategic plan is usually developed by top management; they are, after all, responsible for the long-range direction of the company. A good strategic plan addresses the needs of the company's constituencies. First and foremost, of course, is the customer, which can be internal and/or external. The customer wants a quality product or service at the lowest possible cost. Meeting the needs of the shareholders is another objective. Shareholders want to maximize their return on investment. Top management has the difficult task of balancing these needs and creating the long-term plan that will accomplish them.

What management needs are specific practices that enable them to install the quality program. That is what this chapter is about, but first we need some terminology. In this context, the term **total quality management (TQM)** refers to a comprehensive approach to improving quality. According to the U.S. Department of Defense, TQM is both a philosophy and a set of guiding principles that comprise the foundation of a continuously improving organization. Other frequently used terms are synonymous to TQM; among them are *continuous quality improvement* (CQI), *quality management* (QM), *total quality control* (TQC), and *companywide quality assurance* (CQA). For example, Ford Motor Company has a **total quality excellence (TQE)** program. This system integrates all aspects of the company and its suppliers in the strategic decisions on marketing, design engineering, manufacturing engineering, production, purchasing, distribution, and service. In this program, quality is measured by satisfying and exceeding customer needs and expectations in product and service quality. Table 3-1 lists the major criteria used in the evaluation process of Ford's TQE program.

This TQE is based on Ford Motor Company's (1989) vision, which is "to be a low-cost producer of the highest quality products and services which provide the best customer value." This vision continues to be used currently. Several core concepts are embraced in this vision. First, as stressed throughout this text, quality is defined by the customer. Second, quality is achieved through prevention of problems, not detection of them. This supports the notions that quality is designed into the product and that doing

TABLE 3-1 Major Criteria for Ford Motor Company's Total Quality Excellence (TQE) Program

Functional Area	*Total Points Possible*
ENGINEERING	
Product Design	
Design engineering	3
Sales engineers	1
System design	3
Component design	4
FMEA	3
Manufacturing feasibility	2
Computer-aided design	2
FEA or other modeling	2
Subtotal	20
Product Development	
Development engineering	4
Product development	3
Material development	3
R&D laboratory	2
Prototype build	5
Prototype tooling build	2
Other	1
Subtotal	20
Product Test	
Test equipment	3
Durability testing	4
Engineering specification	3
Product	3
Material	2
Subtotal	15
Manufacturing Engineering	
Manufacturing engineering	3
Process engineering	2
Industrial engineering	1
Process FMEA	2
Manufacturing feasibility	3
Process sheets	2
Production sign-off	2
Subtotal	15
Processing	
All components manufactured starting with base raw material	3
Material and/or component suppliers provide statistical control data	3
All process contained at surveyed locations	3
Process control points clearly defined	2
Statistical control methods used within the process	4
Subtotal	15
Manufacturing Support	
(Productive Maintenance Program)	
Control system scheduled maintenance	1
Spare parts control	1
Tool/machine rehab	1
Tool/die/mold/gage design	3
Tool room	3
Layout inspection	2
Machine shop	1
Machine design	1
Tool/die/mold build	2
Subtotal	15
Total rating points	100

a job right the first time is the right answer. Third, Ford Motor Company uses the vendor-vendee concept to emphasize that each employee is a customer for work done by other employees or vendors. Fourth, similar to Deming's extended process, *all* employees, suppliers, and dealers are part of the process that creates the product or service for a customer. Fifth, Ford supports Deming's idea that quality improvement is never-ending. Finally, the individual is the driving force behind continuous improvement.

The main functional areas in the TQE program are engineering, delivery, and commercial performance. Engineering includes such components as product design, product development, product testing, manufacturing engineering, processing, and manufacturing support. The delivery function spans areas such as utilization of Ford's supplier communication system, shipment performance in relation to order quantity and due date requirements, overall supplier performance as measured by actions such as just-in-time (JIT) operations and suggestions of quality improvement activities,

TABLE 3-1 *(cont.)*

Description	*Total Points Possible*
DELIVERY	
Utilization of the Ford Motor Company supplier communication system	25
Shipment performance in relation to 3086 release quantities and ship date requirements	25
Overall supplier performance (suggests improvements, supports Ford Motor Co. programs such as JIT, etc.)	17
Timely supplier response to problems	13
Frequency of overshipments	10
Timely reconciliation of cumulative shipments	5
Frequency and dollar volume of premium freight charges	5
Total rating points	100
COMMERCIAL PERFORMANCE	
I. Worldwide cost competitiveness	
1. First year cost (annualized piece price plus tool cost)	15
2. Ongoing productivity commitment	15
3. Engineering change costs	5
4. Cost savings contributions	5
Subtotal	40
II. Capability	
1. Management depth	8
2. Financial resources	7
3. Manufacturing flexibility	5
4. Manufacturing technology	5
Subtotal	25
III. Responsiveness to business issues	
1. Quotations (timeliness and thoroughness, including feasibility assessments)	8
2. New part launch performance (meets sample promise dates, supports timing and quantities)	13
3. Adaptability to changing environment (long-term contracts, JIT, SPC, CAD/CAM, etc.)	6
4. Responsiveness to buyer in problem-solving situations (quality, delivery, strike protection, service to Ford Motor Co. plant, etc.) and on day-to-day business issues	8
Subtotal	35
Total rating points	100

Source: Adapted from the "Total Quality Excellence Program" by Ford Motor Company, 1989, p. 8.

timely response to supplier problems, and frequency of overshipments. Commercial performance includes worldwide cost competitiveness, as measured by initial cost and ongoing productivity commitment, capability as a function of technical and financial resources, and responsiveness to business issues, which includes the ability to adapt to a changing environment and be responsive to the needs of the buyer. Note that extreme commitment by senior management is a precondition to the achievement of the TQE program. The TQE program embraces the systems approach. An example at Ford is the Design Approval Process, where input is obtained from various cross-functional areas. These include the design staff, controller's office, sales and marketing, timing and control, advance product planning, vehicle engineering office, body and assembly purchasing, body and assembly manufacturing, powertrain and chassis operations, and product planning.

A partnership between the manufacturer and supplier is a driving force in maintaining excellence in quality. To address this issue, Ford Motor Company developed its **Q101 Preferred Quality Award,** which is awarded to its vendors who have shown a high degree of excellence and commitment to continuous improvement. Ford elected to take a radical approach here: Vendors appraise themselves to determine their level of compliance. If deemed satisfactory, a vendor can request an on-site evaluation. For its first-tier suppliers, Ford has established a set of standards referred to as Q-101. The different areas addressed by Q-101 are shown in Table 3-2.

Management plays a major role in achieving Q-101: The planning activities *all* fall under the domain of management. Depending on the time frame in which planning takes place, both top and middle management will be involved. Process control is emphasized, as is customer feedback on unsatisfactory performance or on returned products, which is essential to improve the process. Maintenance of the quality system

TABLE 3-2 Ford Motor Company's Q-101 Quality System Standards

1.	*Introduction*
1.1	Ford Motor Company's Quality Expectations
1.2	Quality System Evidence Requirements
1.3	How to Use This Document
2.	*Planning for Quality*
2.1	Process Flow Chart
2.2	Feasibility
2.3	Failure Mode and Effects Analysis (FMEA)
2.4	Control Plans
2.5	Gage Planning
2.6	Preliminary Process Capability
2.7	Process Monitoring and Control Instructions
2.8	Packaging Planning
2.9	Initial Sample Approval Requirements
2.10	Prototype Part Quality Initiatives
2.11	Incoming Raw Material and Parts Control
2.12	Planning for Ongoing Quality
3.	*Achieving Process and Product Quality*
3.1	Ongoing Process Capability
3.2	Statistical Process Control (SPC)
3.3	Product Qualification and Monitoring
3.4	Measuring and Test Equipment
3.5	Engineering Specification (ES) Test Performance Requirements
3.6	Indication of Product Status
3.7	Verification of New Set-ups
3.8	Reference Samples
3.9	Reworked Products
3.10	Returned Product Analysis
3.11	Problem Solving Methods
3.12	Scheduled Preventive Maintenance
3.13	Heat-Treated Parts
3.14	Lot Traceability
3.15	Continuous Improvement
4.	*Documenting Quality*
4.1	Procedures
4.2	Records
4.3	Drawings and Change Control
4.4	Part/Process Modification Control
4.5	Changes in Manufacturing Processes
5.	*Special Requirements for Control Item Products*
5.1	Quality Planning
5.2	Documentation
6.	*System for Initial Sample Approvals*
6.1	Requirements for New Products
6.2	Plant Quality Surveys
6.3	Review and Approval of Initial Samples

Source: Adapted from the "Ford Worldwide Quality System Q-101," by Ford Motor Company, 1990.

requires careful documentation, and a management-developed documentation system exists for design and change control. Documentation provides consistency and standardization in decision making.

Total Quality Management

Total quality management revolves around three main themes: the customer, the process, the people. Figure 3-1 shows some basic features of a TQM model. At its core are the company vision and mission and management commitment. They bind the customer, the process, and the people into an integrated whole. A company's vision is quite simply what the company wants to be. The mission lays out the company's strategic focus. Every employee should understand the company's vision and mission so that individual efforts will contribute to the organizational mission. When employees do not understand the strategic focus, individuals and even departments pursue their own goals, rather than those of the company, and the company's goals are inadvertently sabotaged. The classic example is maximizing production with no regard to quality or cost.

Management commitment is another core value in the TQM model. It must exist at all levels for the company to succeed in implementing TQM. Top management

FIGURE 3-1 Features of a TQM model.

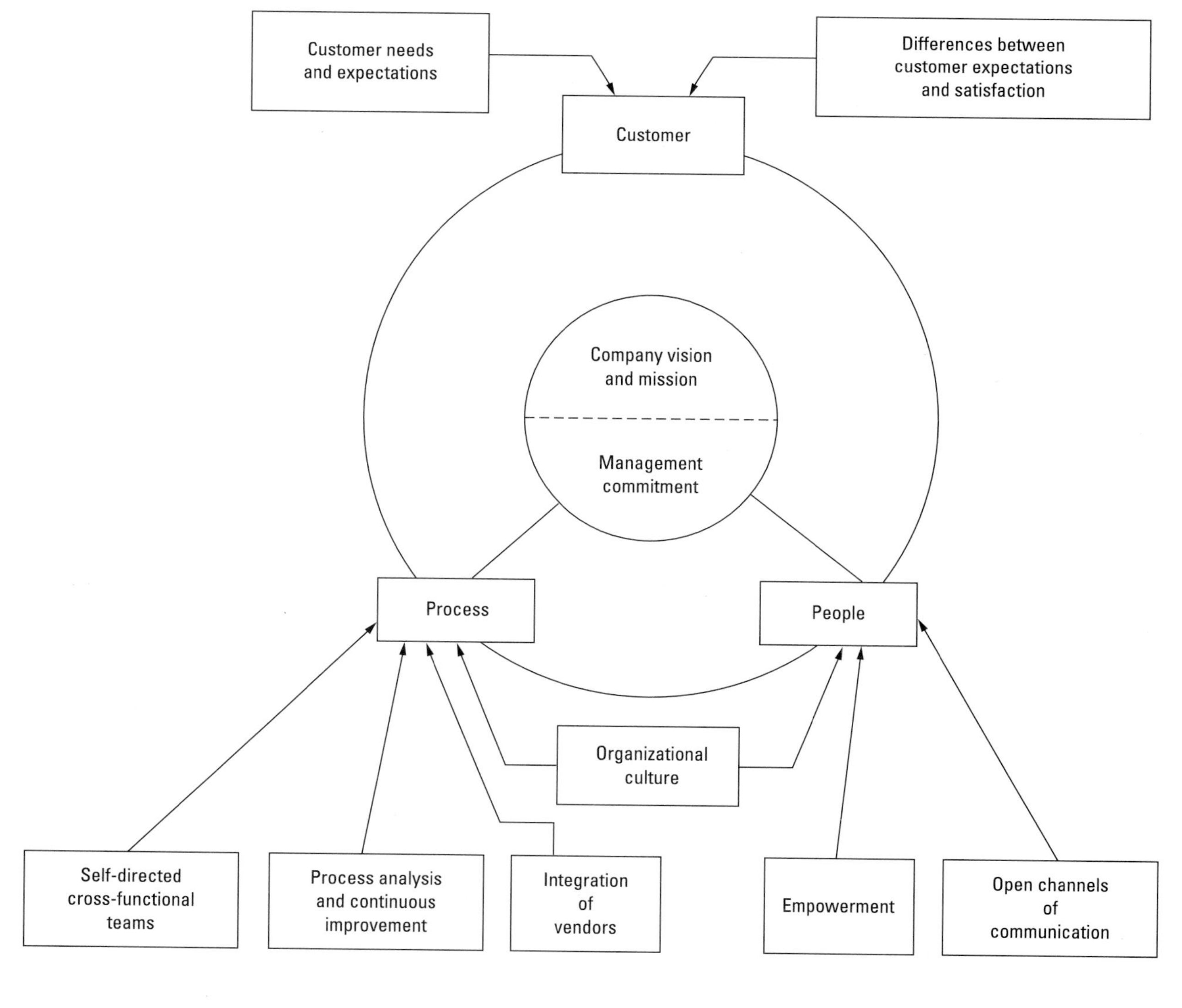

envisions the strategy and creates policy. Middle management works on implementation. At the operational level, appropriate quality management tools and techniques are used.

Satisfying customer needs and expectations is a major theme in TQM—in fact, it is the driving force. Without satisfied customers, market share will not grow, and revenue will not increase. Management should not second-guess the customer. For example, commercial builders should construct general merchandise stores only after they have determined there is enough consumer interest to support them. If consumers prefer specialty stores, then specialty stores should be constructed. Direct feedback using a data-driven approach is the best way to identify customer expectations and needs. A company's strategic plan must conform to these needs.

A key principle in quality programs is that customers are both internal and external. The receiving department of a processed component is a customer of that processing unit. Feedback from such internal customers identifies problem areas before the product reaches its finished stage, thus reducing the cost of scrap and rework.

Customer expectations can, to some extent, be managed by the organization. Factors such as the quality of products and services and warranty policies offered by the competitor directly influence customer expectations. The company through truthful advertising can shape the public's expectations. For example, if the average life of a lawnmower under specified operating conditions is 15 years, there is no reason to exaggerate it. In service operations, customers know which companies are responsive and friendly. This doesn't need advertising. Customer surveys can help management determine discrepancies between expectations and satisfaction. Taking measures to eliminate discrepancies is known as **gap analysis.**

The second theme in TQM is the process. Management is responsible for analyzing the process to continuously improve it. In this framework, vendors are part of the extended process, as advocated by Deming. As discussed in previous chapters, integrating vendors into the process improves the vendors' products, which leads to better final products. Because problems can and do span functional areas, self-directed cross-functional teams are important for generating alternative feasible solutions—the process improves again. Technical tools and techniques along with management tools come in handy in the quest for quality improvement. Self-directed teams are given the authority to make decisions and make appropriate changes in the process.

The third theme deals with people. Human "capital" is an organization's most important asset. **Empowerment**—involving employees in the decision-making process so that they take ownership of their work and of the process—is key in TQM. It is individuals who find better ways to do a job, and this is no small source of pride. With pride comes motivation. There is a sense of pride in making things better through the elimination of redundant or nonvalue-added tasks or combining operations. In TQM, managing is empowering.

Barriers restrict the flow of information. Thus, open channels of communication are imperative, and management had better maintain these. For instance, if marketing fails to talk to product design, a key input on customer needs will not be incorporated into the product. Management must work with its human resources staff to empower people to break down interdepartmental barriers. From the traditional role of management of coordinating and controlling has developed the paradigm of coaching and caring. Once people understand that they, and only they, can improve the state of affairs and once they are given the authority to make appropriate changes, they will do the job that needs to be done. There originates an intrinsic urge from within to do things better. Such an urge has a force that supersedes external forms of motivation.

Linking the human element and the company's vision is the fabric we call organizational culture. Culture is the beliefs, values, norms, and rules that prevail within an organization. How is business conducted? How does management behave? How are employees treated? What gets rewarded? How does the reward system work? How is input sought? How important are ethics? What is the social responsibility of the company? The answers to these and many other questions define an organization's culture. One culture may embrace a participative style of management that empowers its employees and delights its customers with innovative and timely products. Another culture may choose short-term profit over responsibility to the community at large. Consider, for example, the social responsibility adopted by General Electric (GE) Company. The company and its employees made enormous contributions to support education, the arts, the environment, and human services organizations worldwide.

Vision and Quality Policy

The company's **vision** is about its values and beliefs. Their vision is what they want it to be, and it is a message that every employee should not only hear, but should also believe in. Visions, carefully articulated, give a coherent sense of purpose. Visions are about the future. Effective ones are simple and inspirational. Finally, it must be motivational so as to evoke a bond that creates unison in the efforts of individuals working toward a common organizational goal. From the vision emanates a **mission statement** for the organization that is more specific and goal oriented.

Visions, of course, are about quality. For example, the vision of the College of Business at Auburn University is "to promote quality, professional business education for the people of the State of Alabama and the Region." Its mission is "to be a respected business school with a national reputation that promotes cross-disciplinary education through excellence in research, instruction, and outreach. The College, by transmitting technical and professional business knowledge to its constituents, prepares them for rapid placement and advancement in the marketplace." Note that the vision is general and outlines the scope and purpose of the organization, and the mission statement is focused and defines the areas of concentration—namely, research, instruction, and outreach.

Another service organization, IBM Direct, is dedicated to serving U.S. customers who order such IBM products as ES/9000 mainframes, RS/6000 and AS/400 systems, connectivity networks, and desktop software. Their vision for customer service is ". . . create an environment for customers where conducting business with IBM Direct is considered an enjoyable, pleasurable and satisfying experience." This is *what* IBM Direct wants to be. Their mission is ". . . to act as the focal point for post-sale customer issues for IBM Direct customers. We must address customer complaints to obtain timely and complete resolutions. And, through root cause analysis, we must ensure that our processes are optimized to improve our customer satisfaction." Here again, the mission statement gets specific. This is *how* they will get to their vision. Note that no mention is made of a time frame. This issue is usually dealt with in goals and objectives.

Framed by senior management, a **quality policy** is the company's road map. It indicates what is to be done, and it differs from procedures and instructions, which address how it is to be done, where and when it is to be done, and who is to do it. A beacon in TQM leadership, Xerox Corporation is the first major U.S. corporation to regain market share after losing it to Japanese competitors. Xerox attributes its remarkable turnaround to its conversion to TQM philosophy. The company's decision to rededicate itself to quality through a strategy called *Leadership Through Quality* has paid off. Through this process, Xerox created a participatory style of management that focuses on quality improvement while reducing costs. It encouraged teamwork, sought more customer feedback, focused on product development to target key markets, encouraged greater employee involvement, and began competitive benchmarking. Greater customer satisfaction and enhanced business performance are the driving forces in their quality program, the commitment to which is set out in the Xerox Quality Policy—"quality is the basic business principle at Xerox."

Another practitioner of TQM, Eastman Chemical Company, manufactures and markets over 400 chemicals, fibers, and plastics for over 7000 customers around the world. A strong focus on customers is reflected in its vision, "to be the world's preferred chemical company." A similar message is conveyed in its quality goal, "to be the leader in quality and value of products and services." Its vision, values, and goals define Eastman's quality culture. The company's quality management process is set out in four directives: "focus on customers; establish vision, mission, and indicators of performance; understand, stabilize, and maintain processes; and plan, do, check, act for continual improvement and innovation."

Eastman Chemical Company encourages innovation and provides a structured approach to generating new ideas for products. Cross-functional teams help the company understand the needs of both its internal and external customers. The teams define and improve processes, and they help build long-term relationships with vendors and customers. Through the Eastman Innovative Process, a team of employees from various areas—design, sales, research, engineering, and manufacturing—guides an idea from inception to market. People have ownership of the product and of the process. Customer needs and expectations are addressed through the process and are carefully validated. One outcome of the TQM program has been the drastic reduction—almost 50%—of the time required to launch a new product. Through a program called Quality First, employees team with key vendors to improve the quality and value of purchased materials, equipment, and services. Over 70% of Eastman's worldwide customers have ranked the company as their best supplier. Additionally, Eastman has received an outstanding rating on five factors that customers view as most important: product quality, product uniformity, supplier integrity, correct delivery, and reliability. Extensive customer surveys led the company to institute a no-fault return policy on its plastic products. This policy, believed to be the only one of its kind in the chemical industry, allows customers to return any product for any reason for a full refund.

Performance Standards

One intended outcome of a quality policy is a desirable level of performance—that is, a defect-free product that meets or exceeds customer needs. Even though current performance may satisfy customers, organizations cannot afford to be complacent. Continuous improvement is the only way to stay abreast of changing needs of the customer. The tourism industry, for instance, has seen dramatic changes in recent years; options have increased and customer expectations have risen. Top-notch facilities and a room filled with amenities are now the norm and don't necessarily impress the customer. Meeting and exceeding consumer expectations is no small challenge. Hyatt Hotels Corporation has met this challenge head-on. Their "In Touch 100" quality assurance initiative provides the framework for their quality philosophy and culture. Quality at Hyatt means consistently delivering products and services 100% of the time. The In Touch 100 program sets high standards—standards derived from guest and employee feedback—and specifies the pace that will achieve these standards every day. The core components of their quality assurance initiative are standards, technology, training, measurement, recognition, communication, and continuous improvement (Buzanis 1993).

Six-Sigma Quality

While a company may be striving toward an ultimate goal of zero defects, numerical standards for performance measurement should be avoided. Setting numerical values that may or may not be achievable can have an unintended negative emotional impact. Not meeting the standard, even though the company is making significant progress, can be demoralizing for everyone. Numerical goals also shift the emphasis to the short term, as long-term benefits are sacrificed for short-term gains.

So, the question is, how *do* we measure performance? By making continuous improvement the goal and then measuring the *trend* (not the numbers) in improvement. This is also motivational. Another effective method is benchmarking; this involves identifying high-performance companies or intracompany departments and using their performance as the improvement goal. The idea is that, although the goals may be difficult to achieve, others have shown it can be done.

Quantitative goals do have their place, however, as Motorola, Inc. has shown with its concept of **six-sigma quality.** Sigma (σ) stands for the standard deviation, which is a measure of variation in the process. Assuming that the process output is represented by a normal distribution, about 99.73% of the output is contained within bounds that are 3 standard deviations (3σ) from the mean. As shown in Figure 3-2, these are represented as the lower and upper tolerance limits (LTL and UTL). The normal distribution is characterized by two parameters: the mean and the standard deviation. The mean is a measure of the location of the process. Now, if the product specification limits are 3 standard deviations from the mean, the proportion of nonconforming product is about 0.27%, which is approximately 2700 parts per million (ppm); that is, the two tails, each 1350 ppm, add to 2700 ppm. On the surface, this appears to be a good process, but appearances can be deceiving. When we realize that most products and services consist of numerous processes or operations, reality begins to dawn. Even though a single operation may yield 97.73% good parts, the compounding effect of out-of-tolerance parts will have a marked influence on the quality level of the finished product. For instance, for a product that contains 1000 parts or has 1000 operations, an average of 2.7 defects per product unit is expected. The probability that a product contains no defective parts is only 6.72%! This means that only about 7 units in 100 will go through the entire manufacturing process without a defect—not a desirable situation.

For a product to be built virtually defect-free, it must be designed to tolerance limits that are significantly *more* than $\pm 3\sigma$ from the mean. In other words, the process spread as measured by $\pm 3\sigma$ has to be significantly less than the spread between the upper and lower specification limits (USL and LSL). Motorola's answer to this problem is six-sigma quality; that is, process variability must be so small that the specification limits are 6 standard deviations from the mean. Figure 3-3 demonstrates this concept. If the process distribution is stable—that is, it remains centered between the specification limits—the proportion of nonconforming product should be only about 0.001 ppm on each tail.

In real-world situations, the process distribution will not always be centered between the specification limits; process shifts to the right or left are not uncommon. It can be shown that even if the process mean shifts by as much as 1.5 standard deviations from the center, the proportion nonconforming will be about 3.4 ppm. Comparing this to a three-sigma capability of 2700 ppm demonstrates the improvement in the expected level of quality from the process. If we consider the previous example for a product containing 1000 parts and we design it for six-sigma capability, then an average of

FIGURE 3-2 Process output represented by a normal distribution.

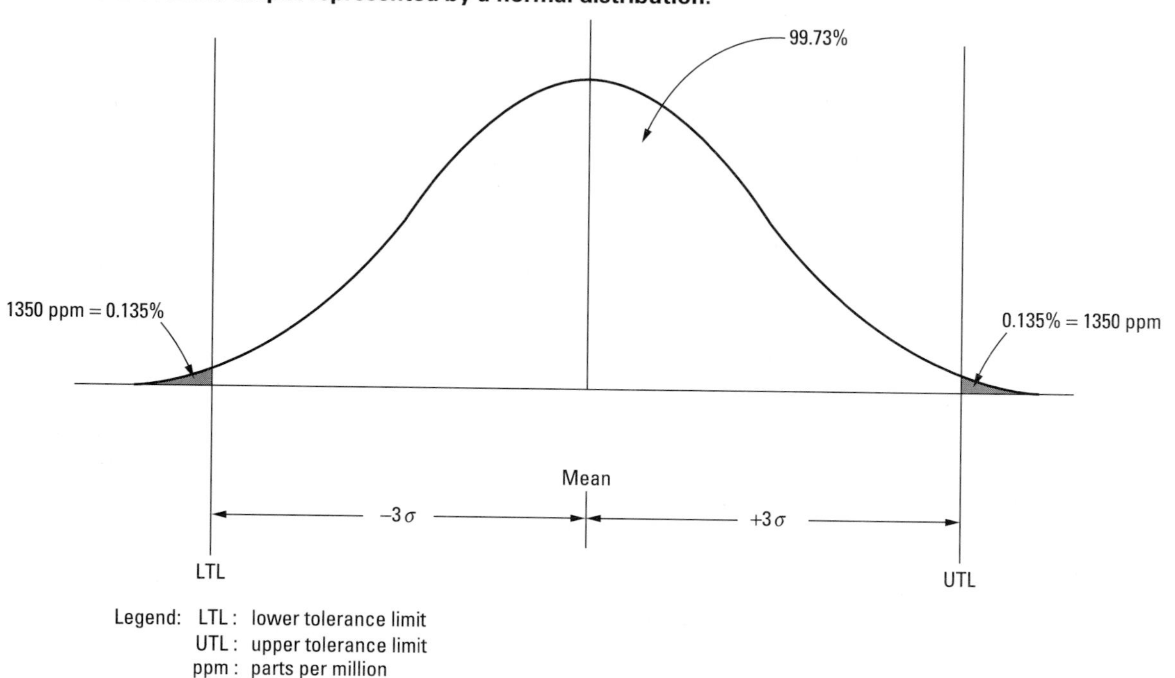

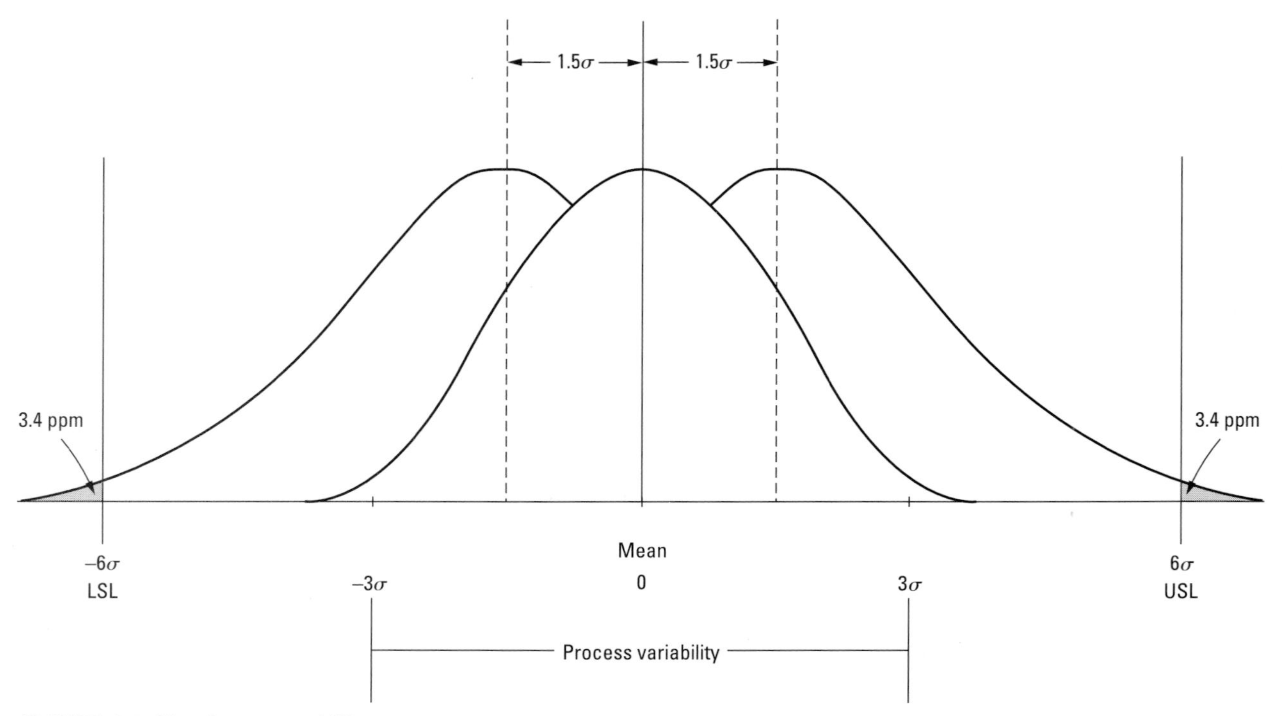

FIGURE 3-3 Six-sigma capability.

0.0034 defect per product unit (3.4 ppm) is expected, instead of the 2.7 defects expected with three-sigma capability. The cumulative yield from the process will thus be about 99.66%—a vast improvement over the 6.72% yield in the three-sigma case.

Establishing a goal of three-sigma capability is acceptable as a starting point, however, because it allows the organization to set a baseline for improvement. As management becomes more process oriented, higher goals such as "six-sigma capability" become possible. Such goals may require fundamental changes in management philosophy and the organizational culture.

3-3 QUALITY FUNCTION DEPLOYMENT

Quality function deployment (QFD) is a planning tool that focuses on designing quality into a product or service by incorporating customer needs. It is a systems approach that involves cross-functional teams (whose members are not necessarily from product design) that looks at the complete cycle of product development. This quality cycle starts with creating a design that meets customer needs and continues on through conducting detailed product analyses of parts and components to achieve the desired product, identifying the processes necessary to make the product, developing product requirements, prototype testing, final product or service testing, and finishing with after-sales troubleshooting.

QFD is customer driven and translates customer needs into appropriate technical requirements in products and services. It is proactive in nature. Also identified by other names—the house of quality, matrix product planning, customer-driven engineering, and decision matrix—it has several advantages. It evaluates competitors from two perspectives—the customer's perspective and a technical perspective. The customer's view of competitors provides the company with valuable information on the market potential of its products. The technical perspective, which is a form of benchmarking, provides information on the relative performance of the company with respect to industry leaders. This analysis identifies the degree of improvements needed in products and processes and serves as a guide for resource allocation.

QFD reduces the product development cycle time in each functional area—from product inception and definition to production and sales. By considering product and part design along with manufacturing feasibility and resource restrictions, QFD cuts down on time that would otherwise be spent on product redesign. Midstream design changes are minimized, along with concerns on process capability and postintroduction problems of the product. This results in significant benefits for products with long lead times such as automobiles. Thus, QFD has been vital for Ford Motor Company and General Motors in their implementation of total quality management.

Companies use QFD to create training programs, select new employees, establish supplier development criteria, and improve service. Cross-functional teams have also used QFD to show the linkages between departments and thereby have broken down existing barriers of communication. While the advantages of QFD are obvious, its success requires a significant commitment of time and human resources because a large amount of information is necessary for its start-up.

QFD Process

Figure 3-4 shows a QFD matrix, also referred to as the **house of quality.** The objective statement delineates the scope of the QFD project, thereby focusing the team effort. For a space shuttle project, for example, the objective could be to identify critical safety features. Only one task is specified in the objective. Multiple objectives are split into separate QFDs in order to keep a well-defined focus.

The next step is to determine customer needs and wants. These are listed as the "Whats" and represent the individual characteristics of the product or service. For example, in credit-card services, the "Whats" could be attributes such as a low interest rate, error-free transactions, no annual fee, extended warranty at no additional cost, customer service 24 hours a day, and a customers' advocate in billing disputes. The list of "Whats" is kept manageable by grouping similar items. On determination of the "Whats" list, a customer importance rating that prioritizes the "Whats" is assigned to each item. Typically, a scale of 1 to 5 is used with 1 being the least important and 5 being the most important. Multiple passes through the list may be necessary to arrive at ratings that are acceptable to the team. The ratings serve as weighting factors and are used as multipliers for determining the technical assessment of the "Hows." The focus is on attributes with high ratings because they maximize customer satisfaction. Let's suppose we have rated attributes for credit-card services as shown in Table 3-3. Our ratings thus imply that our customers consider error-free transactions to be the most important attribute and the least important to be charging no annual fee.

The customer plays an important role in determining the relative position of the organization with respect to that of its competitors for each requirement or "What." Such a comparison is entered in the section on "customer assessment of competitors." Thus, customer perception of the product or service is verified, which will help identify strengths and weaknesses of the company. Different focus groups or surveys should be

TABLE 3-3 Importance Rating of Customer Requirements

Customer Requirements ("Whats")	*Importance Rating*
Low interest rate	2
Error-free transactions	5
No annual fee	1
Extended warranty at no additional cost	3
Customer service 24 hours a day	4
Customers' advocate in billing disputes	4

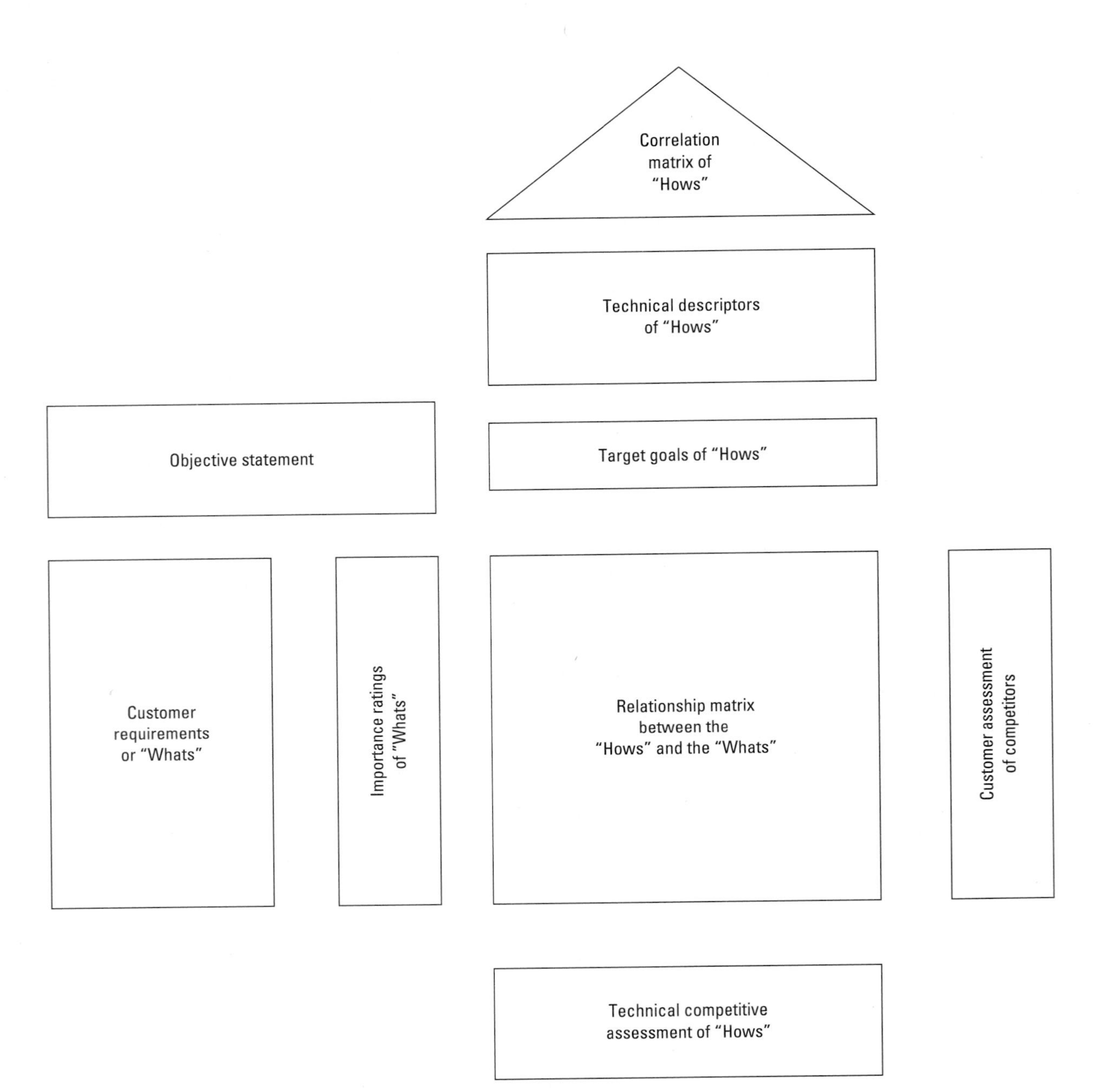

FIGURE 3-4 Quality function deployment matrix, the house of quality.

used to attain statistical objectivity. One outcome of the analysis might be new customer requirements, which would then be added to the list of "Whats," or the importance ratings might change. Results from this analysis will indicate what dimensions of the product or service the company should focus on. The same rating scale that is used to denote the importance ratings of the customer requirements is used in this analysis.

Consider, for example, the customer assessment of competitors shown in Table 3-4, where A represents our organization. The ratings are average scores obtained from various samples of consumers. The three competitors (companies B, C, and D) are our company's competition, so the rating scores in each "What" will serve as benchmarks, and thus the acceptable standard that we will shoot for. For instance, company C has a rating of 4 in the category "customer service 24 hours a day," as compared to our 2 rating; we

TABLE 3-4 Customer Assessment of Competitors

Customer Requirements ("Whats")	*Competitive Assessment of Companies* A	B	C	D
Low interest rate	3	2	④	2
Error-free transactions	4	⑤	3	3
No annual fee	⑤	⑤	2	3
Extended warranty at no additional cost	2	2	1	④
Customer service 24 hours a day	2	2	④	3
Customers' advocate in billing disputes	④	2	3	3

are not doing so well in this "What." We have identified a gap in a customer requirement that we consider important. To close this gap, we could study company C's practices and determine whether we can adopt some of them. We will conduct similar analyses with the other "Whats," gradually implementing improved services. Our goal is to meet or beat the circled values in Table 3-4, which represent best performances in each customer requirement. That is, our goal is to become the benchmark.

Coming up with a list of technical descriptors—the "Hows"—that will enable our company to accomplish the customer requirements is the next step in the QFD process. Multidisciplinary teams whose members originate in various departments will brainstorm to arrive at this list. Departments such as product design and development, marketing, sales, accounting, finance, process design, manufacturing, purchasing, and customer service are likely to be represented on the team. The key is to have a breadth of disciplines in order to "capture" all feasible "Hows." To improve our company's ratings in the credit-card services example, the team might come up with these "Hows": software to detect errors in billing, employee training on data input and customer services, negotiations and agreements with major manufacturers and merchandise retailers to provide extended warranty, expanded scheduling (including flextime) of employee operational hours, effective recruiting, training in legal matters to assist customers in billing disputes, and obtaining financial management services.

Target goals are next set for selected technical descriptors or "Hows." Three symbols are used to indicate target goals: ↑ (maximize or increase the attained value), ↓ (minimize or decrease the attained value), and ⊙ (achieve a desired target value). Table 3-5 shows how our team might define target goals for the credit-card services example. Seven "Hows" are listed, along with their target goals. As an example, for How #1, creating a software to detect billing errors, the desired target value is zero—that is, no billing errors. For How #2, it is desirable to maximize or increase the effect of employee training to reduce input errors and interact effectively with customers. Also, for

TABLE 3-5 Target Goals of Technical Descriptors

"Hows"	*1*	*2*	*3*	*4*	*5*	*6*	*7*
Target Goals	⊙	↑	↑	⊙	↑	↑	↑

<u>*Legend*</u>

Number	Technical Descriptors or "Hows"
1	A software to detect billing errors
2	Employee training on data input and customer services
3	Negotiations with manufacturers and retailers (vendors)
4	Expanded scheduling (including flextime) of employees
5	Effective recruiting
6	Legal training
7	Financial management services

Symbol	Target Goal
↑	Maximize or increase attained value
↓	Minimize or decrease attained value
⊙	Achieve a target value

How #4, the target value is to achieve customer service 24 hours a day. If measurable goals cannot be established for a technical descriptor, it should be eliminated from the list and the inclusion of other "Hows" considered.

The correlation matrix of the relationship between the technical descriptors is the "roof" of the house of quality. In the correlation matrix shown in Figure 3-5, four levels of relationship are depicted: strong positive, positive, negative, and strong negative. These indicate the degree to which the "Hows" support or complement each other or are in conflict. Negative relationships may require a trade-off in the objective values of the "Hows" when a technical competitive assessment is conducted. In Figure 3-5, which correlates the "Hows" for our credit-card services example, How #1, creating a software to detect billing errors, has a strong positive relationship (++) with How #2, employee training on data input and customer services. The user friendliness of the software will have an impact on the type and amount of training needed. A strong positive relationship indicates the possibility of synergistic effects. Note that How #2 also has a strong positive relationship with How #5; this indicates that a good recruiting program where desirable skills are incorporated into the selection procedure will form the backbone for a successful and effective training program.

Following this, a technical competitive assessment of the "Hows" is conducted along the same lines as the customer assessment of competitors we discussed previously. The difference is that, instead of using customers to obtain data on the relative position of the company's "Whats" with respect to those of the competitors, the technical staff of the company provides the input on the "Hows." A rating scale of 1 to 5, as used in Table 3-4, may be used. Table 3-6 shows how our company's technical staff has

FIGURE 3-5 Correlation matrix of "Hows."

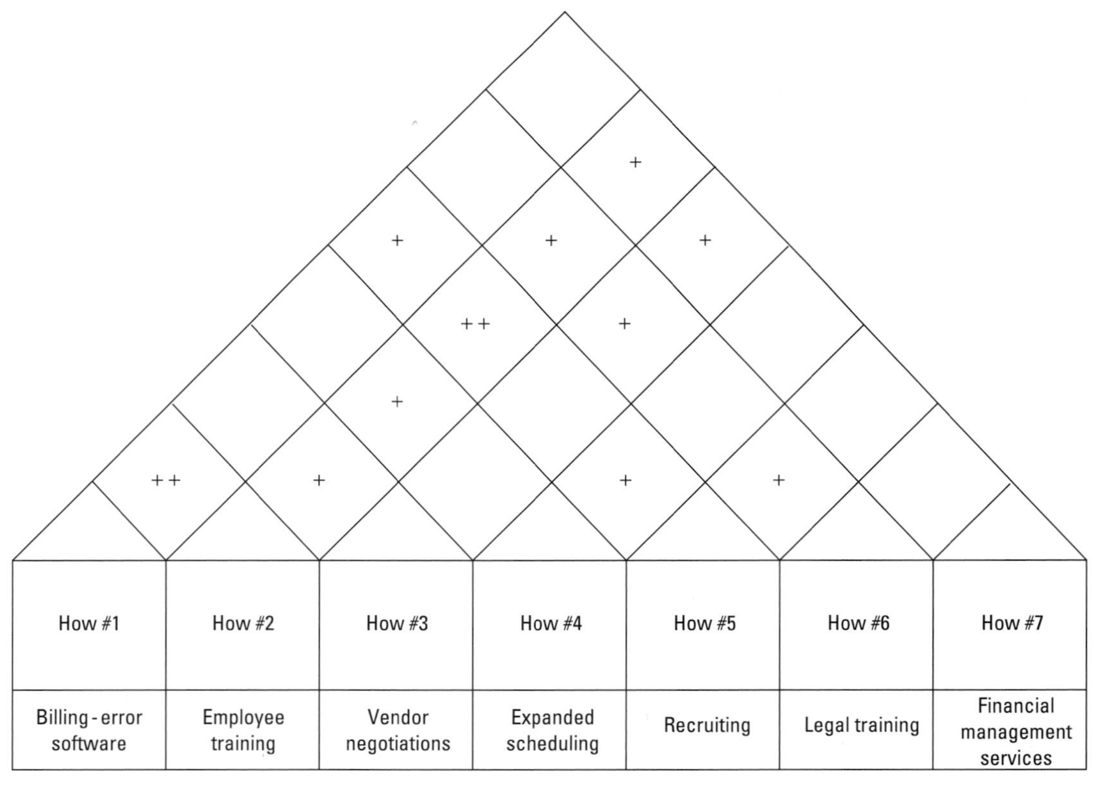

TABLE 3-6 Technical Competitive Assessment of "Hows"

Company	*Technical Descriptors ("Hows")* 1	2	3	4	5	6	7
A	4	3	2	3	4	④	⑤
B	⑤	3	1	④	1	2	3
C	3	⑤	2	2	⑤	3	2
D	2	2	④	1	3	3	4

assessed technical competitiveness for the "Hows" in the credit-card services example. Our three competitors, companies B, C, and D, are again considered. For How #1 (creating a software to detect billing errors), our company is doing relatively well, with a rating of 4, but company B, with its rating of 5, is doing better; company B is therefore the benchmark against which we will measure our performance. Likewise, company C is the benchmark for How #2; we will look to improve the quality and effectiveness of our training programs. The other assessments reveal that we have room to improve in Hows #3, #4, and #5, but in Hows #6 and #7, we are the benchmarks. The circled values in Table 3-6 represent the benchmarks for each "How."

The analysis shown in Table 3-6 can also assist in setting objective values, denoted by the "How Muches," for the seven technical descriptors. The achievements of the highest-scoring companies are set as the "How Muches," which represent the minimum acceptable achievement level for each "How." For example, for How #4, since company B has the highest rating, its achievement level will be the level that our company (company A) will strive to match or exceed. Thus, if company B provides customer service 16 hours a day, this becomes our objective value. If we cannot achieve these levels of "How Muches," we should not consider entering this market because our product or service will not be as good as the competition's.

In conducting the technical competitive assessment of the "Hows," the probability of achieving the objective value (the "How Muches") is incorporated in the analysis. Using a rating scale of 1 to 5, 5 representing a high probability of success, the absolute scores are multiplied by the probability scores to obtain weighted scores. These weighted scores now represent the relative position within the industry and the company's chances of becoming the leader in that category.

The final step of the QFD process involves the relationship matrix located in the center of the house of quality (see Figure 3-4). It provides a mechanism for analyzing how each technical descriptor will help in achieving each "What." The relationship between a "How" and a "What" is represented by the following scale: 0 ≡ No relationship; 1 ≡ Low relationship; 3 ≡ Medium relationship; 5 ≡ High relationship. Table 3-7 shows the relationship matrix for the credit-card services example. Consider, for instance, How

TABLE 3-7 Relationship Matrix of Absolute and Relative Scores

Customer Requirements ("Whats")	*Importance Ratings*	*Technical Descriptors ("Hows")* 1	2	3	4	5	6	7
Low interest rate	2	0 (0)	0 (0)	5 (10)	0 (0)	0 (0)	0 (0)	5 (10)
Error-free transactions	5	5 (25)	5 (25)	0 (0)	3 (15)	5 (25)	0 (0)	0 (0)
No annual fee	1	0 (0)	0 (0)	3 (3)	0 (0)	0 (0)	0 (0)	5 (5)
Extended warranty	3	0 (0)	1 (3)	5 (15)	0 (0)	0 (0)	3 (9)	3 (9)
Customer service 24 hours a day	4	1 (4)	3 (12)	0 (0)	5 (20)	5 (20)	3 (12)	0 (0)
Customers' advocate in billing disputes	4	1 (4)	3 (12)	5 (20)	0 (0)	3 (12)	5 (20)	1 (4)
Absolute score		33	52	48	35	57	41	28
Relative score		6	2	3	5	1	4	7
Technical competitive assessment		5	5	4	4	5	4	5
Weighted absolute score		165	260	192	140	285	164	140
Final relative score		4	2	3	6.5	1	5	6.5

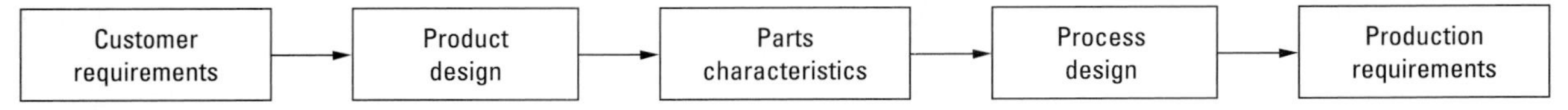

FIGURE 3-6 Phases of use of QFD.

#2 (employee training on data input and customer services). Our technical staff believes that this "How" has a high relationship with providing error-free transactions, and so a score of 5 is assigned. Furthermore, this "How" has a moderate relationship with providing customer service 24 hours a day and serving as customers' advocate in billing disputes, so a score of 3 is assigned for these relationships. Similar interpretations are drawn from the other entries in the table. "Hows" that have a large number of zeros do not support meeting the customer requirements and should be dropped from the list.

The cell values, shown in parentheses in Table 3-7, are obtained by multiplying the rated score by the importance rating of the corresponding customer requirement. The absolute score for each "How" is calculated by adding the values in parentheses. The relative score is merely a ranking of the absolute scores, with 1 representing the most important. It is observed that How #5 (effective recruiting), is most important because its absolute score of 57 is highest.

The analysis can be extended by considering the technical competitive assessment of the "Hows." Using the rating scores of the benchmark companies for each technical descriptor—that is, the objective values (the "How Muches") from the circled values in Table 3-6—our team can determine the importance of the "Hows." The weighted absolute scores in Table 3-7 are found by multiplying the corresponding absolute scores by the technical competitive assessment rating. The final scores demonstrate that the relative ratings of the top three "Hows" are the same as before. However, the rankings of the remaining technical descriptors have changed. How #4 and How #7 are tied for last place, each with an absolute score of 140, and a relative score of 6.5 each. Management may consider the ease or difficulty of implementing these "Hows" in order to break the tie.

Our example QFD exercise illustrates the importance of teamwork in this process. An enormous amount of information must be gathered, all of which promotes cross-functional understanding of the product or service design system. Target values of the technical descriptors or "Hows" are then used to generate the next level of house of quality diagram where they will become the "Whats." The QFD process proceeds by determining the technical descriptors for these new "Whats." We can therefore consider the implementation of the QFD process in different phases. As Figure 3-6 depicts, QFD facilitates the translation of customer requirements into a product whose features meet these requirements. Once such a product design is conceived, QFD may be used at the next level to identify specific characteristics of critical parts that will help in achieving the designed product. The next level may address the design of a process in order to make parts with the identified characteristics. Finally, the QFD process identifies production requirements for operating the process under specified conditions. Use of quality function deployment in such a multiphased environment requires a significant commitment of time and resources. However, the advantages—the spirit of teamwork, cross-functional understanding, and an enhanced product design—offset this commitment.

3-4 INNOVATIVE ADAPTATION AND PERFORMANCE EVALUATION

The goal of continuous improvement forces an organization to look for ways to improve operations. Be it a manufacturing or service organization, the organization must be aware of the best practices in its industry and its relative position in the industry. Such information will set the priorities for areas that need improvement.

Organizations benefit from innovation. Innovative approaches cut costs, reduce lead time, improve productivity, save capital and human resources, and, ultimately, lead to increased revenue. They constitute the breakthroughs that push product or process to new levels of excellence. However, breakthroughs do not happen very often. Visionary ideas are few and far between. Still, when improvements come, they are dramatic and memorable. The development of the computer chip is a prime example. Its ability to store enormous amounts of information in a fraction of the space that was previously required has revolutionized our lives. Figure 3-7 shows the impact of innovation on a chosen quality measure over time. At times a and b innovations occur as a result of which steep increases in quality from x to y and y to z are observed.

Continuous improvement, on the other hand, leads to a slow but steady increase in the quality measure. Figure 3-7 shows that, for certain periods of time, a process with continuous improvement performs better than one that depends only on innovation. Of course, once an innovation takes place, the immense improvement in the quality measure initially outperforms the small improvements that occur on a gradual basis. This can be useful in gaining market share, but it is also a high-risk strategy because innovations are rare. A company must carefully assess how risk averse it is. If its aversion to risk is high, continuous improvement is its best strategy. A process that is guaranteed to improve gradually is always a wise investment.

One way to promote continuous improvement is through innovative adaptation of the best practices in the industry. An organization can incorporate information on the companies perceived to be the leaders in the field to improve their operations.

FIGURE 3-7 Impact of innovation and continuous improvement.

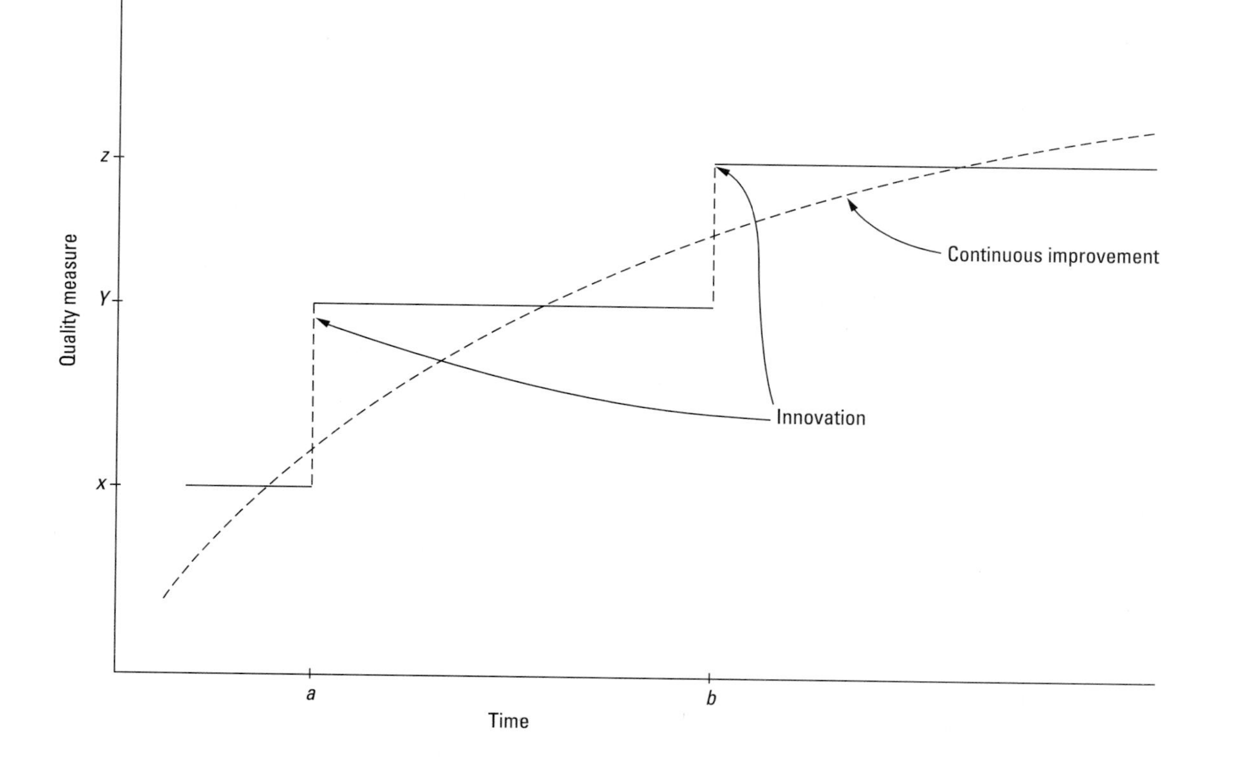

Depending on the relative position of the company with respect to the industry leader, gains will be incremental or dramatic. Incorporating such adaptations on an ongoing basis provides a framework for continuous improvement.

Benchmarking

As discussed earlier, the practice of identifying best practices in industry and thereby setting goals to emulate them is known as **benchmarking.** Companies cannot afford to stagnate; this guarantees a loss of market share to the competition. Continuous improvement is a mandate for survival, and such fast-paced improvement is facilitated by benchmarking. This practice enables an organization to accelerate its rate of improvement. While innovation allows an organization to "leap-frog" its competitors, it does not occur frequently and thus cannot be counted on. Benchmarking, on the other hand, is doable. To adopt the best, innovatively adapt it, and consequently reap improvements is a strategy for success.

Specific steps for benchmarking vary from company to company, but the fundamental approach is the same. One company's benchmarking may not work at another organization because of different operating concerns. Successful benchmarking reflects the culture of the organization, works within the existing infrastructure, and is harmonious with the leadership philosophy. Motorola, Inc., winner of the Malcolm Baldrige Award for 1988, uses a five-step benchmarking model: (1) Decide what to benchmark; (2) select companies to benchmark; (3) obtain data and collect information; (4) analyze data and form action plans; and (5) recalibrate and start the process again.

AT&T, which has two Baldrige winners among its operating units, uses a nine-step model: (1) Decide what to benchmark; (2) develop a benchmarking plan; (3) select a method to collect data; (4) collect data; (5) select companies to benchmark; (6) collect data during site visit; (7) compare processes, identify gaps, make recommendations; (8) implement recommendations; and (9) recalibrate benchmarks.

A primary advantage of the benchmarking practice is that it promotes a thorough understanding of the company's own processes—the company's current profile is well understood. Intensive studies of existing practices often lead to identification of nonvalue-added activities and plans for process improvement. Secondly, benchmarking enables comparisons of performance measures in different dimensions, each with the best practices for that particular measure. It is not merely a comparison of the organization with a selected company, but a comparison with several companies who are the best for the chosen measure. Some common performance measures are return on assets, cycle time, percentage of on-time delivery, percentage of damaged goods, proportion of defects, and time spent on administrative functions. The spider chart shown in Figure 3-8 is used to compare multiple performance measures and gaps between the host company and industry benchmark practices. Six performance measures (PM) are being considered here. The scales are standardized—say between 0 and 1, 0 being at the center and 1 at the outer circumference, which represents the most desired value. Best practices for each performance measure are indicated, along with the companies that achieve them. The current performance level of the company performing the benchmarking (company A) is also indicated in the figure. The difference between company A's level and that of the best practice for that performance measure is identified as the gap. The analysis that focuses on methods and processes to reduce this gap and thereby improve the company's competitive position is known as **gap analysis.**

Another advantage of benchmarking is its focus on performance measures and processes, not on the product. Thus, benchmarking is not restricted to the confines of the industry in which the company resides. It extends beyond these boundaries and identifies organizations in other industries that are superior with respect to the chosen measure. It is usually difficult to obtain data from direct competitors. However, companies outside the industry are more likely to share such information. It then becomes the task of management to find ways to innovatively adapt those best practices within their own environment.

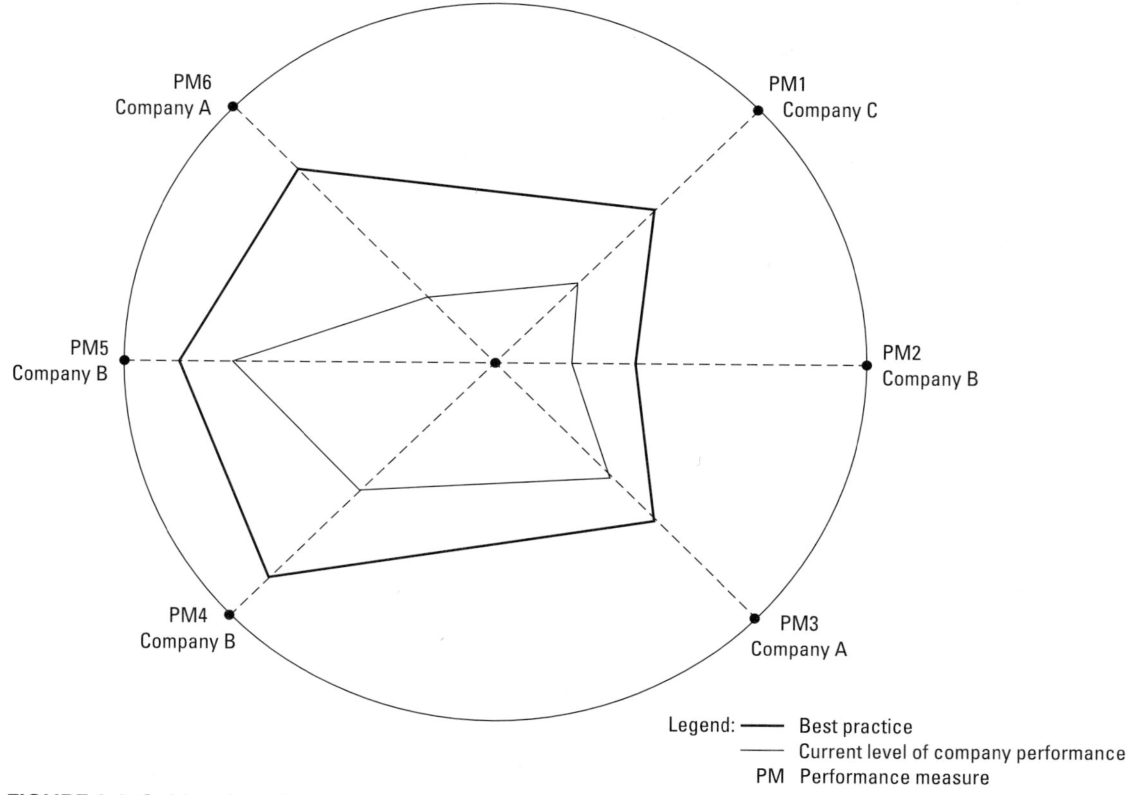

FIGURE 3-8 Spider chart for gap analysis.

In the United States, one of the pioneers of benchmarking is Xerox Corporation. It embarked on this process because its market share eroded rapidly in the late 1970s to Japanese competition. Engineers from Xerox took competitors' products apart and looked at them component by component. When they found a better design, they sought ways to adapt it to their own products or, even better, to improve upon it. Similarly, managers from Xerox began studying the best management practices in the market; this included companies both within and outside the industry. As Xerox explored ways to improve its warehousing operations, it found a benchmark outside its own industry: L. L. Bean, Inc., the outdoor sporting goods retailer.

L. L. Bean has a reputation of high customer satisfaction; the attributes that support this reputation are its ability to fill customer orders quickly and efficiently with minimal errors and to deliver undamaged merchandise. The backbone behind this successful operation is an effective management system aided by state-of-the-art operations planning that addresses warehouse layout, work flow design, and scheduling. Furthermore, the operations side of the processes is backed by an organizational culture of empowerment, management commitment through effective education and training, and a motivational reward system of incentive bonuses.

Figure 3-9 demonstrates how benchmarking brings the "soft" and "hard" systems together. Benchmarking is not merely the identification of the best practices. Rather, it seeks to determine *how* such practices can be adapted to the organization. The real value of benchmarking is accomplished only when the company has successfully integrated the identified best practices into its operation. To be successful in this task, "soft" and "hard" systems must "mesh." The emerging organizational culture should empower employees to make decisions based on the new practice.

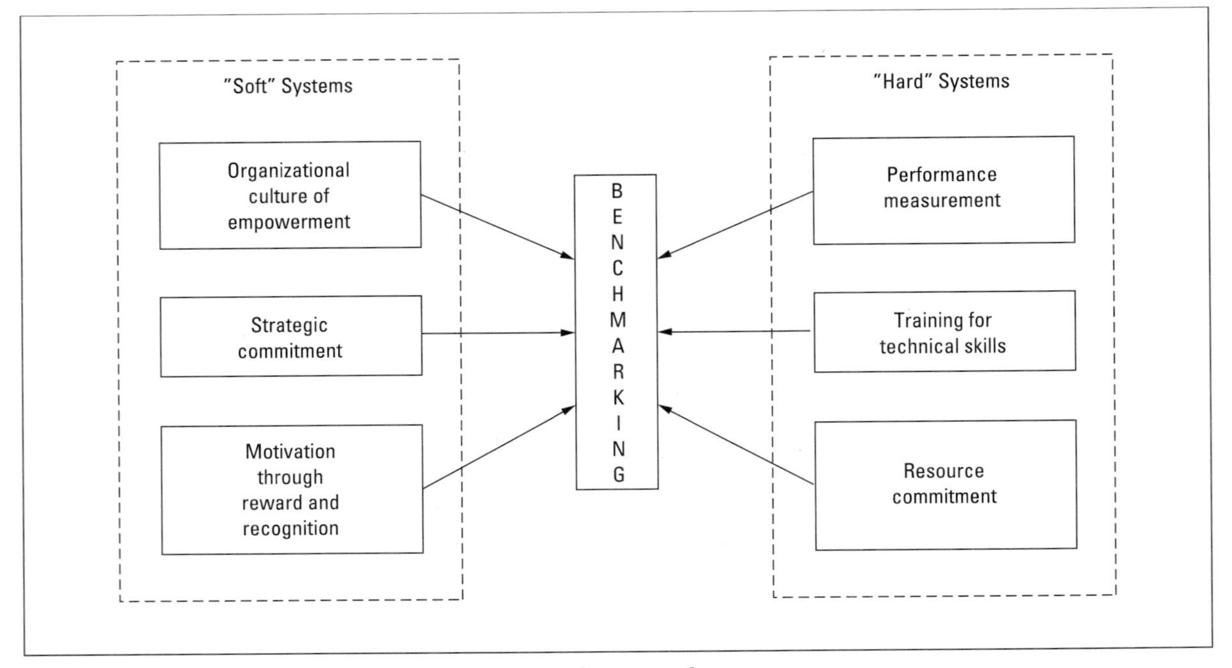

FIGURE 3-9 Role of benchmarking in implementing best practices.

For benchmarking to succeed, management must demonstrate its strategic commitment to continuous improvement and must also motivate employees through an adequate reward and recognition system that promotes learning and innovative adaptation. When dealing with "hard" systems, resources must be made available to allow release time from other activities, access to information on best practices, and installation of new information systems to manage the acquired information. Technical skills, required for benchmarking such as flowcharting and process mapping, should be provided to team members through training sessions. The team must also identify performance measures for which the benchmarking will take place. Examples of such measures are return on investment, profitability, cycle time, and defect rate.

Several factors influence the adoption of benchmarking; **change management** is one of them. Figure 3-10 illustrates factors that influence benchmarking and the subsequent outcomes that derive from it. In the current environment of global competition, change is a given. Rather than react haphazardly to change, benchmarking provides an effective way to manage it. Benchmarking provides a road map for adapting best practices, a major component of change management. These are process-oriented changes. In addition, benchmarking facilitates cultural changes in the organization. These deal with overcoming resistance to change. This is a people-oriented approach, the objective being to demonstrate that change is not a threat but an opportunity.

The ability to reduce process time and create a mode of quick response is important to all organizations. The concept of **time-based competition** is linked to reductions in cycle time. **Cycle time** can be defined as the interval between the beginning and the ending of a process, which may consist of a sequence of activities. From the customer's point of view, cycle time is the elapsed time between placing an order and having it satisfactorily fulfilled. Reducing cycle time is strongly correlated with performance measures such as cost, market share, and customer satisfaction. Detailed flowcharting of the process can identify bottlenecks, decision loops, and nonvalue-added activities. Reducing decision and inspection points, creating electronic media systems for dynamic flow of information, standardizing procedures and reporting forms, and consolidating purchases are examples of tactics that reduce cycle time. Motorola, Inc., for example, reduced its corporate auditing process over a three-year period from an average of seven weeks to five days.

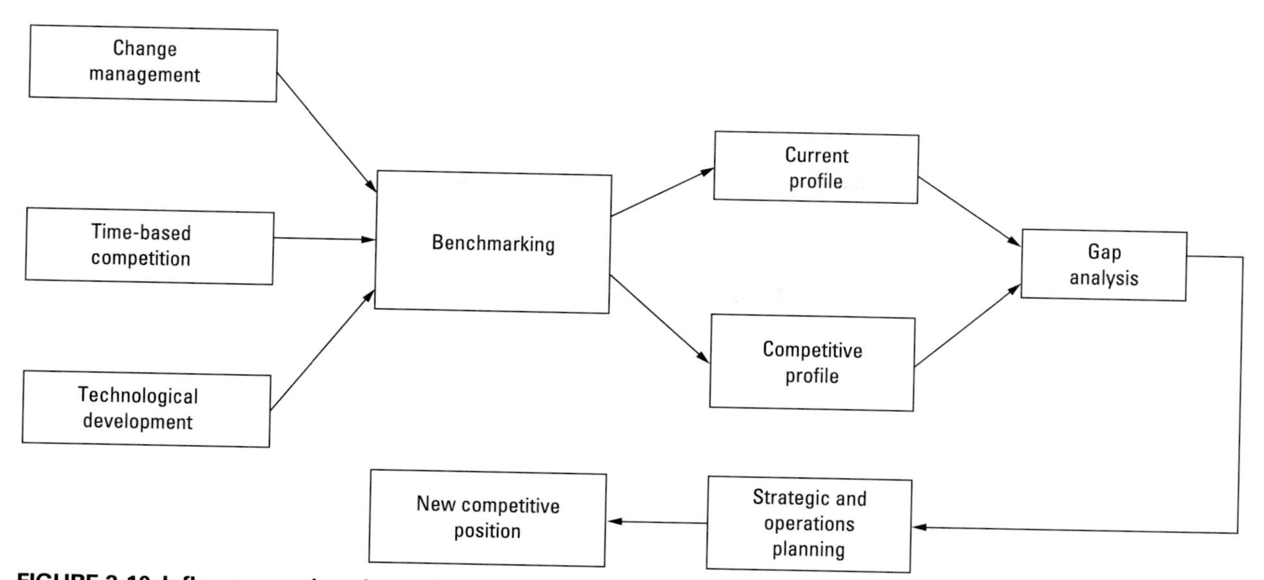

FIGURE 3-10 Influences on benchmarking and its outcomes.

Technological development is another impetus for benchmarking. Consider the microelectronics industry. Its development pace is so rapid that a company has no choice but to benchmark. Falling behind the competition in this industry means going out of business. In this situation, benchmarking is critical to survival.

Quality Auditing

The effectiveness of management control programs may be examined through the practice known as quality auditing. One reason management control programs are implemented is to prevent problems. In spite of such control, however, problems can and do occur. So, **quality audits** are undertaken to identify problems.

In any quality audit, three parties are involved. The party that requests the audit is known as the *client*. The party that conducts the audit is the *auditor,* while the party being audited is the *auditee.* Auditors can be of two types, internal or external. An internal auditor is an employee of the auditee. External auditors are not members of the auditee's organization. An external auditor may be a single individual or a member of an independent auditing organization.

Quality audits fulfill two major purposes. The first purpose, performed in the **suitability quality audit,** deals with an in-depth evaluation of the quality program against a reference standard, usually predetermined by the client. Reference standards are set by several organizations, including the American National Standards Institute/American Society for Quality Control (ANSI/ASQC), International Standards Organization (ISO), and British Standards Institute (BSI). Several ISO standards are discussed later in this chapter. The entire organization may be audited, or specific processes, products, or services may be audited. The second purpose, performed in the **conformity quality audit,** deals with a thorough evaluation of the operations and activities within the quality system and the degree to which they conform to the defined quality policies and procedures.

Quality audits may be categorized as one of three types. The most extensive and inclusive type is the **system audit.** This entails an evaluation of the quality program documentation (including policies, procedures, operating instructions, defined accountabilities and responsibilities to achieve the quality function) with a reference standard. It also includes an evaluation of the activities and operations that are implemented to accomplish the desired quality objectives. Such audits therefore explore conformance of quality management standards and their implementation to specified norms. They encompass the evaluation of the phases of planning, implementation, evaluation, and comparison. An example of a system audit is a preaward survey, which typically evaluates the ability of a potential vendor to provide a desired level of product or service.

A second type of quality audit (not as extensive as the system audit) is the **process audit,** which is an in-depth evaluation of one or more processes in the organization. All relevant elements of the identified process are examined and compared to specified standards. Because a process audit takes less time to conduct than a system audit, it is more focused and less costly. If management has already identified a process that needs to be evaluated and improved, the process audit is an effective means of verifying compliance and suggesting places for improvement. A process audit can also be triggered by unexpected output from a process. For industries that use continuous manufacturing processes, such as chemical industries, a process audit is the audit of choice.

The third type of quality audit is the **product audit,** which is an assessment of a final product or service on its ability to meet or exceed customer expectations. This audit may involve conducting periodic tests on the product or obtaining information from the customer on a particular service. The objective of a product audit is to determine the effectiveness of the management control system. Such an audit is separate from decisions on product acceptance or rejection and is therefore not part of the inspection system used for such processes. Customer or consumer input plays a major role in the decision to undertake a product audit. For a company producing a variety of products, a relative comparison of product performance that indicates poor performers could be used as a guideline for a product audit.

Audit quality is heavily influenced by the independence and objectivity of the auditor. For the audit to be effective, the auditor must be independent of the activities being examined. Thus, whether the auditor is internal or external may have an influence on audit quality. Consider the assessment of an organization's quality documentation. It is quite difficult for an internal auditor to be sufficiently independent to perform this evaluation effectively. For such suitability audits, external auditors are preferable. System audits are also normally conducted by external auditors. Process audits can be internal or external, as can product audits. An example of an internal product audit is a "dock audit," where the product is examined prior to shipment. Product audits conducted at the end of a process line are also usually internal audits. Product audits conducted at the customer site are typically external audits.

Vendor audits are external. They are performed by representatives of the company that is seeking the vendor's services. Knowledge of product and part specifications, contractual obligations and their secrecy, and purchase agreements often necessitate a second-party audit where the client company sends personnel from its own staff to perform the audit. Conformity quality audits may be carried out by internal or external auditors so long as the individuals are not directly involved in the activities being audited.

Methods for conducting a quality audit are of two types. One approach is to conduct an evaluation of all quality system activities at a particular location or operation within an organization, known as a *location-oriented quality audit.* This audit examines the actions and interactions of the elements in the quality program at that location and may be used to interpret differences between locations. The second approach is to examine and evaluate activities relating to a particular element or function within a quality program at all locations where it applies before moving on to the next function in the program. This is known as a *function-oriented quality audit.* Successive visits to each location are necessary to complete this latter audit. It is helpful in evaluating the overall effectiveness of the quality program and also useful in tracing the continuity of a particular function through the locations where it is applicable.

The utility of a quality audit is derived only when remedial actions in deficient areas, exposed by the quality audit, are undertaken by company management. A quality audit does not necessarily prescribe actions for improvement; it typically identifies areas that do not conform to prescribed standards and therefore need attention. If several areas are deficient, a company may prioritize those that require immediate attention. Only on implementation of the remedial actions will a company improve its competitive position. Tools that help identify critical areas, find root causes to problems, and propose solutions include cause-and-effect diagrams, flow charts, and Pareto charts; these will be discussed later in this chapter.

An example of the various functions of a quality system audit based on standards specified in *ISO 9001—Quality Systems—Model for Quality Assurance in Design/Development, Production, Installation, and Servicing* is illustrated in Figure 3-11. The main functional areas of the system audit are management, marketing, purchasing, design, production, inspection, and quality. Subsections that should be addressed for each functional area in the audit are shown. Management, for example, should have a defined quality policy that everyone in the organization knows, understands, follows, and supports. There must be well-defined areas of responsibility that clearly state the person responsible for a functional area, along with channels of authority. Audit responsibilities should not be assigned to personnel who have direct control over audited activities in a particular function. Management also maintains a review system that examines its role and responsibilities, provides feedback, and takes corrective action when there is a deviation between the observed and desired output. Managerial and technical training are responsibilities of management. A quality system audit examines and evaluates the degree to which each of these activities meets adopted standards.

A main function of marketing is contract review. Detailed knowledge of the issues and contractual obligations that exist between the company and vendor is necessary. A mechanism should be in place that monitors constant compliance with stipulated contracts. A quality audit will identify the effectiveness of this mechanism. As changes in contracts are dynamic in nature, the monitoring mechanism must capture such changes on a real-time basis. The functional area of purchasing includes activities such as vendor selection, verification of incoming product, and nonconforming product control. A vendor quality survey report describes the quality status of the vendor's products discrepancies that contribute to inferior quality, and recommends actions that will correct these discrepancies.

A major activity of design is document control. As changes are made in the design, it becomes critical to ensure that such changes are made in all appropriate documents and communicated to other functional areas that are influenced by these changes. Production is another major function of a quality systems audit. Document control and process control are key activities under this function. The topic of process control, through control charts, is treated in depth in subsequent chapters. Specific procedures must exist for handling and storage and packaging and delivery of products and components. Special care must be taken to indicate existing methods for nonconforming material control. What actions are taken when a nonconforming component is found? How is it kept distinct from conforming components? How is it rectified, or how is it disposed? Such questions are addressed in the audit process.

FIGURE 3-11 Functions of a quality systems audit based on ISO 9001 standards.

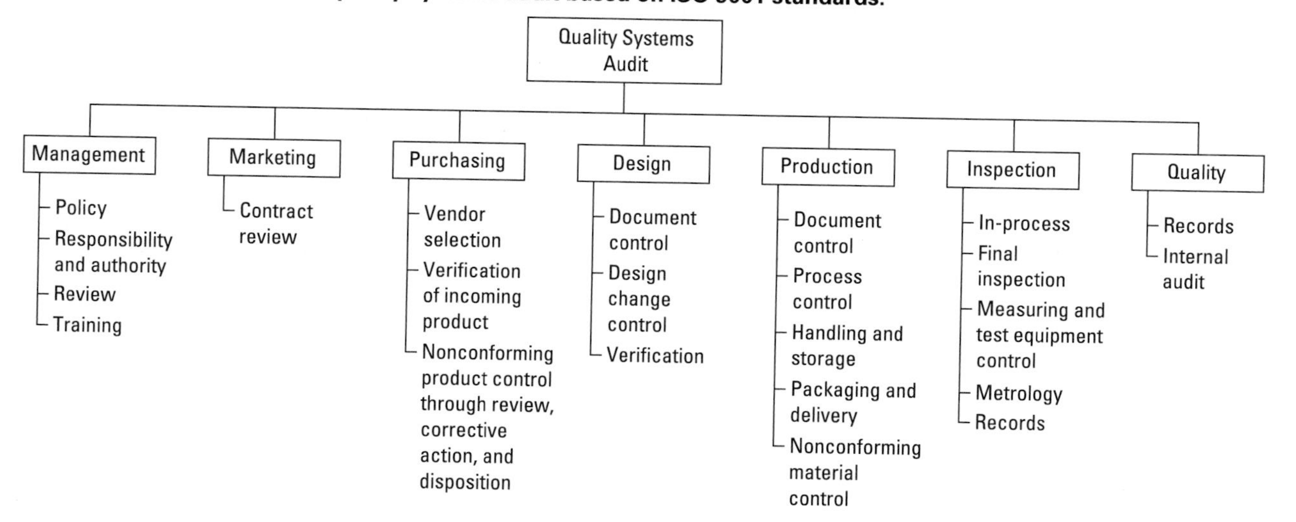

The function of inspection includes in-process and final inspection and the maintenance of adequate records that demonstrate the disposition of the product. Measuring and test equipment control are important activities under this function. What methods are used to calibrate equipment? How often are they calibrated? What processes exist to determine whether equipment needs to be reset or calibrated? Are there discrepancies in the existing operating and environmental conditions of this equipment versus those specified in the standard? Is the equipment's precision adequate? These are typical issues that are considered in the system audit. In the functional area of quality, records control and internal audits are activities of interest. Maintaining adequate and complete records that demonstrate the suitability of the process and the product to meet specified standards is a must. Such records provide a way to trace nonconforming products and associated problems with the process. Quality audits thus identify areas that deserve the attention of management.

Vendor Selection and Certification Programs

As discussed in Chapter 2, the modern trend is to establish long-term relationships with vendors. In an organization's pursuit of continuous improvement, the purchaser (customer) and vendor must be integrated in a quality system that serves the strategic missions of both companies. The vendor must be informed of the purchaser's strategies for market-share improvement, advance product information (including changes in design), and delivery requirements. The purchaser, on the other hand, should have access to information on the vendor's processes and also be advised of their unique capabilities.

Cultivating a partnership between purchaser and vendor has several advantages. First, it is a win-win situation for both. To meet unique customer requirements, a purchaser can then redesign products or components collaboratively with the vendor. The vendor, who makes those particular components, has intimate knowledge of the components and the necessary processes that will produce the desired improvements. The purchaser is thus able to design its own product in a cost-effective manner and can be confident that the design will be feasible to implement. Alternatively, the purchaser may give the performance specification to the vendor and entrust them with design, manufacture, and testing. The purchaser thereby reduces design and development costs, lowers internal costs, and gains access to proprietary technology through its vendor, technology that would be expensive to develop internally. Through such partnerships, the purchaser is able to focus on its areas of expertise, thereby maintaining its competitive edge. Vendors gain from such partnerships by taking ownership of the product or component from design to manufacture; they can meet specifications more effectively because of their involvement in the entire process. They also gain an expanded insight into product and purchaser requirements, through linkage with the purchaser; this helps them better meet those requirements. This, in turn, strengthens the vendor's relationship with the purchaser.

Partnership Between Purchaser and Vendor

Effective partnership is facilitated by open channels of communication between the vendor and vendee (purchaser) and is based on mutual trust and respect. The American Society of Quality Control (ASQC) Vendor-Vendee Technical Committee has developed an informal code of ethics for this partnership, which is summarized as follows:

1. *Personal behavior.* Contacts between the quality functions of the purchaser (customer) and the vendor should credit both parties and avoid compromising relationships.
2. *Objectivity.* The purchaser and vendor should fulfill all contractual obligations with a moral obligation to achieve a satisfactory end product through a fair and equitable cost distribution. So, if the purchaser imposes additional requirements, the costs for achieving these should be borne, to some extent, by the purchaser.
3. *Product definition.* The purchaser should transmit all specifications and workmanship requirements in a clear and complete manner to the vendor. Any help that is sought on interpretation of the requirements should be readily provided to the vendor.

4. *Mutual understanding.* Direct communications should take place between the quality functions of the purchaser and the vendor.
5. *Quality evaluation.* The purchaser must fairly evaluate the vendor's quality performance and communicate the results of such an evaluation to the vendor. The vendor should be advised of the rating system, if any, that is used.
6. *Product quality.* The vendor should fairly inform the purchaser of the quality status of the delivered items. For items that are delivered, it is implied that contractual obligations are met, unless otherwise specified. If any deviations from the contract specifications arise, the purchaser must be notified.
7. *Corrective action.* Such action, when necessary, should be pursued jointly by the purchaser and the vendor. Joint effort at solving problems bonds the partnership between the two parties.
8. *Technical aid.* On request from the vendor, the purchaser should provide technical support either from its own staff or from an independent company. The vendor still bears responsibility for the product quality that it delivers.
9. *Integrity.* The usage of facilities and services by the visiting party should be according to the guidelines stated in the contract, unless offered on a voluntary basis by the vendor or the purchaser. So, if contractual obligations require the purchaser to observe inspections or test at the vendor's site, these must be accommodated.
10. *Rewards.* It is preferable to use only qualified vendors who have met certain performance criteria. Those vendors that consistently meet conformance requirements should be rewarded.
11. *Proprietary information.* To maintain ethical and moral standards, the purchaser and vendor should not divulge proprietary information to other parties.
12. *Safeguarding reputation.* False or unsupported statements should not be made by either party. An open and honest relationship must be maintained by the purchaser and vendor.

Vendor Rating and Selection

Maintaining data on the continual performance of vendors requires an evaluation scheme. **Vendor rating** based on established performance measures facilitates this process. There are several advantages in monitoring vendor ratings. Since the quality of the output product is a function of the quality of the incoming raw material or components, procured through vendors, it makes sense to establish long-term relationships with vendors that consistently meet or exceed performance requirements. Analyzing the historical performance of vendors enables the company to select vendors that deliver their goods on time. Vendor rating goes beyond reporting on the historical performance of the vendor. It ensures a disciplined material control program. Rating vendors also helps reduce quality costs by optimizing purchased material costs.

Measures of vendor performance, which comprise the rating scheme, address the three major categories of quality, cost, and delivery. Under quality, some common measures are percent defective as expressed by defects in parts per million, process capability, product stoppages due to poor quality of vendor components, number of customer complaints, and average level of customer satisfaction. The category of cost includes such measures as scrap and rework cost, return cost, incoming-inspection cost, life-cycle costs, and warranty costs. The vendor's maintenance of delivery schedules is important to the purchaser in order to meet customer-defined schedules. Some measures in this category are percent of on-time deliveries, percent of late deliveries, percent of early deliveries, percent of underorder quantity, and percent of overorder quantity.

Which measures should be used are influenced by the type of product or service, the customer's expectations, and the level of quality systems that exists in the vendor's organization. For example, the Federal Express Corporation, winner of the 1990 Malcolm Baldrige National Quality Award in the service category, is *the* name in fast and reliable delivery. FedEx tracks its performance with such measures as late delivery, invoice adjustment needed, damaged packages, missing proof of delivery on invoices, lost packages, and missed pickups. For incoming material inspection, defectives per shipment, inspection costs, and cost of returning shipment are suitable measures. For vendors with statistical process control systems in place, measuring process capability is also useful. Customer satisfaction indices can be used with those vendors that have extensive companywide quality systems in place.

Vendor performance measures are prioritized according to their importance to the purchaser. Thus, a weighting scheme similar to that described in the house of quality (Figure 3-4) is often used. Let's consider a purchaser that uses rework and scrap cost, price, percent of on-time delivery, and percent of underorder quantity as its key performance measures. Table 3-8 shows these performance measures and the relative weight assigned to each one. This company, as shown in the table, feels that rework and scrap costs are most important, with a weight of 40. Note that price is not the sole determinant; in fact, it received the lowest weighting.

Table 3-8 shows the evaluation of vendors A, B, and C. For each performance measure, the vendors are rated on a 1 to 5 scale, with 1 representing the least desirable performance. A weighted score is obtained by adding the products of the weight and the assigned rating for each performance measure (weighted rating). The weighted scores are then ranked, with 1 denoting the most desirable vendor. From Table 3-8, we can see that vendor B, with the highest weighted score of 350, is the most desirable.

Accurate vendor ratings require reliable data. As ratings are communicated to vendors, subsequent discussions may identify means through which the vendor can improve its ratings, if they are deficient. If particular problems are suspected, product or process audits can be undertaken. The outcome of the audits should be followed by corrective actions. The vendor can then notify the purchaser that remedial action has been taken, and the previously assigned ratings can be updated.

Vendor evaluation in quality programs is quite comprehensive. Even the vendor's culture is subject to evaluation as the purchaser seeks to verify the existence of a quality program. Commitment to customer satisfaction as demonstrated by appropriate actions is another attribute the purchaser will examine closely. The purchaser will measure the vendor's financial stability; the purchaser obviously prefers vendors that are going to continue to exist so the purchaser will not be visited with the problems that follow from liquidity or bankruptcy. The vendor's technical expertise relating to product and process design is another key concern as vendor and purchaser work together to solve problems and to promote continuous improvement.

Vendor Certification

Vendor certification occurs when the vendor has reached the stage such that it consistently meets or exceeds the purchaser's expectations. Consequently, there is no need for the purchaser to perform routine inspections of the vendor's product. Certification motivates vendors to improve their processes and, consequently, their products and services. A vendor must also demonstrate a thorough understanding of the strategic quality goals of the customer such that its own strategic goals are in harmony with those of the customer. Improving key processes through joint efforts strengthens the relationship between purchaser and vendor. The purchaser should therefore assess the vendor's capabilities on a continuous basis and provide adequate feedback.

TABLE 3-8 Prioritizing Vendor Performance Measures Using a Weighting Scheme

		Vendor A		*Vendor B*		*Vendor C*	
Performance Measure	*Weight*	*Rating*	*Weighted Rating*	*Rating*	*Weighted Rating*	*Rating*	*Weighted Rating*
Price	10	4	40	2	20	3	30
Rework and scrap cost	40	2	80	4	160	3	120
Percent of on-time delivery	30	1	30	3	90	2	60
Percent of underorder quantity	20	2	40	4	80	5	100
Weighted Score			190		350		310
Rank			3		1		2

A vendor goes through several levels of acceptance before being identified as a long-term partner. Typically, these levels are an approved vendor, a preferred vendor, and finally a certified vendor—that is, a "partner" in the quality process. To move from one level to the next, the quality of the vendor's product or service must improve. The certification process usually transpires in the following manner. First, the process is documented; this defines the roles and responsibilities of the involved personnel of both organizations. Performance measures, previously described, are chosen, and measurement methods are documented. An orientation meeting occurs at this step.

The next step is to gain commitment from the vendor. The vendor and purchaser establish an environment of mutual respect. This is important because they must share vital and sometimes sensitive information in order to improve process and product quality. A quality system survey of the vendor is undertaken. In the event that the vendor is certified or registered by a third party, the purchaser may forego its own survey and focus instead on obtaining valid performance measurements. At this point, the purchaser sets acceptable performance standards on quality, cost, and delivery and then identifies those vendors that meet these standards. These are the **approved vendors.**

Following this step, the purchaser decides on what requirements it will use to define its **preferred vendors.** Obviously, these requirements will be more stringent than for approved vendors. For example, the purchaser may give the top 20% of its approved vendors preferred vendor status. Preferred vendors may be required to have a process control mechanism in place that demonstrates its focus on problem prevention (as opposed to problem detection).

At the next level of quality—the **certified vendor**—the criteria entail not only quality, costs, and delivery measures but also technical support, management attitude, and organizational quality culture. The value system for the certified vendor must be harmonious with that of the purchaser. An analysis of the performance levels of various attributes is undertaken, and vendors that meet the stipulated criteria are certified. Finally, a process is established to ensure vendor conformance on an ongoing basis. Normally, such reviews are conducted annually.

Several certification programs—industry-specific and organization-specific—exist. An example of an organization-specific certification program is Ford Motor Company's Q101 Total Quality Excellence program discussed earlier. Q101 is an extremely comprehensive quality audit program of vendors that examines such functional areas as product design, development, test, manufacturing engineering, processing, manufacturing support, delivery, cost competitiveness, and quality capability. Details are listed in Table 3-1.

General Motors (GM) has a vendor certification program known as the Purchased Input Concept Optimization System (PICOS). The thrust of PICOS is to identify the vendor's capability to meet performance criteria set by GM so that GM can meet its own customer requirements. GM certifies vendors only if they share the same vision of continuous improvement. Chrysler also uses the partnership concept; indeed, Chrysler has worked with its vendors on concurrent engineering projects that have produced significant improvements in key assemblies and parts.

3M Company, as part of its vendor management process, uses five categories to address increasing levels of demonstrated quality competence to evaluate its vendors. The first category is the *new vendor.* Their performance capabilities are unknown initially. Interim specifications would be provided to them on an experimental basis. The next category is the *approved vendor,* where agreed-upon specifications are used and a self-survey is performed by the vendor. To qualify at this level, vendors need to have a minimum performance rating of 90% and must also maintain a rating of no less than 88%. Following this is the *qualified vendor.* To enter at this level, the vendor must demonstrate a minimum performance rating of 95% and must maintain a rating of at least 93%. Furthermore, the vendor must show that it meets ISO 9001, 9002, or 9003 standards, *or* be approved by the Food and Drug Administration, *or* pass a quality system survey conducted by 3M. The next category is the *preferred vendor.* To enter this category, the vendor must demonstrate a minimum performance rating of 98% and must maintain a rating of at least 96%. The preferred vendor demonstrates continuous improvement in the process and constantly meets 3M standards. Minimal or no incoming inspection is performed. The highest level of achievement is the *strategic vendor* category. These are typically high-volume, critical-item or equipment vendors that have entered into strategic partnerships with the company. They share their own strategic plans and cost data, make available their plants and processes for study by representatives from 3M, and are open to joint ventures, where they pursue design and process innovations with 3M. The strategic vendor has a long-term relationship with the company.

Other certification criteria are based on accepted norms set by various agencies. The *International Standards Organization (ISO)* is one such organization that has prepared a set of five standards, ISO 9000–9004. The *American National Standards Institute (ANSI)* and the *American Society for Quality Control (ASQC)* have established standards ANSI/ASQC Q9000–Q9004, which are equivalent to the ISO 9000–9004 standards. We will discuss these standards later in the chapter. Certification through these standards sends a message to the purchaser that the vendor has a documented quality system in place.

Another form of certification is attained through meeting the criteria set in the *Malcolm Baldrige National Quality Award,* which is administered by the U.S. Department of Commerce's National Institute of Standards and Technology. If vendors can demonstrate that the various criteria set out in the award are met, without necessarily applying for or winning the award, that by itself is an accomplishment that purchasers look at carefully. Details of the Malcolm Baldrige Award are described later in this chapter.

3-5 TOOLS FOR CONTINUOUS QUALITY IMPROVEMENT

To make rational decisions using data obtained on the product, or process, or from the consumer, organizations use certain graphical tools. We will explore the technical details of these tools in later chapters, but for now a brief description will suffice.

Check Sheets

Check sheets facilitate systematic record keeping or data collection. Observations are recorded as they happen, which reveals patterns or trends. Table 3-9 is a check sheet for an organization's computer-related problems. A tally was kept of each problem type on a weekly basis. A glance shows that the majority of problems are related to e-mail.

Pareto Diagrams

Pareto diagrams help prioritize problems by arranging them in decreasing order of importance. In an environment of limited resources, these diagrams help companies decide on the order in which they should address problems. Figure 3-12 shows a Pareto diagram of reasons for airline customer dissatisfaction. Delays in arrival is the major reason, as indicated by 40% of the customers. Thus, this is the problem that the airlines should address first.

TABLE 3-9 Check Sheet for Computer-Related Problems

	Week				
Problems	*1*	*2*	*3*	*4*	*Total*
Software access	IIII	I	II	I	8
Network problems	III	II	~~IIII~~	III	13
Insufficient memory	II	II	III	II	9
e-mail	IIII	~~IIII~~ I	~~IIII~~ ~~IIII~~ II	~~IIII~~ IIII	31
Server problems	I		II	I	4

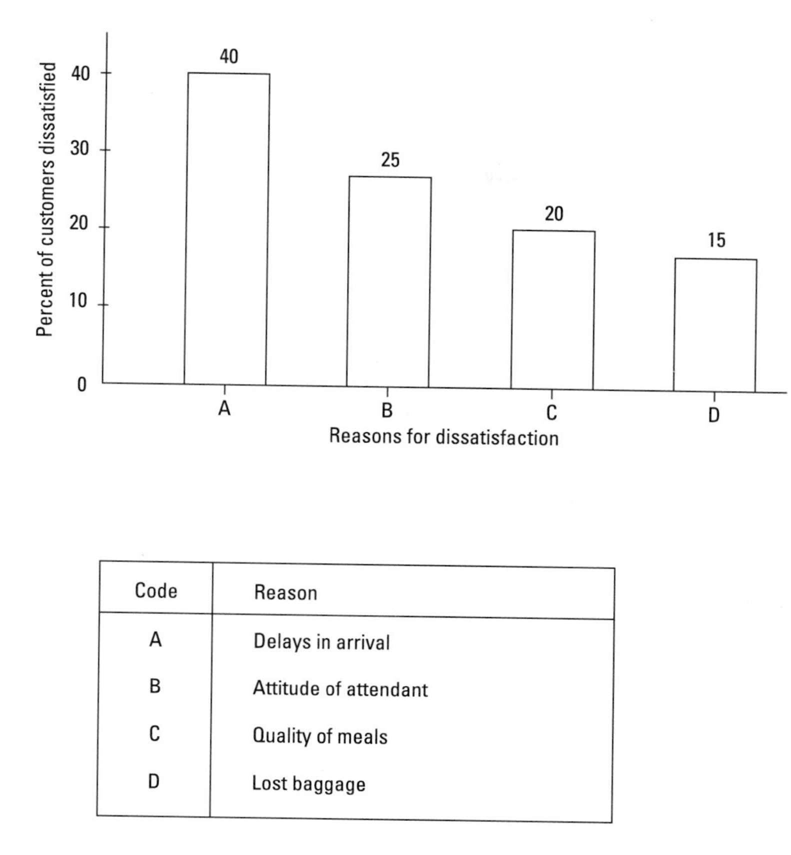

Code	Reason
A	Delays in arrival
B	Attitude of attendant
C	Quality of meals
D	Lost baggage

FIGURE 3-12 Pareto diagram for dissatisfied airline customers.

Flow Charts

Flow charts show the sequence of events in a process. They are used for manufacturing and service operations. Flow charts are often used to diagram operational procedures to simplify the system. They can identify bottlenecks, redundant steps, and nonvalue-added activities. A realistic flow chart can be constructed by using the knowledge of the personnel who are directly involved in the particular process. Valuable process information is usually gained through the construction of flow charts. Figure 3-13 shows a flow chart for processing incoming orders. The flow chart identifies where delays can occur—for example, when a purchase order has not been received. A more detailed flow chart would allow pinpointing of key problem areas.

Cause-and-Effect Diagrams

Cause-and-effect diagrams explore possible causes of problems, with the intention being to discover the root cause(s). Figure 3-14 is a generic view of how equipment, material, methods, and people—the causes—can impact a quality characteristic—the effect. To move closer to the root of the problem, each cause would be broken down into subcauses, which would probably be broken down again. As an example, the strength of a cement beam for construction of bridges could be a quality characteristic representing the effect. The proportions of the various ingredients used to make the cement are causes, in the category of material, that have an impact on beam strength. We will examine cause-and-effect diagrams in detail in Chapter 5.

Histograms

Histograms display large amounts of data that are difficult to interpret in their raw form. By providing a visual summary of the data, histograms reveal whether the process is centered around a target value, the degree of variation in the data, and

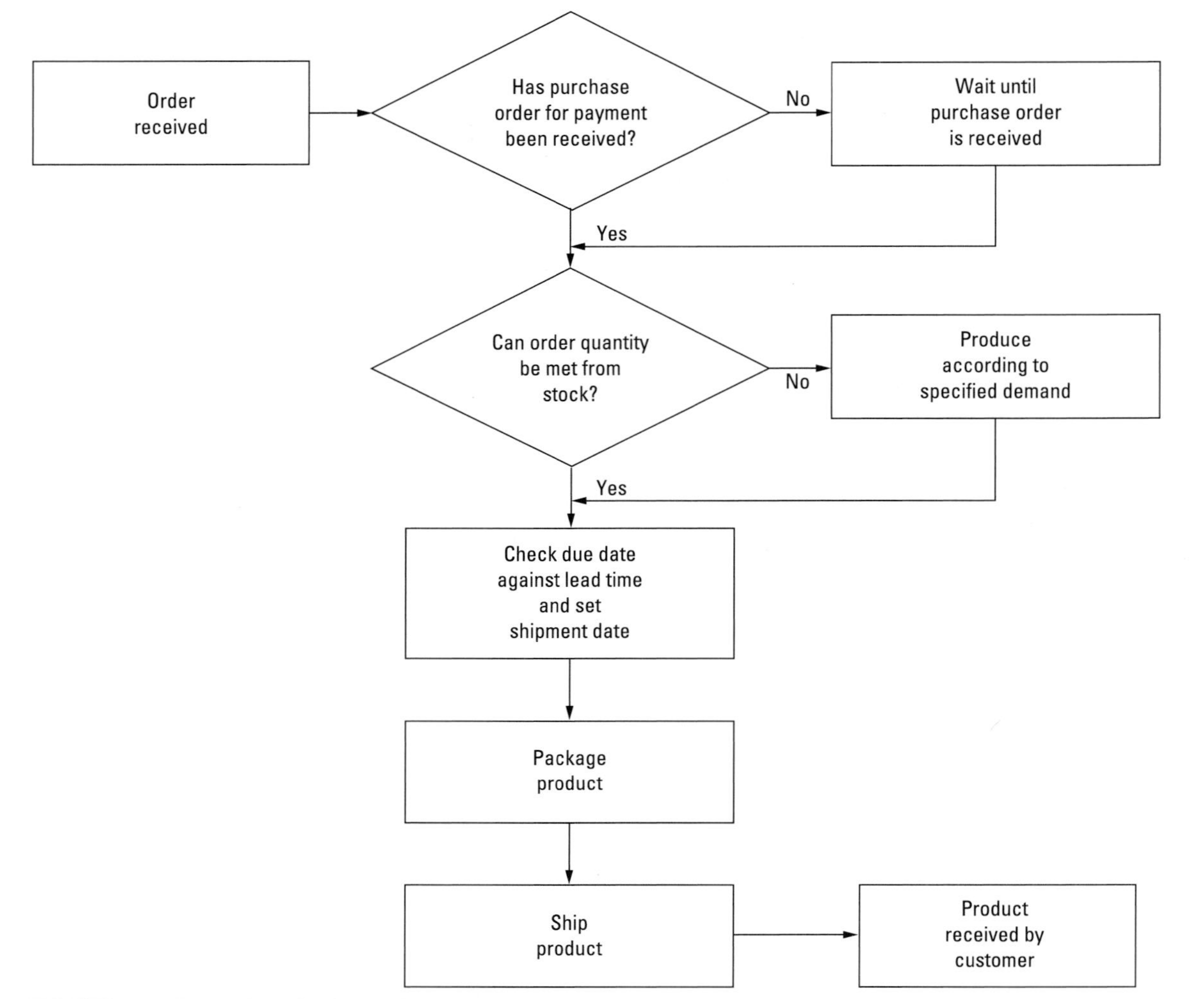

FIGURE 3-13 Flow chart for incoming order processing.

whether the data meet specifications. Thus, a histogram could help in identifying process capability relative to customer requirements. For example, Figure 3-15 is a histogram of the thickness of steel tubes required for an assembly where the specifications on thickness are 9.5 ± 0.5 mm. The histogram reveals that the process is not centered in the specification's target range. Furthermore, at its current setting, a proportion of the product is above the upper specification limit and thus is nonconforming. The histogram also reveals that the process variability is quite small, compared to the spread between the specifications. If the process setting is recentered at the nominal value of the specification, the product will be conforming.

Control Charts

Control charts distinguish special (assignable) causes of variation from common causes of variation. They are used to monitor and control a process on an ongoing basis. A typical control chart plots a selected quality characteristic, found from subgroups of observations, as a function of a sample number (which represents the time sequence of events). Characteristics such as sample average, sample range, and sample proportion of nonconforming units are plotted. The *center line* on a control chart represents the

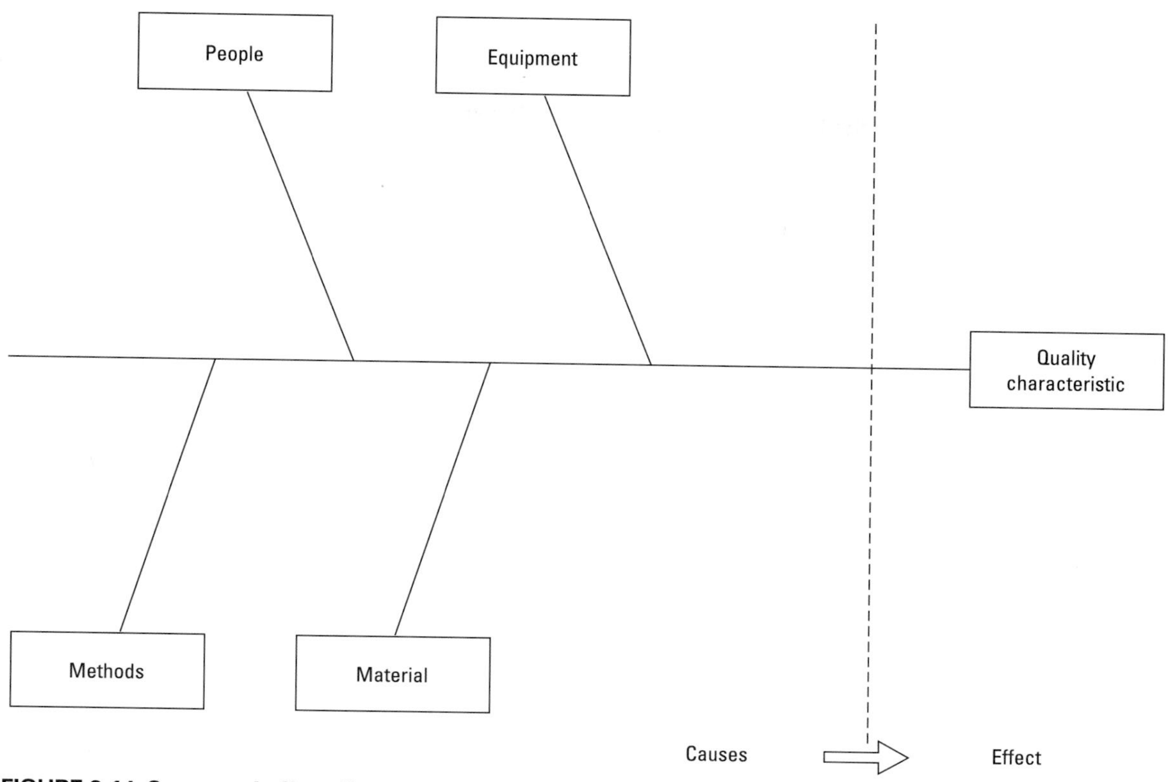

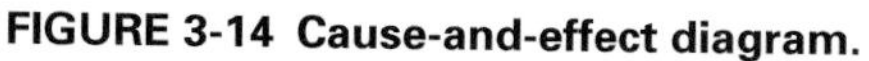
FIGURE 3-14 Cause-and-effect diagram.

average value of the characteristic being plotted. Two limits, known as the *upper control limit* (UCL) and *lower control limit* (LCL), are also shown on control charts. These limits are constructed so that, if the process is operating under a stable system of chance causes, the probability of an observation falling outside these limits is quite small. Thus, the control limits are the triggers that signal when the process is out of control. Figure 3-16 is a control chart for the proportion of nonconforming items. Note that two observations above the upper control limit have been recorded. This signals

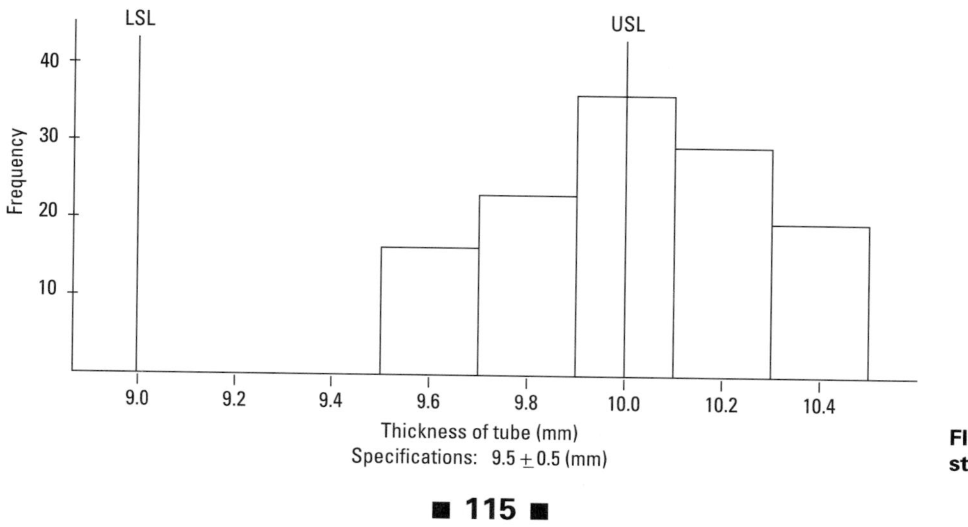

FIGURE 3-15 Histogram of steel tube thickness.

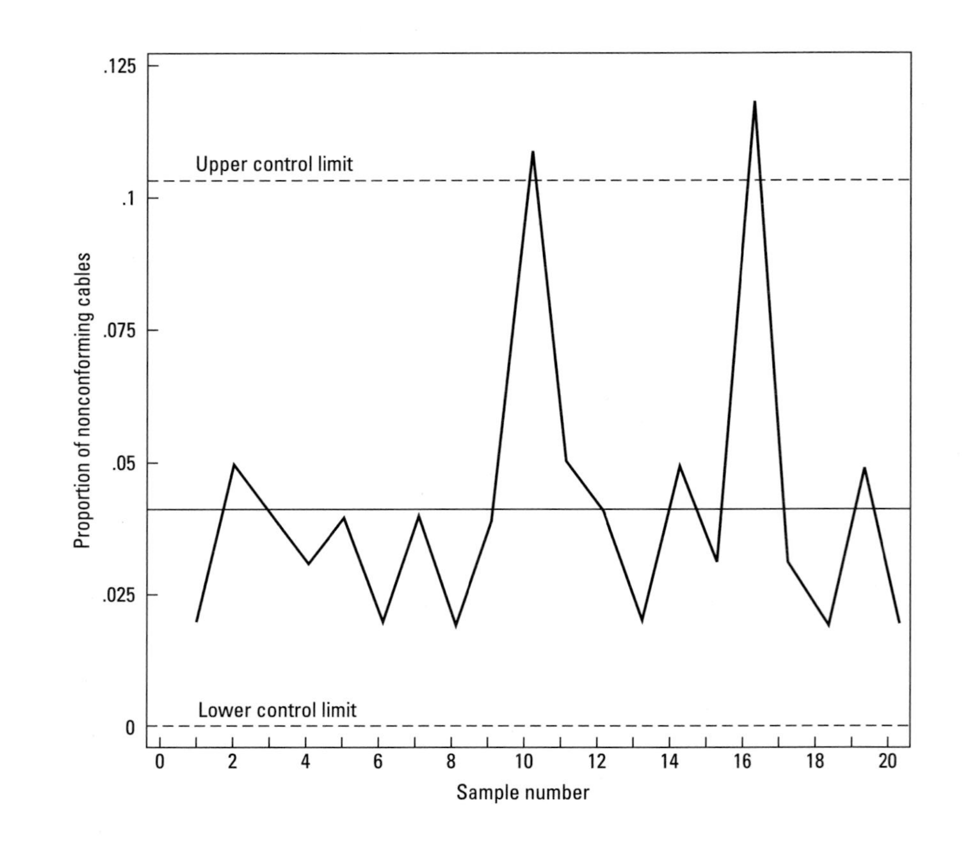

FIGURE 3-16 Control chart for proportion of nonconforming items.

FIGURE 3-17 Scatter plot of sales versus advertising expenditure.

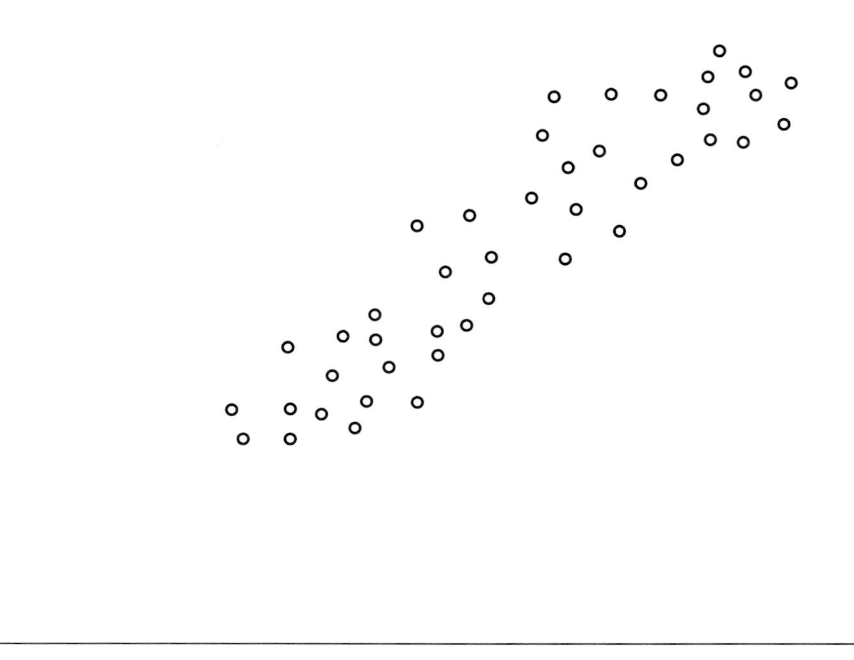

that workers should look for a special cause so they can take remedial action to bring the process back into control. Control charts are an important part of statistical process control, and we will discuss them extensively in Chapters 6, 7, and 8.

Scatter Plots

Scatter plots show the relationship between two variables. They are often used as follow-ups to a cause-and-effect analysis to determine whether a stated cause truly does impact the quality characteristic. Figure 3-17 plots advertising expenditure against company sales and indicates a strong positive relationship between the two variables. As the level of advertising expenditure increases, sales tend to increase.

3-6 INTERNATIONAL STANDARDS ISO 9000–9004*

Quality philosophies have revolutionized the way business is conducted. It can be argued that without quality programs the global economy would not exist because quality programs have been so effective in driving down costs and increasing competitiveness. Total quality systems are no longer an option—they are required. Companies without quality programs are at risk. The emphasis on customer satisfaction and continuous quality improvement has necessitated a system of standards and guidelines that support the quality philosophy. To address this need, the International Standards Organization (ISO) developed a set of five standards, **ISO 9000–9004.** The American National Standards Institute and the American Society for Quality Control brought their standards in line with ISO standards when they developed *ANSI/ASQC Q90–Q94* in 1987. The American standards are technically equivalent to the ISO 9000–9004 series, the difference being that they incorporate American English usage and spelling. As the marketplace has gone global and with the imminent consolidation of the European Economic Community, meeting international standards such as ISO 9000–9004 is a necessity. The ISO 9000–9004 standards were revised in 1994, with the corresponding ANSI/ASQC standards revised and renumbered to ANSI/ASQC Q9000–Q9004.

Features of ISO 9000 and ANSI/ASQC Q9000-1994

ISO 9000 and ANSI/ASQC Q9000-1994 facilitate the selection and use of the following two groups of standards:

1. ISO 9004 or ANSI/ASQC Q9004-1994, which guide organizations in quality management. Activities aimed at providing evidence that the intended quality is being achieved are often referred to as *internal quality assurance.*

   These standards also provide guidance on the technical, administrative, and human factors affecting the quality of products or services at all stages of the quality loop, from detection of need to customer satisfaction. Throughout these standards, emphasis is on customer satisfaction, establishing functional responsibilities, and the importance of assessing (as far as possible) the potential risks and benefits. These aspects should all be considered in establishing and maintaining an effective quality system.
2. ISO 9001–9003 or ANSI/ASQC Q9001–Q9003-1994, which are used for external quality assurance purposes in contractual situations. Activities aimed at providing confidence to the purchaser that the vendor's quality system will satisfy the purchaser's stated quality requirements are often referred to as *external quality assurance.*
   - ISO 9001—design/development, production, installation, and servicing (most stringent)
   - ISO 9002—production and installation
   - ISO 9003—final inspection and test (least stringent)

---

*Adapted from ASQC (1994), ANSI/ASQC Q9000–Q9004. Reprinted with the permission of ASQC.

Another objective of these standards is to clarify the interrelationships among the following three principal quality objectives:

1. The organization should achieve and sustain the quality of the product or service so as to continually meet the purchaser's stated or implied needs.
2. The organization should provide confidence to its own management that the intended quality is being achieved and sustained.
3. The organization should provide confidence to the purchaser that the intended quality is being, or will be, achieved in the delivered product or service. When contractually required, this provision of confidence may involve agreeing upon demonstration requirements.

In this context, the standards make the following definitions:

- *Quality policy* The overall quality intentions and direction of an organization concerning quality, as formally expressed by top management. The quality policy forms one element of the corporate policy and is authorized by top management.
- *Quality management* That aspect of the overall management function that determines and implements the quality policy. The attainment of desired quality requires the commitment and participation of all members of the organization, whereas the responsibility for quality management belongs to top management. Quality management includes strategic planning, allocation of resources, and other systematic activities for quality, such as quality planning, operations, and evaluations.
- *Quality system* The organizational structure, responsibilities, procedures, processes, and resources for implementing quality management. The quality system should be only as comprehensive as needed to meet the quality objectives. For contractual, mandatory, and assessment purposes, demonstration of the implementation of identified elements in the system may be required.
- *Quality control* The operational techniques and activities used to fulfill requirements for quality. To avoid confusion, care should be taken to include a modifying term when referring to a subset of quality control such as "manufacturing quality control" or when referring to a broader concept such as "companywide quality control." Quality control involves operational techniques and activities aimed both at monitoring a process and at eliminating causes of unsatisfactory performance at relevant stages of the quality loop in order to result in economic effectiveness.
- *Quality assurance* All those planned and systematic actions necessary to provide adequate confidence that a product or service will satisfy given requirements for quality. Unless given requirements fully reflect the needs of the user, quality assurance will not be complete. To be effective, quality assurance usually requires a continuing evaluation of factors that affect the adequacy of the design or specification for intended applications as well as verifications and audits of production, installation, and inspection operations. Providing confidence may involve producing evidence. Within an organization, quality assurance serves as a management tool. In contractual situations, quality assurance also serves to provide confidence in the supplier.

The relationship of these concepts is shown in Figure 3-18, which should not be interpreted as a rigid model. The elements that comprise a quality system are shown in Table 3-10, which is found in ANSI/ASQC Q9000 and does not form an integral part of the standard; it is given for information purposes only. Corresponding paragraphs or subsections dealing with the various topics of quality assurance can be found in each of the standards Q9001–Q9004, as noted in Table 3-10. It serves as an index that aids users in their search for specific topics in the standards.

For external quality assurance, the standards rate is as follows: ISO 9001 and Q9001 (most stringent); then ISO 9002 and Q9002; and ISO 9003 and Q9003 (least stringent). Note that the ISO 9002 and ISO 9003 standards are effectively subsets of the ISO 9001 standard because they deal with fewer topics. Internal quality assurance is dealt with in ISO 9004 and its equivalent Q9004. Since these standards overlap, we will examine only one—ISO 9001 and its equivalent ANSI/ASQC Q9001.

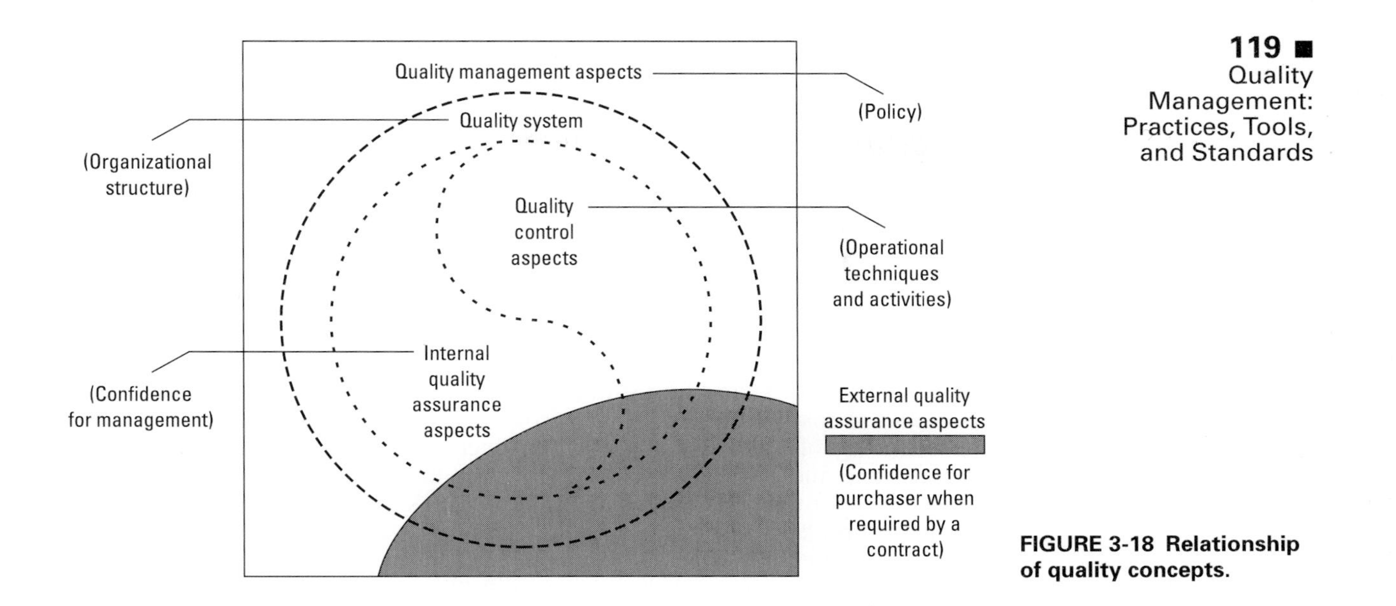

FIGURE 3-18 Relationship of quality concepts.

TABLE 3-10 Cross-Reference List of Quality System Elements* Adapted from ASQC (1994), ANSI/ASQC Q9000–Q9004. Reprinted by permission of ASQC.

Paragraph (or Subsection) No. in Q9004	*Title***	*Corresponding Paragraph (or Subsection) Nos. in*		
		Q9001	*Q9002*	*Q9003*
4	Management Responsibility	4.1 ●	4.1 ◐	4.1 ○
5	Quality System Principles	4.2 ●	4.2 ●	4.2 ◐
5.4	Auditing the Quality System (Internal)	4.17 ●	4.16 ◐	—
6	Economics—Quality-Related Cost Considerations	—	—	—
7	Quality in Marketing (Contract Review)	4.3 ●	4.3 ●	—
8	Quality in Specification and Design (Design Control)	4.4 ●	—	—
9	Quality in Procurement (Purchasing)	4.6 ●	4.5 ●	—
10	Quality in Production (Process Control)	4.9 ●	4.8 ●	—
11	Control of Production	4.9 ●	4.8 ●	—
11.2	Material Control and Traceability (Product Identification and Traceability)	4.8 ●	4.7 ●	4.4 ◐
11.7	Control of Verification Status (Inspection and Test Status)	4.12 ●	4.11 ●	4.7 ◐
12	Product Verification (Inspection and Testing)	4.10 ●	4.9 ●	4.5 ◐
13	Control of Measuring and Test Equipment (Inspection, Measuring, and Test Equipment)	4.11 ●	4.10 ●	4.6 ◐
14	Nonconformity (Control of Nonconforming Product)	4.13 ●	4.12 ●	4.8 ◐
15	Corrective Action	4.14 ●	4.13 ●	—
16	Handling and Post-Production Functions (Handling, Storage, Packaging, and Delivery)	4.15 ●	4.14 ●	4.9 ◐
16.2	After-Sales Servicing	4.19 ●	—	—
17	Quality Documentation and Records (Document Control)	4.5 ●	4.4 ●	4.3 ◐
17.3	Quality Records	4.16 ●	4.15 ●	4.10 ◐
18	Personnel (Training)	4.18 ●	4.17 ◐	4.11 ○
19	Product Safety and Liability	—	—	—
20	Use of Statistical Methods (Statistical Techniques)	4.20 ●	4.18 ●	4.12 ◐
—	Purchaser-Supplied Product	4.7 ●	4.6 ●	—

Key: ● full requirement; ◐ less stringent than ANSI/ASQC Q9001, ○ less stringent than ANSI/ASQC Q9002; — element not present.

*Note that the quality system elements in Q9001, Q9002, and Q9003 are in many cases, but not in every case, identical.

**The paragraph (or subsection) titles quoted are taken from Q9004; the titles given in parentheses are taken from the corresponding paragraphs and subsections in Q9001, Q9002, and Q9003.

Features of ISO 9001 (and ANSI/ASQC Q9001-1994)

The ISO 9001 standard specifies quality system requirements for use when a contract between two parties requires the demonstration of a vendor's capability to design and supply the product or service. The standard is aimed at *preventing nonconformity* at all stages from design to servicing (reinforcing the notion that quality assurance efforts should be directed toward defect prevention rather than defect detection).

ISO 9001 requires that the vendor describe its *management's responsibilities* in all aspects of the vendor's quality system. The vendor's management should define a policy that documents goals and objectives for attaining quality. In creating an organizational structure for a quality system, the responsibilities, distribution of authority, and interrelationship of all employees who manage, perform, and verify work affecting quality must be defined. The vendor should provide adequate resources for such verification activities as inspection, testing, monitoring of the design, production, installation, and servicing of the process and/or product. The employees responsible for design reviews and audits of the quality system must be distinct from those having direct responsibility for the work performed. The vendor's management should conduct reviews at appropriate intervals to ensure the effectiveness of the quality system.

The documentation of a *quality system* must ensure product conformance to specified requirements. In the creation of a quality system, standards should be developed for the following: acceptability, measurement procedures, design compatibility, production processes, installation, inspection and test procedures, preparation of quality plans and a quality manual, and record keeping.

Quality in marketing is maintained through *contract reviews.* The vendor is expected to review each contract to ensure that the requirements are adequately defined and documented. Also, the vendor should make sure that the contractual requirements can be met.

An important section of the standard deals with *design control.* Here the vendor is expected to establish and maintain procedures to control and verify the design to ensure that the requirements are met. To achieve this control, the vendor should draw up plans that identify responsibilities for each design activity. Design input requirements should be reviewed, as should design output. The output is expected to meet the design input requirements, conform to appropriate regulatory requirements, and identify those characteristics of the design that are crucial to the safe and proper functioning of the product.

The *document control* section of ISO 9001 ensures the availability of pertinent information from appropriate documents at all locations where such documents are essential for conducting operations. Obsolete documents should promptly be removed.

The *purchasing* section of the standard requires the vendor to ensure that the products the vendor purchases conform to the requirements. Selection of subcontractors should be based on their ability to meet contractual requirements, as demonstrated through past performance. Purchasing documents should contain data describing the product ordered, the type (or other form of identification), the requirements for approval or qualification of product, the procedures, the process equipment, and the personnel. The purchaser or purchaser's representative should be given the right to verify the purchased product at the source or upon receipt of the product. Such verification, however, cannot be used by the vendor as evidence of effective quality control for the subcontractors.

The vendor is expected to establish and maintain procedures for *identifying the product* from applicable drawings or other documents during all stages of production, delivery, and installation. If *traceability* is a requirement, an individual product or batch should have a unique identification.

Guidelines for *process control* are laid out in the standard. They require the vendor to identify and plan production and installation processes and to operate them under controlled conditions. The control section requires documenting work instructions that define the process and its installation in addition to monitoring suitable product and process characteristics. Product and process control should be monitored continuously to ensure that specified requirements are met.

Product verification through inspection and testing may involve in-process as well as final inspection. The vendor should ensure that the incoming product conforms to requirements. The amount and nature of receiving inspection will be influenced by the control exercised at the source and by documented evidence of quality conformance. In-process product conformance can be verified by process monitoring and control methods. Nonconforming products should be identified. Final inspection and testing should be conducted according to the quality plan, and documented procedures and appropriate records should be maintained.

Measuring and test equipment should be appropriate to the task and should be carefully calibrated; this is the responsibility of the vendor. Inspection, measuring, and test equipment should be capable of achieving the required accuracy and precision. The environmental conditions under which calibrations, inspections, and measurements are carried out should be adequate. Furthermore, handling, preservation, and storage of such test equipment should be such that the accuracy and fitness for use are maintained.

The standard requires that products should be *inspected and tested* to indicate conformance or nonconformance. These procedures must be maintained throughout production and installation to ensure the delivery of conforming products only. *Control of nonconforming products* should be established by the vendor through identification, documentation, evaluation, segregation (or disposal), and notification of the relevant departments. Nonconforming products should be examined to determine whether they should be reworked, regraded for alternative applications, or rejected.

Maintaining procedures for *corrective action* is critical. The vendor's procedures should investigate the cause of nonconforming products and should stipulate corrective actions to prevent recurrence. Analysis of processes, work operations, quality records, service reports, and customer complaints is crucial to detecting and eliminating potential causes of nonconforming products. The vendor should implement and record changes in procedures that result from corrective action.

The vendor is responsible for establishing, documenting, and maintaining *procedures for postproduction handling functions* such as storage, packaging, and delivery. The vendor should provide means of handling that prevent damage or deterioration. Areas for storage should be secure, with appropriate methods stipulated for authorizing receipt and dispatch. Packing and preservation processes should conform to requirements. The protection of quality, where contractually specified, should be extended to include delivery to destination.

An important feature of the standard makes the vendor responsible for establishing and maintaining procedures for identification, collection, indexing, filing, storage, maintenance, and disposition of *quality records.* These records should be maintained to demonstrate achievement of the required quality and the effective operation of the quality system. Pertinent subcontractor quality records should be included in this data. Retention times for quality records should be established and recorded. Where agreed upon contractually, quality records should be made available for a specified period for evaluation by the purchaser or purchaser's representative.

To verify whether quality activities comply with planned arrangements and to determine the effectiveness of the quality system, the vendor must carry out a comprehensive system of planned and documented *internal quality audits.* Scheduling for these audits should be based on the status and importance of the activity. The results of the audits should be documented and brought to the attention of the personnel responsible for the area audited. Appropriate management personnel should then take timely corrective action on the deficiencies found by the audit.

The vendor is responsible for establishing and maintaining procedures that identify the *training needs* and for providing training to all personnel who perform activities that affect quality. Task assignments should be based on such qualifications as appropriate education, training, and experience, as required. Appropriate records of training activities should be maintained.

In the event that *servicing* is specified in the contract, the vendor should establish and maintain procedures for performing and verifying that servicing meets the specified requirements.

Finally, where appropriate, the vendor should establish procedures for identifying adequate *statistical techniques* to verify the acceptability of process capability and product characteristics.

For a detailed discussion of each of these areas, the interested reader may consult the standards.

3-7 MALCOLM BALDRIGE NATIONAL QUALITY AWARD*

In the United States, the strategic importance of quality control and improvement has been formally recognized through the **Malcolm Baldrige National Quality Award,** which was created by Public Law 100-107 and signed into effect on August 20, 1987. The findings of Public Law 100-107 revealed that poor quality has cost companies as much as 20% of sales revenues nationally and that an improved quality of goods and services goes hand in hand with improved productivity, lower costs, and increased profitability. Furthermore, improved management understanding of the factory floor, worker involvement in quality, and greater emphasis on statistical process control can lead to dramatic improvements in the cost and quality of products and services. Quality improvement programs must be management-led and customer oriented. All of these findings are consistent with the quality philosophies discussed in Chapter 2.

The U.S. government felt that a national quality award would motivate American companies to improve quality and productivity. The idea was that national recognition of achievements in quality would provide an example to others and would establish guidelines and criteria that could be used by business, industrial, governmental, and other organizations to evaluate their own quality improvement efforts. Furthermore, detailed information on the quality practices of the winning organizations would be made available to others so that they in turn could adopt the desirable features. In 1991, 106 companies applied for the award, whereas in 1995, only 47 companies did so. Pursuing the criteria, regardless of whether a company applies for the award, is of major benefit as it stimulates a companywide quality effort.

Award Eligibility Criteria and Categories

Responsibility for the award is assigned to the National Institute of Standards and Technology (NIST), an agency of the U.S. Department of Commerce. The American Society for Quality Control assists in administering the award program under contract to NIST. Awards are made annually to U.S. companies that excel in quality management and quality achievement; the first award was given in 1988.

Any for-profit business located in the United States or its territories is eligible to apply. Local, state, or national government agencies may not apply. Four restrictions, called *eligibility criteria,* apply; they are as follows:

1. A company or its subsidiary is eligible only if the quality practices of its major business functions are inspectable in the United States or its territories.
2. At least 50% of a subsidiary's customer base (dollar volume for products and services) must be free of direct financial and line organization control by the parent company. For example, a subsidiary is not eligible if its parent company or another subsidiary of the parent company is the customer for more than one-half of its total products or services.
3. Individual units or partial aggregations of units of "chain" organizations (such as hotels, retail stores, banks, or restaurants) are not eligible.
4. Subsidiaries performing any of the business support functions of the company are not eligible. Examples of business support functions include sales/marketing/distribution, customer service, finance and accounting, human resources, purchasing, legal services, and research and development.

---

*Adapted from "1997 Criteria for Performance Excellence—Malcolm Baldrige National Quality Award," U.S. Department of Commerce, National Institute of Standards and Technology, Gaithersburg, Md. 20899.

Multiple-application restrictions exist in that a subsidiary and its parent company may not both apply for awards in the same year. Furthermore, if a company receives an award, the company and all its subsidiaries are ineligible to apply for another award for a period of five years.

There are three *award eligibility categories:* manufacturing, service, and small business. Up to two awards may be given in each of the three categories. Briefly, the categories are defined as follows:

1. *Manufacturing.* Companies or subsidiaries that produce and sell manufactured products or manufacturing processes or produce agricultural, mining, or construction products. For a complete list of manufacturing and products, the *Standard Industrial Classification (SIC)* codes may be consulted. Typical examples are agriculture (crops and livestock), forestry, building and heavy construction contractors, food and tobacco products, apparel, chemicals, petroleum refining, machinery/computer equipment, electrical/electronic equipment, and transportation equipment.

2. *Service.* Companies or subsidiaries that sell services. The list of SIC codes may be consulted for a complete listing of services. Some examples are agricultural services, transportation (air, railroad, buses, water), communications, utilities (electricity, gas), automobile dealers and service stations, banking and insurance, real estate, hotels and lodging places, personal and business services, health and legal services, and amusement and recreation centers.

3. *Small business.* Complete businesses with no more than 500 full-time employees. Business activities may include manufacturing and/or service. A small business must be able to document that it functions independently of any other businesses that are equity owners.

Criteria for Evaluation

Applying for the award is a two-step process. First, potential applicants must establish their eligibility in one of the three award categories. Second, an application form and a report must be completed. These forms are reviewed in a four-stage process. In the first stage, a review is conducted by at least five members of a board of examiners. At the conclusion of the first stage, the panel determines which applications should be referred for consensus review. In the second stage, the application report is reviewed by at least four members of a board of examiners led by a senior examiner. After the second stage, the panel determines which applications should receive the site visits. The third stage involves an on-site verification of the application report. The site-visit review team develops a report for the panel of judges, leading to the fourth stage involving the judges' final review. The panel of judges develops a set of recommendations for the National Institute of Standards and Technology. Awards are traditionally presented by the President of the United States.

The award criteria are built on the following core values.

Customer-driven quality: **The customer is the final judge of quality, so the company emphasizes product and service features that contribute value to customers and lead to customer satisfaction, preference, and retention, which are dynamic in nature and continually changing.**

Leadership: **The senior leaders set company strategies, which are oriented to meeting and exceeding customer needs, and prescribe methods to accomplish stated goals and objectives. Senior leadership serves as a role model and encourages participation, learning, and creativity by all employees.**

Continuous improvement and learning: **Continuous improvement and learning are ongoing activities. Problems are eliminated at the source.**

Employee participation and development: **Employees are encouraged to grow through education and training. The practices of selection, performance, training, and career advancement are integrated and aligned with business strategy.**

Fast response: **Cycle times for new or improved products or services are continually being reduced. To create a faster response time requires process analysis, simplification, and elimination of redundant or nonvalue-added activities.**

Design quality and problem prevention: **Quality is designed into the product or service. The design stage plays an important role in building an efficient production and delivery system and reducing costs of correcting problems that occur downstream. The design-to-introduction cycle time of a product is short because the company captures relevant information in the design stage.**

Long-range view of the future: **Management has a long-term commitment to its stakeholders; they include customers, employees, suppliers, stockholders, the public, and the community.**

Management by fact: **Data and information are used to arrive at various performance measures relating to product and process, customer, market, supplier, employee-related, cost, and financial objectives. Such information is used to identify factors that may lead to improvement in the company operations and better customer satisfaction.**

Partnership development: **Strategic goals are accomplished through internal and external partnership alliances. Thus, the company promotes internal alliances through good labor relations and interdepartmental networking, and it also seeks external partnerships with its customers and vendors.**

Company responsibility and citizenship: **The company maintains proper business ethics in all its dealings and protects public health, safety, and the environment. The company emphasizes resource conservation and waste reduction and shoulders its responsibilities in the community at large by improving education and health care.**

Results focus: **The company attempts to meet the needs of all stakeholders, including customers, employees, suppliers, and stockholders. While the needs of these constituencies may change and may conflict, the company should use a balanced composite of performance measures to monitor performance.**

Seven categories incorporate the core values and concepts. They determine the framework for evaluation. Table 3-11 shows the award criteria categories, subcategories, and point values of the *1997 Criteria for Performance Excellence*. Briefly, the categories may be described as follows:

1. *Leadership.* The leadership category examines senior executives' *personal* leadership and involvement in creating and sustaining a customer focus and clear and visible quality values. The way in which these quality values are integrated into the company's management system and the way in which the company addresses its public responsibilities are also examined.

2. *Strategic planning.* This category examines the manner in which the company sets strategic directions to define and strengthen its competitive position. How do the company's key action plans link to their performance?

3. *Customer and market focus.* The manner in which the company determines requirements and expectations of customers and markets is addressed in this category. Also of importance is the process through which customer satisfaction is enhanced and assessed. Does the company continually monitor those factors that influence customer preference and loyalty?

4. *Information and analysis.* This category deals with how the company manages its information. These procedures aid in sustaining company performance. The company's selection, use, and management of information and data affect its process management. Thus, both financial and nonfinancial data management techniques are examined.

5. *Human resource development and management.* This category examines how the company develops and realizes the full potential of its workforce in pursuing the com-

TABLE 3-11 The 1997 Malcolm Baldrige Award Criteria Categories

1997 Categories	*Point Value*	
1 Leadership		**110**
1.1 Leadership System	80	
1.2 Company Responsibility and Citizenship	30	
2 Strategic Planning		**80**
2.1 Strategy Development Process	40	
2.2 Company Strategy	40	
3 Customer and Market Focus		**80**
3.1 Customer and Market Knowledge	40	
3.2 Customer Satisfaction and Relationship Enhancement	40	
4 Information and Analysis		**80**
4.1 Selection and Use of Information and Data	25	
4.2 Selection and Use of Comparative Information and Data	15	
4.3 Analysis and Review of Company Performance	40	
5 Human Resource Development and Management		**100**
5.1 Work Systems	40	
5.2 Employee Education, Training, and Development	30	
5.3 Employee Well-Being and Satisfaction	30	
6 Process Management		**100**
6.1 Management of Product and Service Processes	60	
6.2 Management of Support Processes	20	
6.3 Management of Supplier and Partnering Processes	20	
7 Business Results		**450**
7.1 Customer Satisfaction Results	130	
7.2 Financial and Market Results	130	
7.3 Human Resource Results	35	
7.4 Supplier and Partner Results	25	
7.5 Company-Specific Results	130	
Total Points		**1000**

Source: U.S. Department of Commerce, National Institute of Standards and Technology, "1997 Criteria for Performance Excellence—Malcolm Baldrige National Quality Award," Gaithersburg, Md. 20899.

pany's quality and performance objectives. Does the company maintain an environment for excellence that encourages full participation and personal and organizational growth?

6. *Process management.* Key aspects of process management, which include customer-focused design, product and service delivery processes, support processes, and vendor and partnering processes involving all work units, are examined in this category. Does the company incorporate changing customer requirements and technology into its product and service designs? Are production and delivery processes designed to meet operational performance requirements?

7. *Business results.* This category examines the company's performance and improvement in such key business areas as customer satisfaction, financial and marketplace performance, human resources, vendor and partner performance, and operational performance. Performance levels relative to competitors are also considered. Current levels and trends in key measures of customer satisfaction and dissatisfaction are taken into account. Financial and marketplace performance include such measures as aggregate return on investment, market share, business growth, and new markets entered. Human resource results include employee well-being, satisfaction, development, and work system performance.

Note that these evaluation criteria embody the tenets of the quality philosophies presented in Chapter 2. For further details concerning each of the subcategories, consult the relevant reference from the U.S. Department of Commerce (1996).

Profiles of Winners

Table 3-12 lists the *Baldrige Award* winners from 1988, the year the award was first given, to 1996. In recognizing the achievements of these companies, the award promotes awareness and competitiveness. Thus, the winners become models for organizations to emulate in their pursuit of quality.

Motorola, Inc., winner in the manufacturing category in 1988, has a corporate objective of total customer satisfaction. The company's quality goal is "zero defects in everything we do." One of the key initiatives undertaken by Motorola is the concept of six-sigma quality, a performance measure described earlier in this chapter. The company has a target of no more than 3.4 defects per million products. Motorola achieved significant improvement in quality by reducing the total cycle time, which is the time between when a customer places an order and when it is received.

Westinghouse Electric Corporation—Commercial Nuclear Fuel Division (CNFD), another winner in the manufacturing category in 1988, embraced a quality culture with their slogan of doing the right things right the first time. Continuous improvement is achieved through four initiatives: management leadership, product and process leader-

TABLE 3-12 Malcolm Baldrige Award Winners

Year	*Manufacturing Companies*	*Small Business*	*Service Companies*
1988	Motorola, Inc. Westinghouse Electric Corporation—Commercial Nuclear Fuel Division	Global Metallurgical, Inc.	
1989	Milliken & Company Xerox Corporation—Business Products & Systems		
1990	General Motors—Cadillac Motor Car Division IBM Rochester	Wallace Co., Inc.	Federal Express Corporation
1991	Solectron Corporation Zytec Corporation	Marlow Industries	
1992	AT&T Network Systems Texas Instruments, Inc.—Defense Systems & Electronics Group	Granite Rock Company	AT&T Universal Card Services The Ritz-Carlton Hotel Company
1993	Eastman Chemical Company	Ames Rubber Corporation	
1994		Wainwright Industries, Inc.	AT&T Consumer Communications Services GTE Directories Corporation
1995	Armstrong World Industries, Inc., Building Products Operations Corning Telecommunications Products Division		
1996	ADAC Laboratories	Customer Research, Inc. Trident Precision Manufacturing, Inc.	Dana Commercial Credit Corporation

ship, human resource excellence, and customer satisfaction. The system tracks improvements in over 60 key performance areas through statistical techniques and other evaluative tools. The company invests in state-of-the-art technology, such as robotics, supercomputer simulations, expert systems, laser diagnostics, and laser welding. Management commitment through integration of quality into all design, production, and customer service activities along with a motivated workforce facilitated the successful implementation of a total quality approach.

General Motors' Cadillac Motor Car Division, winner in the manufacturing category in 1990, initiated its quality approach with customer satisfaction as a master plan. Top management began implementation of simultaneous engineering, which is an integrated approach to product design and development and process monitoring. This concept anticipates the impact of changes in one functional area on other functional areas. In contrast, in the traditional serial approach of automobile development and manufacturing, individual functional departments operate largely in isolation. At Cadillac, teamwork is a key element in the quality journey. Furthermore, Cadillac developed a partnership with the United Auto Workers (UAW) that has been an important catalyst in defining the quality culture of the organization. Executives and plant managers and union leaders serve on the Divisional Quality Council, which is part of the UAW/GM Quality Network.

Three 1996 *Malcolm Baldrige Award* winners are profiled next.

ADAC Laboratories

ADAC Laboratories, a Silicon Valley–based maker of high-technology healthcare products, initiated a management system based on quality management principles as a way to change the culture of the company after a successful turnaround in the mid-1980s. This customer-focused system, based on the Malcolm Baldrige Award criteria, has helped transform the company into a world leader in markets for diagnostic imaging and healthcare information systems. Founded in 1970, ADAC Laboratories designs, manufactures, markets, and supports products for nuclear-medicine imaging, radiation-therapy planning, and managing healthcare information. Many of the company's products are regulated by the Food and Drug Administration, which requires adherence to strict safety standards. ADAC has installed about 5000 systems at more than 2500 hospitals, clinics, and other sites around the world. These systems are extremely complex, comprising several thousand parts, the vast majority purchased from suppliers. Most of ADAC's 710 employees are based at its corporate headquarters and production facility in Milpitas, California, and at offices in Houston, Texas, headquarters of the company's healthcare information systems business. About 300 ADAC employees work out of their homes or in small field offices in North America and Europe.

ADAC's whole-organization approach to increasing customer satisfaction and improving quality is best illustrated by its novel decision to eliminate the Quality Council, a body composed of executives and managers and charged with overseeing the company's quality management processes. As a result of benchmarking a Baldrige company, ADAC replaced the council with two weekly meetings that are open to all employees as well as customers and suppliers. During these meetings, numerous employees present data on key measures of customer satisfaction, quality, productivity, and operational and financial performance.

The company's corporate planning process—known as DASH—develops strategic plans for the next three to five years and an annual business plan. Consistent with ADAC's primary core value, "Customers come first," the DASH process begins with a thorough, fact-based analysis of customer requirements—today's requirements and tomorrow's requirements. This analysis uses data gathered from a variety of sources, including surveys, lost-order information, interviews conducted by customer-contact employees, logs of service calls, and focus groups. These results are integrated with those from analyses of competitive forces, risks, company capabilities, and supplier capabilities.

Short- and long-term strategies are then distilled into the "vital few"—key business drivers that focus and align plans and continuous improvement efforts over the next year. In turn, each department translates the strategic directions and business drivers into specific requirements and action plans. These are the basis for MITs—"most important tasks"—the top priority improvements set for functional units and for individual employees. Alignment of plans is ensured by cross-functional work sessions at which MITs are presented.

Staying on Course ADAC management recognizes that good decisions begin with good information. The company has made significant investments in data collection systems targeted to key needs and activities, such as tracking design defects and customer calls for support. Most workers participate in highly empowered teams, and all manufacturing employees are members of self-directed work teams. All employees receive training on customer and supplier models, problem solving, and basic statistical analysis. In 1995, each received, on average, more than 60 hours of training.

At quarterly "measurement summits," representatives from all departments review the types of data collected according to the company's three criteria—that is, whether the data support key business drivers; address one of the "five evils" (waste, defects, delays, accidents, or mistakes); or support objective analysis for improvement. Participants also examine whether new categories of data are needed to guide continuous improvement efforts. Benchmarking, an integral element of ADAC's standardized problem-solving process, is used regularly by all continuous improvement teams to set performance goals and to gauge the effectiveness of its management processes.

QUALITY AND BUSINESS PERFORMANCE ACHIEVEMENTS

- Customer focus at ADAC Laboratories is revealed by its core value, "Customers Come First." All executives are expected to spend 25% of their time with customers, personally take customer calls, and invite customers to attend weekly quality meetings. Customer satisfaction results have shown positive and improving trends over a five-year period for postsales technical support (10% increase), customer retention (from 70% to 90%), and service contract renewals (from 85% to 95%).
- One of ADAC's measures of service quality is service cycle time, which determines the total time for getting a system back in operation. Service cycle time is critical to customers because they often cannot treat patients until a problem is fixed. Since the company began tracking this measure in 1990, the average cycle time has declined from 56 to 17 hours.
- ADAC's nuclear medicine market share has grown over the past five years from 10% to approximately 52% in the United States and from 5% to approximately 28% in Europe. Also, its revenue has tripled since 1990—compared to a 50% increase for the industry as a whole.
- ADAC consistently brings products to market faster than its competitors. Time to market leadership is evidenced by three product releases that averaged just over half the development time of competitors for similar products.
- Current levels and trends in key business measures demonstrate positive trends and performance. For example, revenue per employee has risen from about $200,000 in 1990 to almost $330,000 in 1995. On this overall measure of productivity, ADAC has achieved a 65% greater efficiency than its best competitor. Another measure is the number of direct-labor dollars required to build cameras used to detect and diagnose health problems. Through more efficient processes and technology improvements, labor dollars per camera have decreased 40% since 1994.
- ADAC has a strong focus on process that is standardized through training and education of all employees. Over 100 customer and operational measures are reported in twice-weekly quality meetings, which are open to all employees, customers, and suppliers.
- ADAC's business planning process, known as DASH, measures financial, customer, and operational performance and regulatory compliance. At quarterly DASH meetings, progress is checked and, if needed, midcourse changes are made emphasizing recent performance compared with plans and the vision of the future.
- ADAC started an Advanced Clinical Research Program in 1992 to fund research at leading hospitals to improve the quality and efficiency of healthcare. ADAC donates approximately $350,000 annually to the program.

Trident Precision Manufacturing, Inc.

Founded in 1979, privately held Trident Precision Manufacturing, Inc., located in Webster, New York, manufactures precision sheet metal components, electromechanical assemblies, and custom products, mostly in the office-equipment, medical-supply, computer, and defense industries. It has grown from a three-person operation to an employer of 167 people, occupying a modern, 83,000-square-foot facility. In 1995, revenues totaled $14.5 million. Well established as a local supplier, Trident is now diversifying to serve regional and international markets. The company's Senior Executive Team devised and launched "Excellence in Motion," a strategy designed to sharpen Trident's focus on its customers and to instill a commitment to continuous improvement throughout the organization.

Satisfying Customers Trident has established "quality as its basic business plan" to accomplish short- and long-term goals for each of its five key business drivers: customer satisfaction, employee satisfaction, shareholder value, operational performance, and supplier partnerships. All goals, however, contribute to achieving Trident's overarching aim of total customer satisfaction. Each improvement project begins with a thorough analysis of how to meet or exceed customer requirements in four critical areas: quality, cost, delivery, and service. Metrics are designed to ensure that progress toward the customer-targeted improvements can be evaluated. The company's data-collection system provides all personnel with a current record of the company's progress toward its goals. Forty-five networked computers make this information readily accessible. Performance data are reviewed daily in each department and weekly by the Senior Executive Team. Once each month, this team aggregates the data for the entire company and reports on progress toward goals set for each of the five key business drivers.

Beyond tracking its operational and financial performance, Trident also analyzes data collected from a variety of other internal and external sources. These include semiannual surveys of customers, suppliers, and employees; benchmarking studies; discussions with customers; employee forums; market reports; quarterly quality audits; and an independently conducted annual assessment of the company's competitive position within its industry. Regular contact with customers and suppliers is an essential element of Trident's quality strategy. Senior executives meet twice a year with representatives of each customer company for in-depth discussions on Trident's performance as a supplier, while 41 customer-contact personnel interact with these firms on a daily basis. Customers, as well as key suppliers, also participate in Continuous Involvement Meetings, initiated by Trident to gain full understanding of a customer's new or modified product design. Direct feedback flags real and potential problems that can be acted upon immediately, and it alerts Trident to changing customer requirements that can be addressed in short- and long-range planning. Responding to future requirements identified through such discussions, for example, Trident recently raised its goal for manufacturing- process reliability to a level significantly more stringent than now specified by its most demanding customers. Trident also uses technology to strengthen links to customers and suppliers. Electronic data interchange capabilities, for example, permit paperless transactions, while file-exchange capabilities enable customers to send their designs electronically to Trident's computer-aided design and manufacturing equipment.

Fulfilling Employee Expectations Top management takes the lead in planning, setting improvement priorities, and systematically reviewing progress toward quality goals, but executives and managers see their primary role as facilitating the transformation to a continuous improvement culture. Workers, the company believes, are the "source and foundation for quality leadership and competitiveness." Consequently, Trident's human resource strategies emphasize training, involvement through teams, empowerment, and reward and recognition. Organized into functional departmental teams, employees "own" specific processes and are given responsibility for identifying problems and opportunities for improvement. To foster innovation, employees have the authority to modify their process, using the company's documented process improvement procedure, which focuses attention on nonvalue-added activities that can be eliminated.

The company also relies heavily on the contributions of cross-functional teams, and it encourages employees to diversify their work skills and abilities. In fact, 80% of Trident workers are trained in at least two job functions, well on the way to the 1998 goal of 100%. To reinforce worker commitment to continuous improvement, the company regularly acknowledges exemplary performance. Reward and recognition of employees have climbed steadily, from just nine incidents in 1988 to 1201 in 1995.

QUALITY AND BUSINESS PERFORMANCE ACHIEVEMENTS

- On-time delivery indicates a positive trend over time, rising from 87% in 1990 to 99.94% in 1995.
- Trident monitors custom product reliability through defects per 100 machines. Custom products go directly from Trident to the customer's distribution center for shipping to their customers. For the past two years, Trident's custom products have had "zero" defects. As a result, Trident has been able to give its customers a full guarantee against defects.
- Results in achieving full employee involvement are indicated by 100% participation on departmental work teams since 1992, 97% of recommendations for process improvements being accepted in 1995, with over 95% accepted since 1991.
- The percentage of direct-labor hours spent on rework declined from 8.7% in 1990 to 1.1% in 1995.
- Financial indicators are positive: sales per employee rose from $67,000 in 1988 to $116,000 per employee in 1995. Return on assets rose from 7.9% in 1992 to 10% in 1995, compared to similar companies' return on assets of 4.7% in 1992 and 7.8% in 1995.
- Trident's quality rating (based on customer reports and customer rejects) for its major customers shows performance results consistent over time and above 99.8%.
- Customer satisfaction performance is strong: Trident has never lost a customer to a competitor, and sales volume has increased steadily from $4.4 million in 1988 to $14.3 million in 1995. Due to its strong customer focus, Trident has been able to maintain its status as a key supplier to major customers, even after those customers reduced their suppliers by 65% to 75%. Trident is the only supplier of General Dynamics to receive its Supplier Excellence Award.
- To ensure that new work being considered by the company will not introduce toxic or hazardous materials into Trident's manufacturing environment, the company works with potential customers before bids are placed on a project in an effort to ensure that no materials required for the project would pose a health or safety risk. If this cannot be done, Trident will not bid on the job.
- Employee turnover has shown dramatic improvement, falling from 41% in 1988 to 5% in 1994 and 1995. Trident's goal is to have less than 2% turnover.
- Trident's investment in training and education over the last several years is impressive—4.6% of payroll—for a small company. Expenditures in this area have consistently been two or three times the national average for the past seven years.
- Through customer participation in Continuous Improvement Meetings, Trident shares manufacturing plans, methods, times, and costs with its customers in the quotation or concurrent engineering stages. This facilitates process modification and plan changes and allows agreement on performance metrics.

Dana Commercial Credit Corporation

When it comes to customer satisfaction, competitor performance, operational effectiveness, and workforce capabilities, Dana Commercial Credit (DCC) Corporation always wants to know the score. A provider of leasing and financing services to a broad range of commercial customers, the Dana Corporation subsidiary has developed a collection of quality-linked "scoring processes" that assesses how the company is progressing in its pursuit of continuous improvement goals set for all key areas of the business.

Since 1992, when DCC embarked on an effort to improve teamwork and organizational communications, the company has scored gains in the quality of its performance, customer satisfaction, and the percentage of repeat business. DCC competes on the basis of value-added lease products and services, not just financing. Since 1991, the dollar volume of DCC leases has more than tripled, to more than $1 billion.

About DCC Since 1980, when it was started with a $2.5 million investment by its corporate parent, DCC has grown to become the eleventh largest U.S. leasing company among the more than 2000 in business here, with 1995 revenues of nearly $200 million and total assets of $1.5 billion. Headquartered in Toledo, Ohio, DCC consists of seven major product groups, each aligned with a different market segment. These include leases for power generation facilities and real estate properties with values up to $150 million and leases for commercial equipment resellers, manufacturers, and distributors ranging in price from $4000 to $3 million. Unique transactions are DCC's specialty, such as arranging the short-term lease of microcomputers for the television network covering the 1994 Winter Olympic games, providing full-service leasing of on-site photo processing equipment to retail outlets, and helping put a major gas processing facility in the North Sea "on-line."

DCC lease contracts are prepared for distributors and organizations that lease equipment or buildings. However, the company views financial intermediaries, such as investment bankers and equipment manufacturers, as its primary customers, because they are the major source of leasing recommendations and referrals. Most of DCC's 547 employees are located in Ohio, Michigan, Toronto, London, Paris, and Zurich.

Adding Value for Customers DCC aims to be the preferred financial services provider in its selected markets. To achieve this objective, DCC is increasing customer satisfaction through the commitment, skills, and innovativeness of its people and through its quality improvement system. The system provides the strategy, direction, incentives, tools, and resources necessary for continuous improvement, but it is customized by design so that each group concentrates on the particular requirements and expectations of customers in its market niche.

DCC's strategic plan integrates customer, operational, people, supplier, and quality plans into seven guiding plans, one for each product group. Product group improvement goals are translated into actions that address the company's key business drivers—customer satisfaction, knowledgeable people, quality processes, and profit for the shareholder. In all groups, action plans are linked directly to anticipated improvements in meeting four key customer requirements, which are determined by the Division Operating Committee but are adapted to each market. Customer-related performance metrics are established for each process and each improvement project.

Measurements are tracked closely. Each month, scorecards are compiled to inform all DCC people of progress toward reaching goals for customer satisfaction, human resources, and key processes. A monthly competitor scorecard is also prepared to compare DCC performance on key customer satisfaction measures. In 1995, DCC piloted a customer expectation scorecard, compiled largely from information gathered by cross-functional sales teams that work closely with customers in the design of leasing arrangements and new products. Now deployed companywide, this scorecard alerts DCC to changing customer requirements and indicates how well the company is responding. DCC's "SWOT" (strengths, weaknesses, opportunities, and threats) analyses compare company performance to benchmark measures. Performance in the key process areas is flagged as a strength or weakness compared to the benchmark, as an opportunity, or as a threat to the business.

Knowledgeable People DCC's mission statement and the Dana Style of Management assert that "people are our most important asset." To promote organizational flexibility and responsiveness to customers, DCC limits the number of management layers within its groups to five or fewer, and its "just do it" policy empowers DCC's people to act on their ideas for improvement without prior approval. The company uses employee education and training to differentiate itself from its competitors. In 1992, it created the Education Group to develop and teach courses in interpersonal communication, quality, and marketing as well as in technical areas needed to structure customized loans. The company also provides 100% reimbursement for successfully completed college courses. Training and education needs and effectiveness are reviewed monthly. Careful attention is paid to further enhancing the skills of people, including senior managers, who have direct contact with customers.

QUALITY AND BUSINESS PERFORMANCE ACHIEVEMENTS

- DCC consistently meets or exceeds key customer requirements, including completing transactions that competitors cannot, closing transactions on time, getting transactions done as agreed and providing customized lease products before the competition. For example, DCC's Capital Markets Group (CMG), which accounts for 50% of new transactions, has closed all of its transactions on time for the past five years. DCC's Dealer Products Group (DPG) U.S., accounting for 20% of new volume, has reduced the time it takes to approve a transaction from about seven hours in 1992 to an hour or less in 1996.
- CMG and DPG U.S. consistently have achieved excellent performance levels with their primary customers. CMG maintains 50% and DPG U.S. has 20% of DCC's lease portfolio. Since 1994, CMG's customers have ranked them between 4 and 5 on a 5-point scale (5 being very satisfied). DPG U.S. ranks 8 to 9 on a 10-point scale (10 is superior). The industry average is about 6.
- Each DCC employee receives an average of 48 hours of education, exceeding the industry average and the average of key competitors. DCC develops and teaches its employees more than 40 classes in areas such as accounting, finance, and law, as well as interpersonal communications, quality, and marketing.
- DCC has a policy of promoting from within, with 100% of senior leadership and 95% of supervisory and management positions filled internally, providing all employees with significant opportunity for advancement and growth.
- Most financial performance measures show sustained improvement and very good comparative performance. Return on equity and return on assets have increased more than 45% since 1991. Return on equity has been at or above 20% since 1992, compared to industry averages of 15% to 18% and exceeding DCC's two largest competitors by 2% and 10%.
- With more than 2000 competitors, DCC ranks as number 11 with 0.6% of the market.
- DCC committed 45% of the money it saved as a result of a tax incentive from the City of Toledo to the Toledo School Board. As a result, the school board will receive 1½ times more revenue than it would have received from DCC through a normal tax distribution. Toledo has adopted this approach as the standard for future tax incentives. This practice has also received positive recognition throughout the state of Michigan as a model of excellence.

3-8 SUMMARY

The chapter has examined the philosophy of total quality management and the role management plays in accomplishing desired organizational goals and objectives. A company's vision describes what it wants to be; the vision molds quality policy. This policy, along with the support and commitment of top management, defines the quality culture that prevails in the organization. Since meeting and exceeding customer needs is a fundamental criteria for the existence and growth of any company, the steps of product design and development, process analysis, and production scheduling have to be integrated into the quality system.

The planning tool of quality function deployment (QFD) is used in an interdisciplinary team effort to accomplish the desired customer requirements. Benchmarking enables a company to understand its relative performance with respect to industry performance measures and thus helps the company improve its competitive position. Adaptation of best practices to the organization's environment also ensures continuous improvement. Vendor quality audits, selection, and certification programs are important because final product quality is influenced by the quality of raw material and components. Since quality decisions are dependent on the collected data and information on products, processes, and customer satisfaction, simple tools for quality improvement that make use of such data have been presented.

Finally, some international standards on quality assurance practices have been depicted. In particular, the standards ISO 9000–9004 and their U.S. equivalents, ANSI/ASQC Q9000–Q9004 are discussed in this chapter. Furthermore, the criteria for the *Malcolm Baldrige National Quality Award* bestowed annually in the United States have been presented. Several profiles of award-winning companies examine the unique strategies through which they accomplished success in the marketplace.

CASE STUDY

Xerox Corporation—Leadership Through Quality

Xerox Corporation is a global company that offers the widest array of document-processing products and consulting services in the industry. Xerox sells its publishing systems, copiers, printers, scanners, fax machines, and document management software, along with related products and services, in more than 130 countries. Xerox products and services are designed to help customers master the flow of information from paper to electronic form and back again. The Xerox customer is anyone who uses documents: Fortune 500 corporations and small companies; public agencies and universities.

Xerox leads the way in digital imaging and what is called distributed publishing. Xerox technology enables the home office to copy, print, scan, and fax documents using a single device; and far-flung enterprises to transmit complicated, multipage documents across networks for copying or printing. Xerox started the office copying revolution with the introduction of its 914 copier in 1959. Today, Xerox stands poised for the continued expansion of the global document-processing market, already enormous at $200 billion a year and growing 10% a year. In 1995, 20% of revenues were in businesses that grew more than 20%: personal copying and printing (29%); document outsourcing (50%); production publishing (24%); and color copying and printing (45%). Including Fuji Xerox, about two-thirds of the $25 billion in revenues are generated outside the United States.

XEROX AND QUALITY

Xerox practices total quality management and is committed to providing its customers with innovative products and services that fully meet their needs. Xerox products are consistently rated among the world's best by independent testing organizations. Since 1980, Xerox has won numerous quality awards, including the world's three most prestigious: the Malcolm Baldrige National Quality Award for Xerox Products and Systems in 1989, the first European Quality Award for Rank Xerox in 1992, and the Deming Prize, Japan's highest quality award for Fuji Xerox in 1980.

Xerox is the first major U.S. corporation to regain market share after losing it to Japanese competitors. The company's decision to rededicate itself to quality explains that accomplishment. In the 1970s, Xerox nearly became a victim of its own success, lulled into complacency by the easy growth of its early years. Market share dropped to less than 50% by 1980, from nearly 100% a few years earlier. Fortunately, Xerox reacted to this challenge with a strategy called "Leadership Through Quality." Using Fuji Xerox in Japan as a model, Xerox created a participatory management style that stressed improving quality while reducing costs. Quality circles flourished and teamwork was fostered. Xerox also sought more customer feedback, changed its approach to product development to target key markets, reduced costs, encouraged greater employee involvement and began competitive benchmarking: the now widely used process of measuring performance against the toughest competitors and against companies recognized as the best in a particular area, such as L.L. Bean for distribution and Toyota for quality control.

LEADERSHIP THROUGH QUALITY

The "Leadership Through Quality" thrust has made quality improvement and, ultimately, customer satisfaction the job of every employee. All have received at least 28 hours of training in problem-solving and quality improvement techniques. The company has invested more than 4 million man-hours and $125 million in educating employees about quality principles. Workers are vested with authority over day-to-day work decisions. And, they are expected to take the initiative in identifying and correcting problems that affect the quality of products or services. Both salaried and hourly personnel have embraced these added responsibilities.

For example, the company's 1989 labor contract (the year they won the Baldrige Award) with the Amalgamated Clothing and Textile Workers' Union pledged employee support to "continuous quality improvement while reducing quality costs through teamwork and the tools and processes of "Leadership Through Quality." This partnership with the union is considered a model in the industry. The phrase "Team Xerox" is not an empty slogan. It accurately reflects the firm's approach to tackling quality issues. Xerox Business Products and Services (BP&S) estimates that 75% of its workers are members of at least one of more than 7000 quality improvement teams. In 1988, teams in manufacturing and development were credited with saving $116 million by reducing scrap, tightening production schedules, and devising other efficiency and quality-enhancing measures.

Teamwork also characterizes the company's relationship with many of its 480 suppliers. Vendors are "process qualified" through a step-by-step procedure to analyze and quantify suppliers' production and control processes. Vendors receive training and follow-up in such areas as statistical process control and total quality techniques; firms credit Xerox with improving their products and operations. For BP&S, increasing reliance on qualified vendors over the last five years has reduced the number of defective parts reaching the production line by 73%.

Planning new products and services is based on detailed analyses of data organized in 375 information management systems, including 175 specific to planning, managing, and evaluating quality improvement. Much of the wealth of data has been amassed through an extensive network of market surveillance and customer feedback, all designed to support systematic evaluation of customer requirements. Over half of the company's marketing-research budget is allocated for this purpose, and each year its Customer Service Measurement System tracks the behavior and preferences of about 200,000 owners of Xerox equipment.

BENCHMARKING SYSTEM

In its quest to elevate its products and services to world-class status, Xerox BP&S devised a benchmarking system that has, in itself, become a model. The company measures its performance in about 240 key areas of product, service, and business performance. Derived from international studies, the ultimate target for each attribute is the level of performance achieved by the world leader, regardless of industry.

Returns from the company's strategy for continuous quality improvement have materialized quickly. Gains in quality over the last five years include a 78% decrease in the number of defects per 100 machines; greatly increased product reliability, as measured by a 40% decrease in unscheduled maintenance; increasing copy quality, which has strengthened the company's position as world leader; a 27% drop (nearly two hours) in service response time; and significant reductions in labor and material overhead. These improvements have enabled Xerox BP&S to take additional steps to distinguish itself from the competition; for instance, it was the first in the industry to offer a three-year product warranty.

The thrust of "Leadership Through Quality" is ongoing. The process of continuous quality improvement is directed toward greater customer satisfaction and enhanced business performance. Such goals illustrate the commitment contained in the Xerox Quality Policy, which states that "quality is the basic business principle at Xerox."

XEROX AND DIVERSITY

Xerox views diversity in the workplace as more than a moral imperative or a business necessity. They see it as a business opportunity. They believe diversity makes them better: People of all ages and with different backgrounds bring fresh ideas, opinions, perspectives, and boundless creativity to the company. Under the company's balanced workforce strategy, senior managers are evaluated on their ability to hire, keep, and promote minorities and women. Even when the company must reduce its ranks, the smaller workforce is expected to mirror the workforce before the reduction, in the percentages of minorities and women. Caucus groups are another aspect of the diversity story at Xerox. These independent groups of Xerox employees date from the 1960s. The caucuses, not labor unions but something akin to self-help groups, help members negotiate the corporate world and work to ensure that their members, like all Xerox employees, have equal opportunities in hiring, promotion, and training.

XEROX AND SOCIAL RESPONSIBILITY

Xerox understands that corporations, like private citizens, have a responsibility to society at large. Civic virtue and community involvement are among the most cherished corporate values. The company carries out much of its philanthropic work through the Xerox Foundations, which in 1995 contributed $14 million in five areas: community affairs; education and workforce preparedness; science and technology; cultural affairs; and national affairs. Xerox also supports employee involvement through two innovative programs, Social Service Leave and the Xerox Community Involvement Program (XCIP).

Under Social Service Leave, employees are granted paid leaves of absence to work on community projects of their choice. More than 400 have taken leaves since 1971. The program is believed to be the oldest of its kind in American business. Through XCIP, groups of Xerox employees can get corporate seed money to work on meeting needs they identify in their communities. In 1995 alone, more than 20,000 Xerox employees took part in nearly 700 XCIP projects.

XEROX AND WORK/FAMILY

Xerox is committed to helping employees balance the demands of professional and personal life. The company believes that by relieving some of the pressures in people's personal lives, Xerox can help them be more focused and productive at work. Xerox offers childcare subsidies; salary redirection for dependent-care expenses; childcare sources and referrals; eldercare consultations and referrals; adoption assistance; leaves of absence; and flexible work arrangements.

In 1993, Xerox introduced LifeCycle Assistance to address the changing needs of a diverse workforce. Money from this program of flexible benefits can be put toward the purchase of a first home, and toward the purchase of health insurance for household members not generally eligible for coverage under the Xerox health plans, such as a domestic partner. LifeCycle Assistance is another important step toward giving employees greater choice in how benefit dollars are spent. Xerox ranked No. 1 on *Money* magazine's 1995 list of U.S. corporations with the best employee benefits.

XEROX AND THE ENVIRONMENT

Xerox is proving that what is good for the environment can also be good for business. Their environmental initiatives have already saved hundreds of millions of dollars while reducing pollution, waste and energy consumption. The motto "Reuse, Remanufacture, Recycle" reflects their goal: To create waste-free products in waste-free factories and offices using what is called Design for the Environment. Most copiers, printers, and multifunction devices are now designed to be remanufactured at the end of their initial life cycles, an approach made possible by the durability and quality of Xerox products and parts. Xerox uses only recyclable and recycled thermoplastics and metals. The company has adopted snap-together designs to facilitate assembly and disassembly, for the cleaning, testing, and reuse of parts. In 1995, Xerox received the Environmental Achievement Award from the National Wildlife Federation, which cited the company's Design for the Environment program.

QUESTIONS FOR DISCUSSION

1. What are some of the unique characteristics of the quality culture at Xerox?
2. Discuss the major strategic goals of Xerox. Compare and contrast this to another company of your choice. List some goals of a company other than profit maximization and discuss the social responsibilities of an organization as promoted by Xerox.
3. Describe the benchmarking efforts of Xerox. Discuss what steps Xerox could have taken to avoid loss of market share prior to its turnaround in the 1980s.
4. Discuss the role played by the management at Xerox in adopting the quality policy.
5. What are some efforts undertaken by Xerox to ensure satisfaction of its employees? How would you monitor such a system to maintain employee commitment?

Key Terms

- approved vendor
- benchmarking
- cause-and-effect diagram
- certified vendor
- change management
- check sheet
- cycle time
- control chart
- empowerment
- flow chart
- gap analysis
- histogram
- house of quality
- ISO 9000–9004
- Malcolm Baldrige National Quality Award
- mission statement
- Pareto diagram

- performance standards
- preferred vendor
- quality audit
  - conformity quality audit
  - process audit
  - product audit
  - suitability quality audit
  - system audit
- quality function deployment
- quality policy
- scatter plot
- six-sigma quality
- time-based competition
- total quality excellence (TQE)
- total quality management (TQM)
- Q101 Preferred Quality Award
- vendor certification
- vendor rating
- vision

Exercises

1. Describe the total quality management philosophy. Choose a company and discuss how its quality culture fits this theme.
2. What are the advantages of creating a long-term partnership with vendors?
3. Discuss the role of top management in setting a road map for the organization. What are some reasons for the failure of a total quality management program?
4. Compare and contrast a company vision, mission, and quality policy. Discuss these concepts in the context of a company of your choice.
5. Describe Motorola's concept of six-sigma quality and explain the level of nonconforming product that could be expected from such a process.
6. What are the advantages of using quality function deployment? What are some key ingredients that are necessary for its success?
7. Consider the airline transportation industry. Develop a house of quality showing customer requirements and technical descriptors.
8. What are the advantages of benchmarking? How does it contribute to the goal of continuous quality improvement?
9. Describe the steps of benchmarking relative to a company of your choice. What is the role of top management in this process?
10. What are the different types of quality audits? Discuss each and identify the context in which they are used.
11. Discuss the role of established standards and third-party auditors in quality auditing. What is the role of ISO 9000 standards in this context?
12. Discuss some criteria for selecting vendors. Describe the selection process for vendors.
13. What is the purpose of vendor certification? Describe typical phases of certification.
14. Discuss the role of national and international standards in certifying vendors.
15. Construct a flow chart for a manufacturing or service operation of your choice. Identify possible measures for process improvement.
16. Use a flow chart to develop an advertising campaign for a new product that you will present to top management.
17. Construct a cause-and-effect diagram of a process that is producing nonconforming items. Discuss some remedial measures.
18. You are asked to make a presentation to senior management outlining the demand for a product. Describe the data you would collect and the tools you would use to organize your presentation.
19. Discuss the emerging role of ISO 9000 standards in the global economy.
20. Select an organization that is registered or is going through the process of being registered to ISO 9000 standards. Describe the key steps of the process that the organization must undertake.
21. What is the role of the Malcolm Baldrige National Quality Award? How is it different from ISO 9000 standards?
22. Explain the obligations of the Malcolm Baldrige Award winners.

References

ANSI/ASQC Q9000-1994 (1994). *American National Standard—Quality Management and Quality Assurance Standards—Guidelines for Selection and Use.* Milwaukee, Wis.: American Society for Quality Control.

____ Q9001-1994 (1994). *American National Standard—Quality Systems—Model for Quality Assurance in Design/Development, Production, Installation, and Servicing.* Milwaukee, Wis.: American Society for Quality Control.

—— Q9002-1994 (1994). *American National Standard—Quality Systems—Model for Quality Assurance in Production and Installation.* Milwaukee, Wis.: American Society for Quality Control.

—— Q9003-1994 (1994). *American National Standard—Quality Systems—Model for Quality Assurance in Final Inspection and Test.* Milwaukee, Wis.: American Society for Quality Control.

—— Q9004-1994 (1994). *American National Standard—Quality Management and Quality System Elements—Guidelines.* Milwaukee, Wis.: American Society for Quality Control.

Bogan, C. E., and M. J. English (1994). *Benchmarking for Best Practices.* New York: McGraw-Hill.

Bossert, J. L. (1991). *Quality Function Deployment—A Practitioner's Approach.* New York: Marcel Dekker.

Bossert, J. L., ed. (1994). *Supplier Management Handbook.* Milwaukee, Wis.: Quality Press, ASQC.

Buzanis, C. H. (1993). "Hyatt Hotels and Resorts: Achieving Quality Through Employee and Guest Feedback Mechanisms." In *Managing Quality in America's Most Admired Companies,* J. W. Spechler, ed. San Francisco: Berrett-Kochler.

Camp, R. C. (1989). *Benchmarking: The Search for Best Practices That Lead to Superior Performance.* Milwaukee, Wis.: Quality Press, ASQC.

Ford Motor Company (1989). "Total Quality Excellence Program."

—— (1990). "Quality System Standards Q101."

ISO 9000 (1994). *Quality Management and Quality Assurance Standards—Guidelines for Selection and Use.* Geneva, Switzerland.

—— 9001 (1994). *Quality Systems—Model for Quality Assurance in Design/Development, Production, Installation, and Servicing.* Geneva, Switzerland.

—— 9002 (1994). *Quality Systems—Model for Quality Assurance in Production and Installation.* Geneva, Switzerland.

—— 9003 (1994). *Quality Systems—Model for Quality Assurance in Final Inspection and Test.* Geneva, Switzerland.

—— 9004 (1994). *Quality Management and Quality System Elements—Guidelines.* Geneva, Switzerland.

Mills, C. A. (1989). *The Quality Audit.* Milwaukee, Wis.: Quality Press, ASQC.

U.S. Department of Commerce (1996). *1997 Criteria for Performance Excellence, Malcolm Baldrige National Quality Award.* Gaithersburg, Md.: National Institute of Standards and Technology.

PART II: Statistical Foundations and Methods of Quality Improvement

CHAPTER

4

Fundamentals of Statistical Concepts and Techniques in Quality Control and Improvement

Chapter Outline

Symbols			
$P(A)$	Probability of event A	$f(x)$	Probability density function for a continuous random variable
μ	Population mean of a quality characteristic	$F(x)$	Cumulative distribution function
$\overline{X}$	Sample average	$E(X)$	Expected value or mean of a random variable X
s	Sample standard deviation	Z	Standard normal random variable
n	Sample size	p	Probability of success on a trial in a binomial experiment
X_i	ith observation in a sample	λ	Mean of a Poisson random variable
N	Population size	λ	Failure rate for an exponential distribution
$T(\alpha)$	α% trimmed mean	γ	Location parameter for a Weibull distribution
σ	Population standard deviation	α	Scale parameter for a Weibull distribution
s^2	Sample variance	β	Shape parameter for a Weibull distribution
σ^2	Population variance	$\Gamma(u)$	Gamma function for the variable u
R	Range	$(1-\alpha)$	Level of confidence for confidence intervals
γ_1	Skewness coefficient		
γ_2	Kurtosis coefficient		
r	Correlation coefficient		
M	Sample median		
s_m	Standard deviation of the sample median		
$p(x)$	Probability distribution (or mass function) for a discrete random variable		

4-1 INTRODUCTION

In this chapter we will build a foundation for the statistical concepts and techniques used in quality control and improvement. Statistics is a subtle science, and it plays an important role in quality programs. Only a clear understanding of statistics will enable you to apply it properly. They are often misused, but a sound knowledge of statistical principles will help you formulate correct procedures in different situations and will also help you interpret the results properly. When we analyze a process, we often find it necessary to study its characteristics individually. Breaking the process down allows us to determine whether some identifiable cause has forced a deviation from the expected norm and whether a remedial action needs to be taken. We may also do this simply to see how we can improve a process. Thus, our objective in this chapter is to review different statistical concepts and techniques. Should you encounter a statistical concept in a later chapter that needs clarification, you can refer to this chapter.

4-2 POPULATION AND SAMPLE

A **population** is the set of all items that possess a certain characteristic of interest.

Example 4-1: Suppose our objective is to determine the average weight of cans of brand A soup processed by our company for the month of July. The population in this case is the set of all cans of brand A soup that are output in the month of July (say, 50,000). Other brands of soup made during this time are not of interest, only the population brand A soup cans.

A **sample** is a subset of a population. Realistically, in many manufacturing or service industries, it is not feasible to obtain data on every element in the population. Measurement, storage, and retrieval of large volumes of data is impractical, and the costs of obtaining such information are high. Thus, we usually obtain data from only a portion of the population—a sample.

Example 4-2: Consider our brand A soup. To save ourselves the cost and effort of weighing 50,000 cans, we randomly select a sample of 500 cans of brand A soup from the July output.

4-3 PARAMETER AND STATISTIC

A **parameter** is a characteristic of a population, something that describes it.

Example 4-3: For our soup example, we will be looking at the parameter average weight of all 50,000 cans processed in the month of July.

A **statistic** is a characteristic of a sample. It is used to make inferences on the population parameters that are typically unknown.

Example 4-4: Our statistic then is the average weight of a sample of 500 cans chosen from the July output. Suppose this value is 12.11 oz; this would then be an *estimate* of the average weight of all 50,000 cans. A statistic is sometimes called an estimator.

4-4 PROBABILITY

Our discussion of the concepts of probability is intentionally brief. For an in-depth look at probability, see the references listed at the end of this chapter. The **probability** of an event describes the chance of occurrence of that event. A probability function is bounded between 0 and 1, with 0 representing the definite nonoccurrence of the event and 1 representing the certain occurrence of the event.

The set of all outcomes of an experiment is known as the **sample space** S.

Relative Frequency Definition of Probability

If each event in the sample space is equally likely to happen, then the probability of an event A is given by

$$P(A) = \frac{n_A}{N} \tag{4.1}$$

where $P(A)$ = probability of event A
n_A = number of occurrences of event A
N = size of the sample space

This definition is associated with the relative frequency concept of probability. It is applicable to situations where historical data on the outcome of interest is available. The probability associated with the sample space is 1 [that is, $P(S) = 1$].

Example 4-5: A company makes plastic storage bags for the food industry. Out of the hourly production of 2000 16-oz bags, 40 were found to be nonconforming. If the inspector randomly chooses a bag from the hour's production, what is the probability of it being nonconforming?

Solution. We define event A as getting a bag that is nonconforming. The sample space S consists of 2000 bags (that is, $N = 2000$). The number of occurrences of event $A(n_A)$ is 40. Thus, if the inspector is equally likely to choose any one of the 2000 bags,

$$P(A) = \frac{40}{2000} = .02$$

Simple Events and Compound Events

Simple events cannot be broken into other events. They represent the most elementary form of the outcomes possible in an experiment. **Compound events** are made up of two or more simple events.

Example 4-6: Suppose an inspector is sampling transistors from an assembly line and identifying them as acceptable or not. Suppose the inspector chooses two transistors. What are the simple events? Give an example of a compound event. Find the probability of finding at least one acceptable transistor.

Solution. Consider the following outcomes:

A_1: Event that the first transistor is acceptable
D_1: Event that the first transistor is unacceptable
A_2: Event that the second transistor is acceptable
D_2: Event that the second transistor is unacceptable

There are four simple events that make up the sample space S:

$$S = \{A_1A_2, A_1D_2, D_1A_2, D_1D_2\}$$

These events may be described as follows:

$E_1 = \{A_1A_2\}$: Event that the first and second transistors are acceptable
$E_2 = \{A_1D_2\}$: Event that the first transistor is acceptable and the second one is not
$E_3 = \{D_1A_2\}$: Event that the first transistor is unacceptable and the second one is acceptable
$E_4 = \{D_1D_2\}$: Event that both transistors are unacceptable

Compound event B is the event that at least one of the transistors is acceptable. In this case, event B consists of the following three simple events: $B = \{E_1, E_2, E_3\}$. Assuming that each of the simple events is equally likely to happen, $P(B) = P(E_1) + P(E_2) + P(E_3) = \frac{1}{4} + \frac{1}{4} + \frac{1}{4} = \frac{3}{4}$. Figure 4-1 shows a Venn diagram, which is a graphical representation of the sample space and its associated events.

Complementary Events

The complement of an event A implies the occurrence of everything but A. If we define A^c to be the complement of A, then

$$P(A^c) = 1 - P(A) \tag{4.2}$$

Figure 4-2 shows the probability of the complement of an event by means of a Venn diagram. Continuing with Example 4-6, suppose we want to find the probability of the event that both transistors are unacceptable. Note that this is the complement of event B, which was defined as at least one of the transistors being acceptable. So

$$P(B^c) = 1 - P(B) = 1 - \tfrac{3}{4} = \tfrac{1}{4}$$

Additive Law

The additive law of probability defines the probability of the union of two or more events happening. If we have two events A and B, then the union of these two implies that A happens *or* B happens *or* both happen. Figure 4-3 shows the union of two

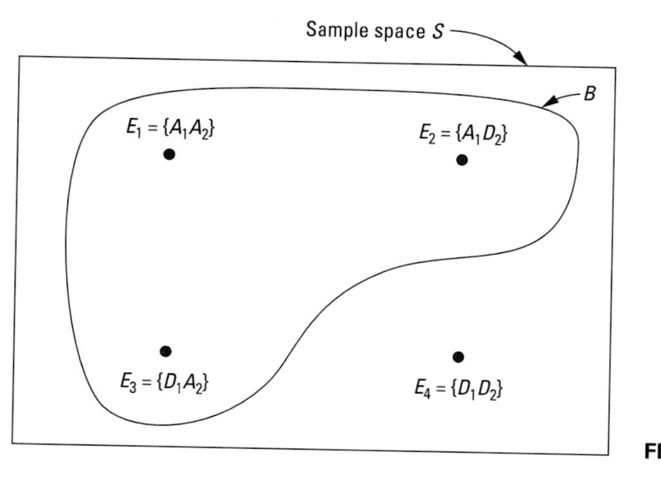

FIGURE 4-1 Venn diagram.

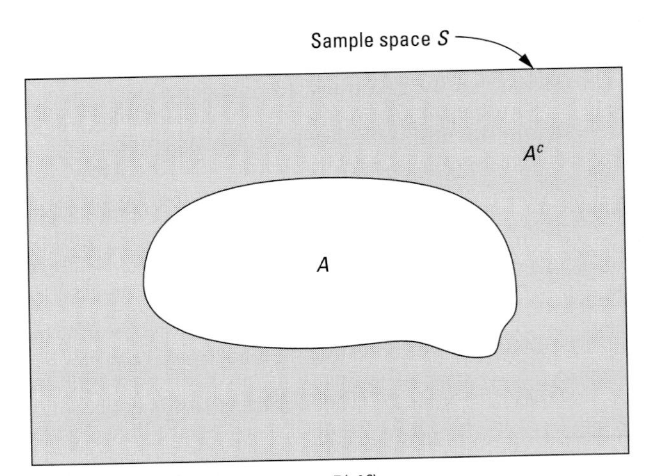

FIGURE 4-2 An event and its complement. *Note:* The shaded area represents $P(A^c)$.

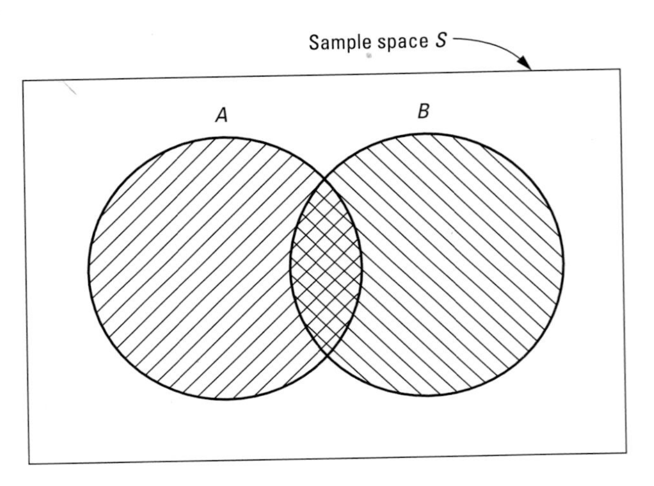

FIGURE 4-3 Union of two events.

events, A and B. The hatched area in the sample space represents the probability of the union of the two events. The **additive law** is as follows:

$$P(A \cup B) = P(A \text{ or } B \text{ or both}) \\ = P(A) + P(B) - P(A \cap B) \quad \textbf{(4.3)}$$

Note that $P(A \cap B)$ represents the probability of the intersection of events A and B—that is, the occurrence of both A and B. The logic behind the additive law can easily be seen from the Venn diagram in Figure 4-3, where $P(A)$ represents the area within the boundary-defining event A. Similarly, $P(B)$ represents the area within the boundary-defining event B. The overlap (cross-hatched) between areas A and B represents the probability of the intersection, $P(A \cap B)$. When $P(A)$ is added to $P(B)$, this intersection is included twice, so eq. (4.3) adds $P(A)$ to $P(B)$ and subtracts $P(A \cap B)$ once.

Multiplicative Law

The **multiplicative law** of probability defines the probability of the intersection of two or more events. Intersection of a group of events means that all the events in that group occur. In general, for two events A and B,

$$P(A \cap B) = P(A \text{ and } B) = P(A)\,P(B \mid A) \\ = P(B)P(A \mid B) \quad \textbf{(4.4)}$$

The term $P(B \mid A)$ represents the conditional probability of B given that event A has happened (that is, the probability that B will occur if A has). Likewise, $P(A \mid B)$ represents the conditional probability of A given that event B has happened. Of the two forms given by eq. (4.4), the problem will dictate which version to use.

Independence and Mutually Exclusive Events

Two events A and B are said to be **independent** if the outcome of one has no influence on the outcome of the other. If A and B are independent, then $P(B \mid A) = P(B)$; that is, the conditional probability of B given that A has happened equals the unconditional probability of B. Similarly, $P(A \mid B) = P(A)$ if A and B are independent. From eq. (4.4), it can be seen that if A and B are independent, the general multiplicative law reduces to

$$P(A \cap B) = P(A \text{ and } B) = P(A)P(B) \qquad \text{if } A \text{ and } B \text{ are independent} \qquad \textbf{(4.5)}$$

Two events A and B are said to be **mutually exclusive** if they cannot happen simultaneously. The intersection of two mutually exclusive events is the null set, and the probability of their intersection is zero. Notationally, $P(A \cap B) = 0$ if A and B are mutually exclusive. Figure 4-4 shows a Venn diagram for two mutually exclusive events. Note that when A and B are mutually exclusive events, the probability of their union is simply the sum of their individual probabilities. In other words, the additive law takes on the following special form:

$$P(A \cup B) = P(A) + P(B) \qquad \text{if } A \text{ and } B \text{ are mutually exclusive}$$

If events A and B are mutually exclusive, what can we say about their dependence or independence? Obviously, if A happens, B cannot happen, and vice versa. Therefore, if A and B are mutually exclusive, they are dependent. If A and B are independent, the additive rule from eq. (4.3) becomes

$$P(A \text{ or } B \text{ or both}) = P(A) + P(B) - P(A)P(B) \qquad \textbf{(4.6)}$$

Example 4-7:

a. In the production of metal plates for an assembly, it is known from past experience that 5% of the plates do not meet the length requirement. Also, from historical records, 3% of the plates do not meet the width requirement. Assume there are no dependencies between the processes that make the length and those that trim the width. What is the probability of producing a plate that meets both the length and width requirements?

Solution. Let A be the outcome that the plate meets the length requirement and B be the outcome that the plate meets the width requirement. From the problem statement, $P(A^c) = .05$ and $P(B^c) = .03$. Then

$$P(A) = 1 - P(A^c) = 1 - .05 = .95$$
$$P(B) = 1 - P(B^c) = 1 - .03 = .97$$

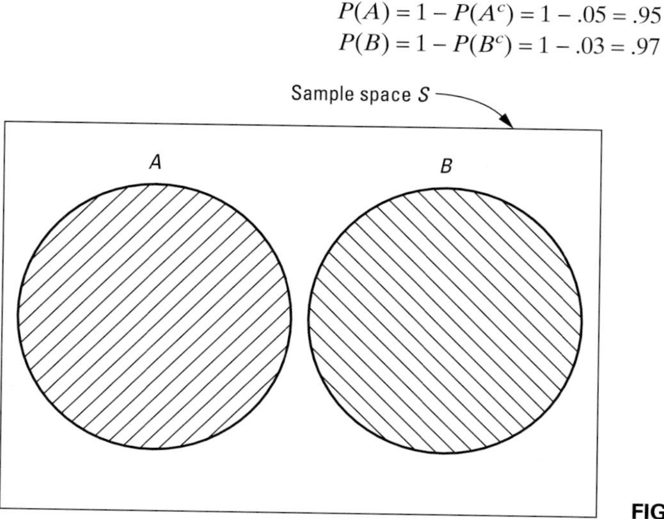

FIGURE 4-4 Mutually exclusive events.

Using the special case of the multiplicative law for independent events, we have

$$\begin{aligned} P(\text{meeting both length } \textit{and} \text{ width requirements}) &= P(A \cap B) \\ &= P(A)P(B) \qquad (\text{since } A \text{ and } B \text{ are independent events}) \\ &= (.95)(.97) = .9215 \end{aligned}$$

b. What proportion of the parts will not meet at least one of the requirements?

Solution. The required probability = $P(A^c \text{ or } B^c \text{ or both})$. Using the additive law, we get

$$\begin{aligned} P(A^c \text{ or } B^c \text{ or both}) &= P(A^c) + P(B^c) - P(A^c \cap B^c) \\ &= .05 + .03 - (.03)(.05) = .0785 \end{aligned}$$

Therefore, 7.85% of the parts will have at least one characteristic (length, width, or both) not meeting the requirements.

c. What proportion of parts will meet neither length nor width requirements?

Solution: We want to find $P(A^c \cap B^c)$.

$$P(A^c \cap B^c) = P(A^c)P(B^c) = (.05)(.03) = .0015$$

Thus 0.15% of the parts will be deficient in meeting both the length and width requirements.

d. Suppose the operations that produce the length and the width are not independent. If the length does not satisfy the requirement, it causes an improper positioning of the part during the width trimming and thereby increases the chances of nonconforming width. From experience, it is estimated that if the length does not conform to the requirement, the chance of producing nonconforming widths is 60%. Find the proportion of parts that will neither conform to the length nor the width requirements.

Solution. The probability of interest is $P(A^c \cap B^c)$. The problem states that $P(B^c \mid A^c) = .60$. Using the general form of the multiplicative law,

$$\begin{aligned} P(A^c \cap B^c) &= P(A^c)P(B^c \mid A^c) \\ &= (.05)(.60) = .03 \end{aligned}$$

So 3% of the parts will meet neither the length nor width requirements. Notice that this value is different from the answer to part c, where the events were assumed to be independent.

e. In part a, are events A and B mutually exclusive?

Solution. We have found $P(A) = .95$, $P(B) = .97$, and $P(A \cap B) = .9215$. If A and B were mutually exclusive, $P(A \cap B)$ would have to be zero; that is, the probability of the plate meeting both the length and width requirements would be zero. However, this is not the case, since $P(A \cap B) = .9215$. So A and B are *not* mutually exclusive.

f. Describe two events in this example setting that are mutually exclusive.

Solution. Events A and A^c are mutually exclusive, to name one instance and $P(A \cap A^c) = 0$, because A and A^c cannot happen simultaneously. This means that it is not possible to produce a part that both meets and does not meet the length requirement!

4-5 DESCRIPTIVE STATISTICS—DESCRIBING PRODUCT OR PROCESS CHARACTERISTICS

Statistics is the science that deals with the collection, classification, analysis, and making of inferences from data or information. Statistics is subdivided into two categories—*descriptive statistics* and *inferential statistics.*

Descriptive statistics describes the characteristics of a product or process using information collected on it. Suppose we have recorded service times for 500 customers in a fast-food restaurant. We can plot this as a frequency histogram where the horizontal axis represents a range of service time values and the vertical axis denotes the number of service times observed in each time range, which would give us some idea of the process condition. Likewise, the average service time for the 500 customers could also tell us something about the process. Procedures used in descriptive statistics are discussed in later sections.

Inferential statistics draws conclusions on unknown process parameters based on information contained in a sample. Let's say that we want to test the validity of a claim that the average service time in the fast-food restaurant is no more than 3 minutes (min). Suppose we find that the sample average service time (based on a sample of 500 people) is 3.5 min. We then need to determine whether this observed average of 3.5 min is significantly greater than the claimed mean of 3 min.

Such procedures fall under the heading of inferential statistics. They help us draw conclusions about the conditions of a process. They also help us determine whether a process has improved by comparing conditions before and after changes. For example, suppose the management of the fast-food restaurant is interested in reducing the average time to serve a customer. They decide to add two people to their service staff. Once this change is implemented, they sample 500 customers and find that the average service time is 2.8 min. The question then is whether this decrease is a statistically significant decrease or whether it is due to random variation inherent to sampling. Procedures that address such problems are discussed later.

Data Collection

To control or improve a process, we need information, or data. Data can be collected in several ways. One of the most common methods is through *direct observation.* Here, a measurement of the quality characteristic is taken by an observer (or automatically by an instrument); for instance, measurements on the depth of tread in automobile tires taken by an inspector are direct observations. On the other hand, data collected on the performance of a particular brand of hair dryer through questionnaires mailed to consumers are *indirect observations.* In this case, the data reported by the consumers have not been observed by the experimenter, who has no control over the data collection process. Thus, the data may be flawed because errors can arise from a respondent's incorrect interpretation of a question, an error in estimating the satisfactory performance period, or an inconsistent degree of precision among responders' answers.

Data on quality characteristics is described by a **random variable** and is categorized as *continuous* or *discrete.*

Continuous Variable

A variable that can assume any value on a continuous scale within a range is said to be **continuous.** Examples of continuous variables are the hub length of lawnmower tires, the viscosity of a certain resin, the specific gravity of a toner used in photocopying machines, the thickness of a metal plate, and the time to admit a patient to a hospital. Such variables are measurable and have numerical values associated with them.

Discrete Variable

Variables that can assume a finite or countably infinite number of values are said to be **discrete.** These variables are counts of an event. The number of defective rivets in an assembly is a discrete random variable. Other examples include the number of paint blemishes in an automobile, the number of operating capacitors in an electrical instrument, and the number of satisfied customers in an automobile repair shop.

Counting events usually costs less than measuring the corresponding continuous variables. The discrete variable is merely classified as being, say, unacceptable or not; this can be done through a go/no-go gage, which is faster and cheaper than finding exact measurements. However, the reduced collection cost may be offset by the lack of detailed information in the data.

Sometimes, continuous characteristics are viewed as discrete to allow easier data collection and reduced inspection costs. For example, the hub diameter in a tire is actually a continuous random variable, but rather than precisely measuring the hub diameter numerically, a go/no-go gage is used to quickly identify the characteristic as either acceptable or not. Hence, the *acceptability* of the hub diameter is a discrete random variable. In this case, the goal is not to know the exact hub diameter but rather to know whether it is within certain acceptable limits.

Accuracy and Precision

The **accuracy** of a data set or a measuring instrument refers to the degree of uniformity of the observations around a desired value such that, on average, the target value is realized. Let's assume that the target thickness of a metal plate is 5.25 mm. Figure 4-5a shows observations spread on either side of the target value in almost equal proportions; these observations are said to be accurate. Even though individual observations may be quite different from the target value, a data set is considered accurate if the average of a large number of observations is close to the target.

For measuring instruments, accuracy is dependent on **calibration.** If a measuring device is properly calibrated, the average output value given by the device for a particular quality characteristic should, after repeated use, equal the true input value.

The **precision** of a data set or a measuring instrument refers to the degree of variability of the observations. Observations may be off the target value but still considered precise, as shown in Figure 4-5b. A sophisticated measuring instrument should show very little variation in output values if a constant value is used multiple times as input. Similarly, sophisticated equipment in a process should be able to produce an output characteristic with as little variability as possible. The precision of the data is influenced by the precision of the measuring instrument. For example, the thickness of a metal plate may be 12.5 mm when measured by callipers; however, a micrometer may yield a value of 12.52 mm, while an optical sensor may give a measurement of 12.523 mm.

Having both accuracy and precision is desirable. Figure 4-5c depicts a situation where not only do the values exhibit small variability, they are also centered around the desired target value. In equipment or measuring instruments, accuracy can usually be altered by changing the setting of a certain adjustment. However, precision is an inherent function of the equipment itself and cannot be improved by changing a setting.

Measurement Scales

Four scales of measurement are used to classify data—the nominal, ordinal, interval, and ratio scales. Notice that each scale builds on the previous scale.

Nominal Scale

The scale of measurement is **nominal** when the data variables are simply labels used to identify an attribute of the sample element. Labels can be "conforming and nonconforming" or "critical, major, and minor." Numerical values are not involved.

Ordinal Scale

The scale of measurement is **ordinal** when the data has the properties of nominal data (that is, labels) and the data ranks or orders the observations. Suppose customers at a clothing store are asked to rate the quality of the store's service. The customers rate the quality according to these responses: 1 (outstanding), 2 (good),

FIGURE 4-5 Accuracy and precision of observations.

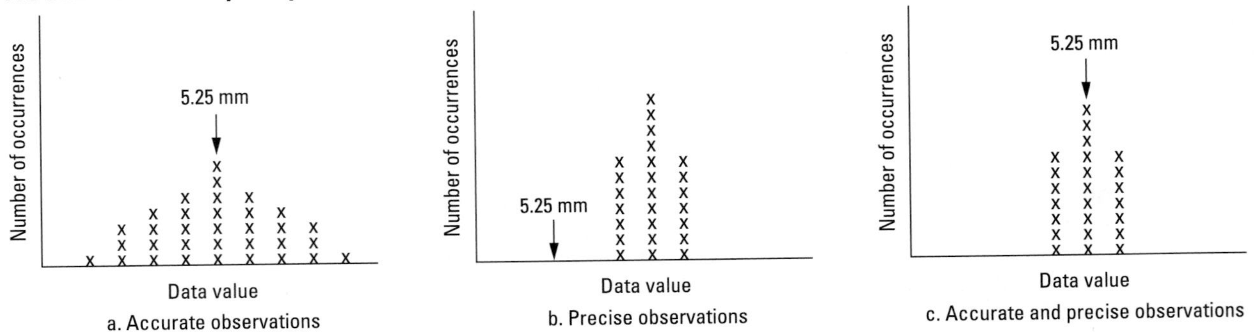

3 (average), 4 (fair), 5 (poor). This is ordinal data. Note that a rating of 1 does not necessarily imply that the service is twice as good as a rating of 2. However, we can say that a rating of 1 is preferable to a rating of 2, and so forth.

Interval Scale

The scale of measurement is **interval** when the data have the properties of ordinal data *and* a fixed unit of measure describes the interval between observations. Suppose we are interested in the temperature of a furnace used in steel smelting. Four readings taken during a 2-hour interval are 2050, 2100, 2150, and 2200°F. Obviously, these data values ranked (like ordinal data) in ascending order of temperature indicating the coolest temperature, the next coolest, and so on. Furthermore, the differences between the ranked values can then be compared. Here the interval between the data values 2050 and 2100 represents a 50°F increase in temperature, as do the intervals between the remaining ranked values.

Ratio Scale

The scale of measurement is **ratio** when the data have the properties of interval data *and* a natural zero exists for the measurement scale. Both the order of, and difference between, observations can be compared and there exists a natural zero for the measurement scale. Suppose the weights of four castings are 2.0, 2.1, 2.3, and 2.5 kg. The order (ordinal) of, and difference (interval) in, the weights can be compared. Thus, the increase in weight from 2 to 2.1 is 0.1 kg, which is the same as the increase from 2.3 to 2.4 kg. Also, when we compare the weights of 2.0 and 2.4 kg, we find a meaningful ratio—casting weighing 2.4 kg is 20% heavier than one weighing 2.0 kg. There is also a natural zero for the scale—0 kg implies no weight.

Measures of Central Tendency

In statistical quality control, collected data needs to be described so that analysts can objectively evaluate the process or product characteristic. This section describes some of the common numerical measures used to derive summary information from observed values. Measures of central tendency tell us something about the locations of the observations and the value about which they cluster and thus help us decide whether the settings of process variables should be changed.

Mean

The **mean** is the simple average of the observations in a data set. In quality control, the mean is one of the most commonly used measures. It is easy to calculate and understand. The mean is used to determine whether, on average, the process is operating around a desirable target value. The **sample mean,** or average (denoted by $\overline{X}$), is found by adding all observations in a sample and dividing by the number of observations (n) in that sample. If the ith observation is denoted by X_i, then the sample mean is calculated as

$$\overline{X} = \frac{\sum_{i=1}^{n} X_i}{n} \tag{4.7}$$

The **population mean** (μ) is found by adding all the data values in the population and dividing by the size of the population (N). It is calculated as

$$\mu = \frac{\sum_{i=1}^{N} X_i}{N} \tag{4.8}$$

The population mean is sometimes denoted as $E(X)$, the expected value of the random variable X. It is also called the mean of the probability distribution of X. Probability distributions are discussed later in this chapter.

Example 4-8: A random sample of five observations of the waiting time of customers in a bank is taken. The times (in minutes) are 3, 2, 4, 1, and 2. The sample average ($\overline{X}$), or mean waiting time, is

$$\overline{X} = \frac{3+2+4+1+2}{5} = \frac{12}{5} = 2.4 \text{ min}$$

The bank can use this information to determine whether the waiting time needs to be improved by increasing the number of tellers.

Median

The **median** is the value in the middle, when the observations are ranked. If there are an even number of observations, the simple average of the two middle numbers is chosen as the median. The median has the property that 50% of the values are less than or equal to it.

Example 4-9:

a. A random sample of 10 observations of piston ring diameters (in millimeters) yields the following values: 52.3, 51.9, 52.6, 52.4, 52.4, 52.1, 52.3, 52.0, 52.5, and 52.5. We first rank the observations:

51.9 52.0 52.1 52.3 52.3
52.4 52.4 52.5 52.5 52.6

The two observations in the middle are 52.3 and 52.4. The median is (52.3 + 52.4)/2 or 52.35.
The median is less influenced by the extreme values in the data set; thus, it is said to be more "robust" than the mean.

b. A department store is interested in expanding its facilities and wants to do a preliminary analysis of the number of customers it serves. Five weeks are chosen at random, and the number of customers served during those weeks were as follows:

3000, 3500, 500, 3300, 3800

The median number of customers is 3300, while the mean is 2820. On further investigation of the week with 500 customers, it is found that a major university, whose students frequently shop at the store, was closed for spring break. In this case, the median (3300) is a better measure of central tendency than the mean (2820) because it gives a better idea of the variable of interest. In fact, had the data value been 100 instead of 500, the median would still be 3300, though the mean would further decrease, another demonstration of the robustness of the median.
Outliers (values that are very large or very small compared to the majority of the data points) can have a significant influence on the mean, which is pulled toward the outliers. Figure 4-6 demonstrates the effect of outliers.

Mode

The **mode** is the value that occurs most frequently in the data set. It denotes a "typical" value from the process.

Example 4-10: A hardware store wants to determine what size of circular saws it should stock. From past sales data, a random sample of 30 shows the following sizes (in millimeters):

80 120 100 100 150 120 80 150 120 80
120 100 120 120 150 80 120 100 120 80
100 120 120 150 120 100 120 120 100 100

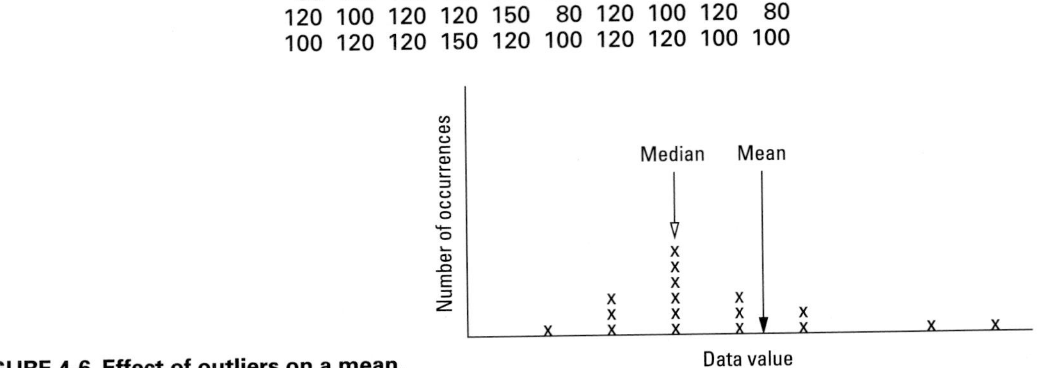

FIGURE 4-6 Effect of outliers on a mean.

A frequency plot of the number of saws of each size sold is shown in Figure 4-7. Note that the mode has the highest frequency. In this case, the mode is 120 (13 is the largest number of occurrences). So, the manager may decide to stock more size-120 saws. A data set can have more than one mode, in which case it is said to be multimodal.

Trimmed Mean

The **trimmed mean** is a robust estimator of the central tendency of a set of observations. It is obtained by calculating the mean of the observations that remain after a proportion of the high and low values have been deleted. The $\alpha\%$ trimmed mean, denoted by $T(\alpha)$, is the average of the observations that remain after trimming (or deleting) $\alpha\%$ of the high observations and $\alpha\%$ of the low observations. This is a suitable measure when it is believed that existing outliers do not represent usual process characteristics. Thus, analysts will sometimes trim extreme observations caused by a faulty measurement process to obtain a better estimate of the population's central tendency.

Example 4-11: The time taken for car tune-ups (in minutes) is observed for 20 randomly selected cars. The data values are as follows:

15 10 12 20 16 18 30 14 16 15
18 40 20 19 17 15 22 20 19 22

To find the 5% trimmed mean [that is, $T(.05)$], first rank the data in increasing order:

10 12 14 15 15 15 16 16 17 18
18 19 19 20 20 20 22 22 30 40

The number of observations (high and low) to be deleted on each side is 20(.05) = 1. Delete the lowest observation (10) and the highest one (40). The trimmed mean (of the remaining 18 observations) is 18.222, which is obviously more robust than the mean (18.9). For example, if the largest observation of 40 had been 60, the 5% trimmed mean would still be 18.222. However, the untrimmed mean would jump to 19.9.

Measures of Dispersion

An important function of quality control and improvement is to analyze and reduce the variability of a process. The numerical measures of location we have described give us indications of the central tendency, or middle, of a data set. They do not tell us much about the variability of the observations. Consequently, sound analysis requires an understanding of measures of dispersion, which provide information on the variability, or scatter, of the observations around a given value (usually the mean).

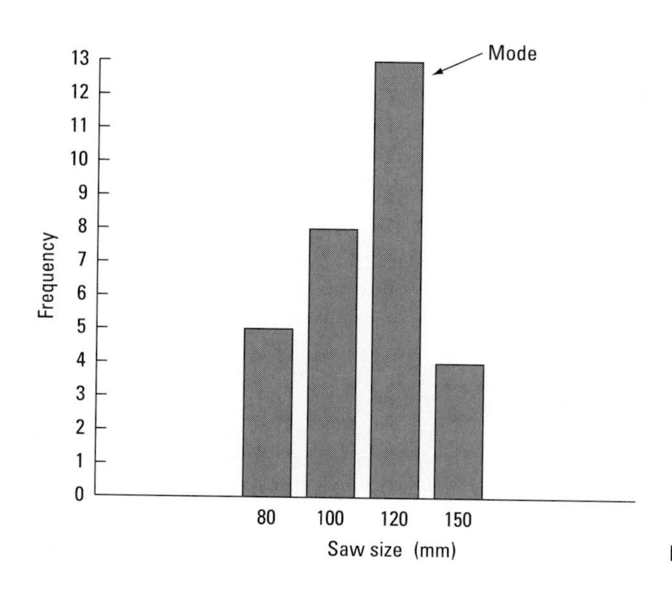

FIGURE 4-7 Frequency plot of saw size.

Range

A widely used measure of dispersion in quality control is the **range,** which is the difference between the largest and smallest values in a data set. Notationally, the range R is defined as

$$R = X_{\mathrm{L}} - X_{\mathrm{S}} \tag{4.9}$$

where X_{L} is the largest observation and X_{S} is the smallest observation.

Example 4-12: The following 10 observations of the time to receive baggage after landing are randomly taken in an airport. The data values (in minutes) are as follows:

15, 12, 20, 13, 22, 18, 19, 21, 17, 20

The range $R = 22 - 12 = 10$ min. This value gives us an idea of the variability in the observations. Management can now decide whether this spread is acceptable.

Variance

The **variance** measures the fluctuation of the observations around the mean. The larger the value, the greater the fluctuation. The population variance σ^2 is given by

$$\sigma^2 = \frac{\sum_{i=1}^{N} (X_i - \mu)^2}{} \tag{4.10}$$

where μ is the population mean and N represents the size of the population. The sample variance s^2 is given by

$$s^2 = \frac{\sum_{i=1}^{n} (X_i - \overline{X})^2}{} \tag{4.11}$$

where $\overline{X}$ is the sample mean and n is the number of observations in the sample. In most applications, the sample variance is calculated rather than the population variance because calculation of the latter is possible only when every value in the population is known.

A modified version of eq. (4.11) for calculating the sample variance is

$$s^2 = \frac{\sum_{i=1}^{n} X_i^2 - \left(\sum_{i=1}^{n} X_i\right)^2 / n}{n-1} \tag{4.12}$$

This version is sometimes easier to use. It involves accumulating the sum of the observations and the sum of squares of the observations as data values become available. On the other hand, with eq. 4.11, we first have to calculate the sample mean $\overline{X}$ using all the observations and then go back to each data value to find the deviation from the sample mean. Using eq. (4.11) requires two passes through the data, whereas using eq. (4.12) requires only one pass. These two equations are algebraically equivalent.

Observe that to calculate either the population variance or the sample variance, we must first obtain the deviation of each observation from the corresponding mean. Only then can we obtain the sum of squares of the deviations from the mean [that is, the numerator of eqs. (4.10) and (4.11). Because of the nature of the mean, the sum of the deviations should equal zero.

Note that in calculating the sample variance, the denominator is $(n - 1)$, whereas for the population variance the denominator is N. Thus, eq. (4.10) can be interpreted as the average of the squared deviations of the observations from the mean, and eq. (4.11) can be interpreted similarly, except for the difference in the denominator, where $(n - 1)$ is used instead of n. This difference can be explained as follows. First, a population variance σ^2 is a parameter, whereas a sample variance s^2 is an estimator, or a statistic. The value of s^2 can therefore change from sample to sample, whereas σ^2 should be constant.

One desirable property of s^2 is that even though it may not equal σ^2 for every sample, on average s^2 does equal σ^2. This is known as the property of unbiasedness, where the mean or expected value of the estimator equals the corresponding parameter. If a denominator of $(n - 1)$ is used to calculate the sample variance, then it can be shown that the sample variance is an unbiased estimator of the population variance. On the other hand, if a denominator of n is used, then on average the sample variance underestimates the population variance.

Figure 4-8 denotes the sampling distribution (that is, the relative frequency) of s^2 calculated using eq. (4.11) or (4.12) over repeated samples. Suppose the value of σ^2 is as shown. If the average value of s^2 is calculated over repeated samples, it will equal the population variance. Technically, the expected value of s^2 will equal σ^2 [that is, $E(s^2) = \sigma^2$]. If a denominator of n is used in eq. (4.11) or (4.12), then $E(s^2)$ will be less than σ^2.

Unlike the range, which uses only the extreme values of the data set, the sample variance incorporates every observation in the sample. Two data sets with the same range can have different variability. As Figure 4-9 shows, data sets A and B have the same range; however, their degree of variability is quite different. The sample variances will thus indicate different degrees of fluctuation around the mean. The units of variance are the square of the units of measurement for the individual values. For example, if the observations are in millimeters, then the units of variance are square millimeters.

Standard Deviation

The **standard deviation,** like the variance, also measures the variability of the observations around the mean. It is equal to the positive square root of the variance. A standard deviation has the same units as the observations and is thus easier to interpret. It is probably the most widely used measure of dispersion in quality control. Using eq. (4.10), the population standard deviation is given by

$$\sigma = \sqrt{\frac{\sum_{i=1}^{N} (X_i - \mu)^2}{}} \qquad \textbf{(4.13)}$$

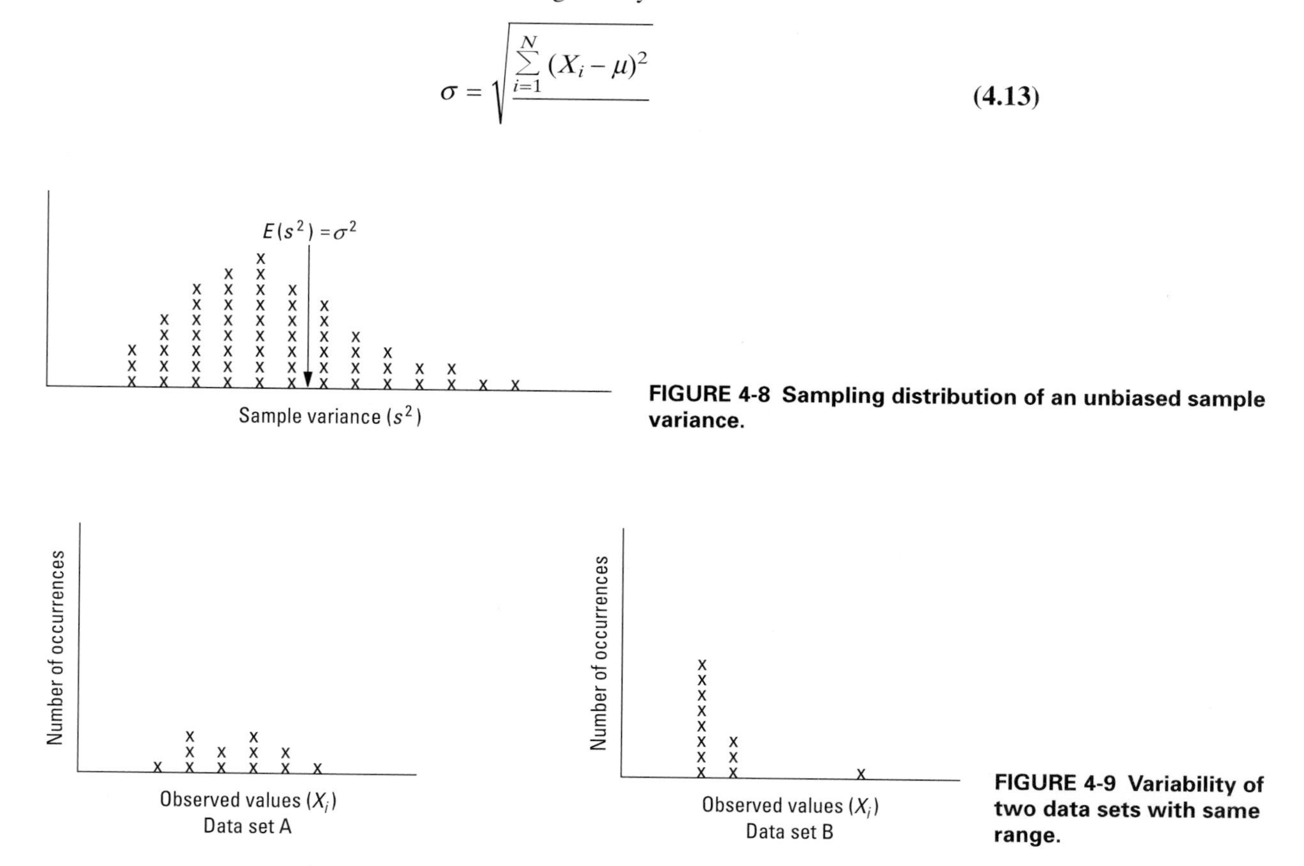

FIGURE 4-8 Sampling distribution of an unbiased sample variance.

FIGURE 4-9 Variability of two data sets with same range.

Similarly, the sample standard deviation s is found using eqs. (4.11) or (4.12) as

$$s = \sqrt{\frac{\sum_{i=1}^{n} (X_i - \overline{X})^2}{n-1}} \tag{4.14}$$

$$= \sqrt{\frac{\sum_{i=1}^{n} X_i^2 - \left(\sum_{i=1}^{n} X_i\right)^2 \Big/ n}{n-1}} \tag{4.15}$$

As with the variance, the data set with the largest standard deviation will be identified as having the most variability about its average. If the probability distribution of the random variable is known—a normal distribution, say—then the proportion of observations within a certain number of standard deviations of the mean can be obtained. Techniques for obtaining such information are discussed later in the section on probability distributions.

Example 4-13: A random sample of 10 observations of the output voltage of transformers is taken. The values (in volts, V) are as follows:

9.2, 8.9, 8.7, 9.5, 9.0, 9.3, 9.4, 9.5, 9.0, 9.1

Using eq. (4.11), the sample mean $\overline{X}$ is

$$\overline{X} = \frac{9.2 + 8.9 + 8.7 + 9.5 + 9.0 + 9.3 + 9.4 + 9.5 + 9.0 + 9.1}{10} = 9.16 \text{ V}$$

Table 4-1 shows the calculations. From Table 4-1, $\sum (X_i - \overline{X})^2 = 0.644$. The sample variance is given by

$$s^2 = \frac{\sum (X_i - \overline{X})^2}{n-1} = \frac{0.644}{9} = 0.0716 \text{ V}^2$$

The sample standard deviation given by eq. (4.14) is

$$s = \sqrt{0.0716} = 0.2675 \text{ V}$$

Next, using eq. (4.12) (the calculations are shown in Table 4-2), the sample variance is given by

$$s^2 = \frac{\sum X_i^2 - (\sum X_i)^2/n}{n-1}$$

$$= \frac{839.70 - (91.60)^2/10}{9}$$

$$= \frac{(839.70 - 839.056)}{9}$$

$$= \frac{0.644}{9} \text{ or } 0.0716 \text{ V}^2$$

The sample standard deviation s is 0.2675 V, as before.

Interquartile Range

The lower quartile Q_1 is the value such that one-fourth the observations fall below it and three-fourths fall above it. The middle quartile is the median—half the observations fall below it and half above it. The third quartile Q_3 is the value such that three-fourths of the observations fall below it and one-fourth above it.

The **interquartile range** IQR is the difference between the third quartile and the first quartile. Thus

$$\text{IQR} = Q_3 - Q_1 \tag{4.16}$$

TABLE 4-1 Calculation of Sample Variance and Standard Deviation Using Eq. (4.11)

X_i	*Deviation from Mean,* $X_i - \overline{X}$	*Squared Deviation,* $(X_i - \overline{X})^2$
9.2	0.04	0.0016
8.9	−0.26	0.0676
8.7	−0.46	0.2116
9.5	0.34	0.1156
9.0	−0.16	0.0256
9.3	0.14	0.0196
9.4	0.24	0.0576
9.5	0.34	0.1156
9.0	−0.16	0.0256
9.1	−0.06	0.0036
	$\sum (X_i - \overline{X}) = 0$	$\sum (X_i - \overline{X})^2 = 0.644$

TABLE 4-2 Calculation of Sample Variance Using Eq. (4.12)

X_i	X_i^2
9.2	84.64
8.9	79.21
8.7	75.69
9.5	90.25
9.0	81.00
9.3	86.49
9.4	88.36
9.5	90.25
9.0	81.00
9.1	82.81
$\sum X_i = 91.60$	$\sum X_i^2 = 839.70$

Note from Figure 4-10 that IQR contains 50% of the observations. The larger the IQR value, the greater the spread of the data. To find the IQR, the data are ranked in ascending order. The first quartile Q_1 is located at rank $0.25(n+1)$, where n is the number of data points in the sample. Likewise, Q_3 is located at rank $0.75(n+1)$.

Example 4-14: A random sample of 20 observations on the welding time (in minutes) of an operation gives the following values:

2.2 2.5 1.8 2.0 2.1 1.7 1.9 2.6 1.8 2.3
2.0 2.1 2.6 1.9 2.0 1.8 1.7 2.2 2.4 2.2

First let's find the locations of Q_1 and Q_3:

$$\text{Location of } Q_1 = 0.25(n+1) = 0.25(21) = 5.25$$
$$\text{Location of } Q_3 = 0.75(n+1) = 0.75(21) = 15.75$$

Now let's rank the data values

Q_1's location = 5.25 $\qquad$ Q_3's location = 15.75

Rank	1	2	3	4	5	6	7	8	9	10	11	12	13	14	15	16	17	18	19	20
Data value	1.7	1.7	1.8	1.8	1.8	1.9	1.9	2.0	2.0	2.0	2.1	2.1	2.2	2.2	2.2	2.3	2.4	2.5	2.6	2.6

$Q_1 = 1.825$ $\qquad$ $Q_3 = 2.275$

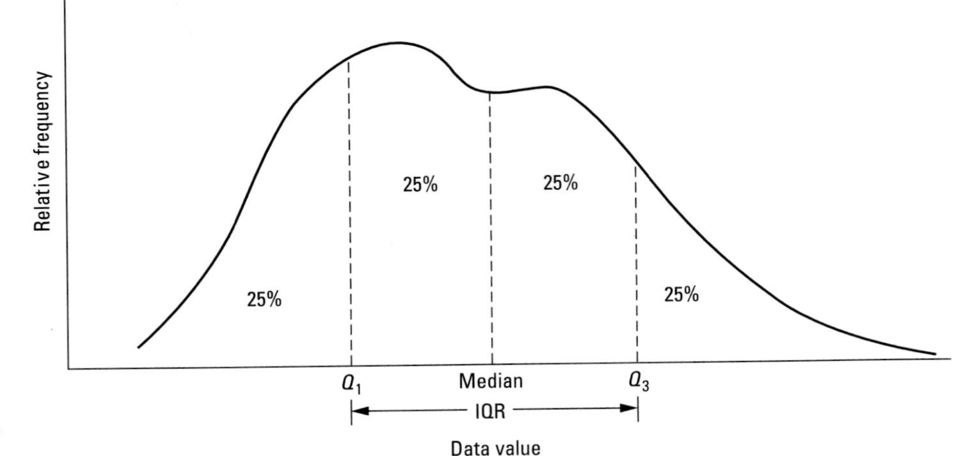

FIGURE 4-10 Interquartile range for a distribution.

Thus, linear interpolation yields a Q_1 of 1.825 and a Q_3 of 2.275. The interquartile range is then

$$\begin{aligned}\text{IQR} &= Q_3 - Q_1 \\ &= 2.275 - 1.825 = 0.45 \text{ min}\end{aligned}$$

Measures of Skewness and Kurtosis

In addition to central tendency and dispersion, two other measures are used to describe data sets: the skewness coefficient and the kurtosis coefficient.

Skewness Coefficient

The **skewness coefficient** describes the asymmetry of the data set about the mean. The skewness coefficient is calculated as follows:

$$\gamma_1 = \left\{ \frac{n\left[\sum_{i=1}^{n} (X_i - \overline{X})^3\right]^2}{\left[\sum_{i=1}^{n} (X_i - \overline{X})^2\right]^3} \right\}^{1/2} \tag{4.17}$$

In Figure 4-11, part a is a negatively skewed distribution (skewed to the left), part b is positively skewed (skewed to the right), and part c is symmetric about the mean. The skewness coefficient is zero for a symmetric distribution, because (as shown in part c) the mean and the median are equal. For a positively skewed distribution, the mean is greater than the median because a few values are large compared to the oth-

FIGURE 4-11 Symmetric and skewed distributions.

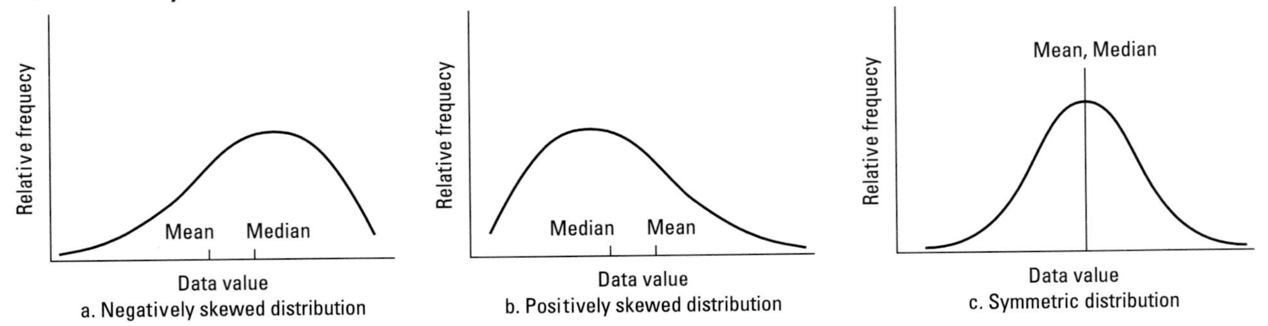

ers; the skewness coefficient will be a positive number. If a distribution is negatively skewed, the mean is less than the median because the outliers are very small compared to the other values, and the skewness coefficient will be negative. The skewness coefficient indicates the degree to which a distribution deviates from symmetry. It is used for data sets that are unimodal (that is, have one mode) and have a sample size of at least 100. The larger the magnitude of the skewness coefficient, the stronger the case for rejecting the notion that the distribution is symmetric.

Kurtosis Coefficient

Kurtosis is a measure of the peakedness of the data set. It is also viewed as a measure of the "heaviness" of the tails of a distribution. The kurtosis coefficient is given by

$$\gamma_2 = \frac{n \sum_{i=1}^{n} (X_i - \overline{X})^4}{\left[\sum_{i=1}^{n} (X_i - \overline{X})^2\right]^2} \tag{4.18}$$

The **kurtosis coefficient** is a relative measure. For a normal distribution (to be discussed in depth later), the kurtosis coefficient is 3. Figure 4-12 shows a normal distribution (mesokurtic), a distribution that is more peaked than the normal (leptokurtic), and one that is less peaked than the normal (platykurtic). For a leptokurtic distribution, the kurtosis coefficient is greater than 3. The more pronounced the peakedness, the larger the value of the kurtosis coefficient. The kurtosis coefficient should only be used to make inferences on a data set when the sample size is at least 100 and the distribution is unimodal.

Example 4-15: A sample of 50 coils to be used in an electrical circuit is randomly selected, and the resistance of each is measured. From the data in Table 4-3, the sample mean $\overline{X}$ is found to be

$$\overline{X} = \frac{1505.5}{50} = 30.11$$

Calculations for the skewness coefficient, γ_1, are also shown in Table 4-3, as are the square and cube of the deviations of each observation from the mean. Table 4-3 gives

$$\sum (X_i - \overline{X})^2 = 727.900$$

$$\sum (X_i - \overline{X})^3 = 736.321$$

The skewness coefficient is

$$\gamma_1 = \sqrt{\frac{50(736.321)^2}{(727.900)^3}} = 0.265$$

The positive skewness coefficient indicates that the distribution of the data points is slightly skewed to the right.

FIGURE 4-12 Distributions with different degrees of peakedness.

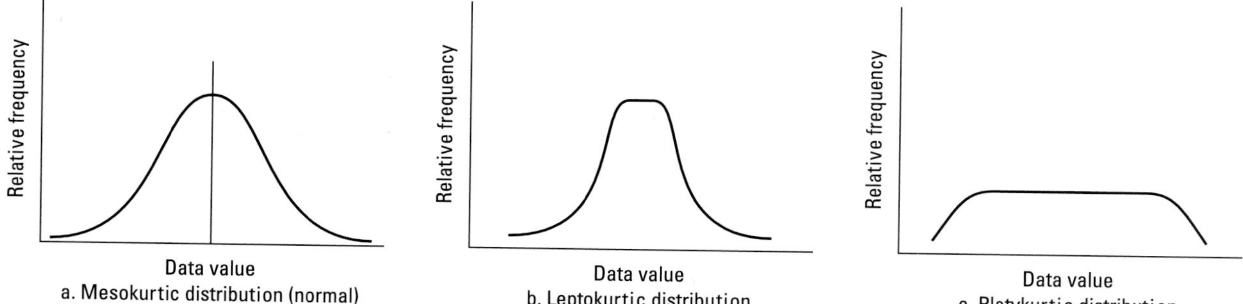

TABLE 4-3 Calculations for Skewness and Kurtosis Coefficients for Example 4-15

Observation i	*Resistance of Coils,* X_i	$(X_i-\overline{X})^2$	$(X_i-\overline{X})^3$	$(X_i-\overline{X})^4$	*Observation i*	*Resistance of Coils,* X_i	$(X_i-\overline{X})^2$	$(X_i-\overline{X})^3$	$(X_i-\overline{X})^4$
1	35.1	24.900	124.251	620.015	26	27.5	6.812	−17.780	46.405
2	35.4	27.984	148.036	783.110	27	26.5	13.032	−47.046	169.836
3	36.3	38.316	237.177	1,468.123	28	26.9	10.304	−33.076	106.174
4	38.8	75.516	656.235	5,702.681	29	26.7	11.628	−39.652	135.213
5	39.0	79.032	702.595	6,246.073	30	27.2	8.468	−24.642	71.709
6	22.5	57.912	−440.711	3,353.811	31	31.8	2.856	4.827	8.157
7	23.7	41.088	−263.375	1,688.232	32	32.1	3.960	7.881	15.682
8	25.0	26.112	−133.433	681.842	33	31.5	1.932	2.686	3.733
9	25.3	23.136	−111.285	535.279	34	31.2	1.188	1.295	1.412
10	25.0	26.112	−133.433	681.842	35	31.4	1.664	2.147	2.769
11	34.7	21.068	96.702	443.865	36	28.5	2.592	−4.173	6.719
12	34.2	16.728	68.418	279.829	37	28.4	2.924	−5.000	8.550
13	34.4	18.404	78.954	338.711	38	27.6	6.300	−15.813	39.691
14	34.7	21.068	96.702	443.865	39	27.6	6.300	−15.813	39.691
15	34.3	17.556	73.560	308.217	40	28.2	3.648	−6.968	13.309
16	26.4	13.764	−51.065	189.450	41	30.8	0.476	0.329	0.227
17	25.5	21.252	−97.972	451.652	42	30.6	0.240	0.118	0.058
18	25.8	18.576	−80.063	345.071	43	30.4	0.084	0.024	0.007
19	26.4	13.764	−51.065	189.450	44	30.5	0.152	0.059	0.023
20	25.6	20.340	−91.734	413.720	45	30.5	0.152	0.059	0.023
21	33.1	8.940	26.731	79.925	46	28.5	2.592	−4.173	6.719
22	33.6	12.180	42.509	148.355	47	30.2	0.008	0.001	0.000
23	32.3	4.796	10.503	23.003	48	30.1	0.000	0.000	0.000
24	32.6	6.200	15.438	38.441	49	30.0	0.012	−0.001	0.000
25	32.2	4.368	9.129	19.080	50	28.9	1.464	−1.772	2.143
Sum		639.112	932.804	25,473.642	Sum		88.788	−196.483	678.250

Next, Table 4.3 gives

$$\sum (X_i-\overline{X})^4 = 26{,}151.892$$

The kurtosis coefficient is found to be

$$\gamma_2 = \frac{50(26{,}151.892)}{(727.900)^2} = 2.468$$

These values imply that the given distribution is slightly less peaked than a normal distribution.

Example 4-16: For the coil resistance data shown in Table 4-3, Microsoft Excel (1994), a software for spreadsheet applications, can be used to obtain descriptive statistics. The Analysis ToolPak provided by Excel includes a feature for creating tables of basic statistics. Click **Tools > Data Analysis,** then click **Descriptive Statistics** in the Analysis Tools list box. Enter the range containing the data and indicate whether the variables are arranged in rows or columns. Also indicate whether the first row (if the data is arranged in columns) or first column (if the data is arranged in rows) contains the variable label. Select the **Summary Statistics** check box to produce the table of descriptive statistics shown in Table 4-4.

The mean and standard deviation of the resistance of coils are 30.11 and 3.854, respectively, while the median is 30.3. The minimum and maximum values are 22.5 and 39.0, respectively. Note that the skewness coefficient in Table 4-4 is 0.273, slightly different from the computed value of 0.265 in Example 4-15, since a different version of the formula is used. The distribution is slightly skewed to the right. Also, the kurtosis coefficient is –0.459, which is different from the computed value of 2.468 in Example 4-15. Excel uses a different version of the formula so that the coefficient will be zero for a normal distribution. Since the calculated kurtosis coefficient is slightly less than zero, it indicates that the distribution is slightly less peaked than a normal distribution.

TABLE 4-4 Descriptive Statistics for Coil Resistance Data Using Excel

Mean	30.11
Standard error	0.545073
Median	30.3
Mode	25
Standard deviation	3.854245
Sample variance	14.8552
Kurtosis	−0.45912
Skewness	0.27339
Range	16.5
Minimum	22.5
Maximum	39
Sum	1505.5
Count	50
Confidence level (95.0%)	1.095364

Measures of Association

Measures of association indicate how two or more variables are related to each other. For instance, as one variable increases, how does it influence another variable? Small values of the measures of association indicate a nonexistent or weak relationship between the variables, and large values indicate a strong relationship.

Correlation Coefficient

A **correlation coefficient** is a measure of the strength of the linear relationship between two variables. If two variables are denoted by X and Y, then the correlation coefficient r of a sample of observations is found from

$$r = \frac{\sum_{i=1}^{n} (X_i - \overline{X})(Y_i - \overline{Y})}{\sqrt{\sum_{i=1}^{n} (X_i - \overline{X})^2} \sqrt{\sum_{i=1}^{n} (Y_i - \overline{Y})^2}} \tag{4.19}$$

where X_i and Y_i denote the coordinates of the ith observation, $\overline{X}$ is the sample mean of the X_i-values, $\overline{Y}$ is the sample mean of the Y_i-values, and n is the sample size. An alternative version for calculating the sample correlation coefficient is

$$r = \frac{\sum X_i Y_i - (\sum X_i)(\sum Y_i)/n}{\sqrt{[\sum X_i^2 - (\sum X_i)^2/n]\,[\sum Y_i^2 - (\sum Y_i)^2/n]}} \tag{4.20}$$

The sample correlation coefficient r is always between −1 and 1. An r-value of 1 denotes a perfect positive linear relationship between X and Y. This means that as X increases, Y increases linearly and that as X decreases, Y decreases linearly. Similarly, an r-value of −1 indicates a perfect negative linear relationship between X and Y. If the value of r is zero, the two variables X and Y are uncorrelated, which implies that if X increases we cannot really say how Y would change. A value of r that is close to zero thus indicates that the relationship between the variables is weak.

Figure 4-13 shows plots of bivariate data with different degrees of strength of the linear relationship. Figure 4-13a shows a perfect positive linear relationship between X and Y. As X increases, Y very definitely increases linearly, and vice versa. In Figure 4-13c, X and Y are positively correlated (say, with a correlation coefficient of 0.8) but not perfectly related. Here, on the whole, as X increases, Y tends to increase, and vice versa. Similar analogies can be drawn for Figures 4-13b and 4-13d, where X and Y are negatively correlated. In Figure 4-13e, note that it is not evident what happens to

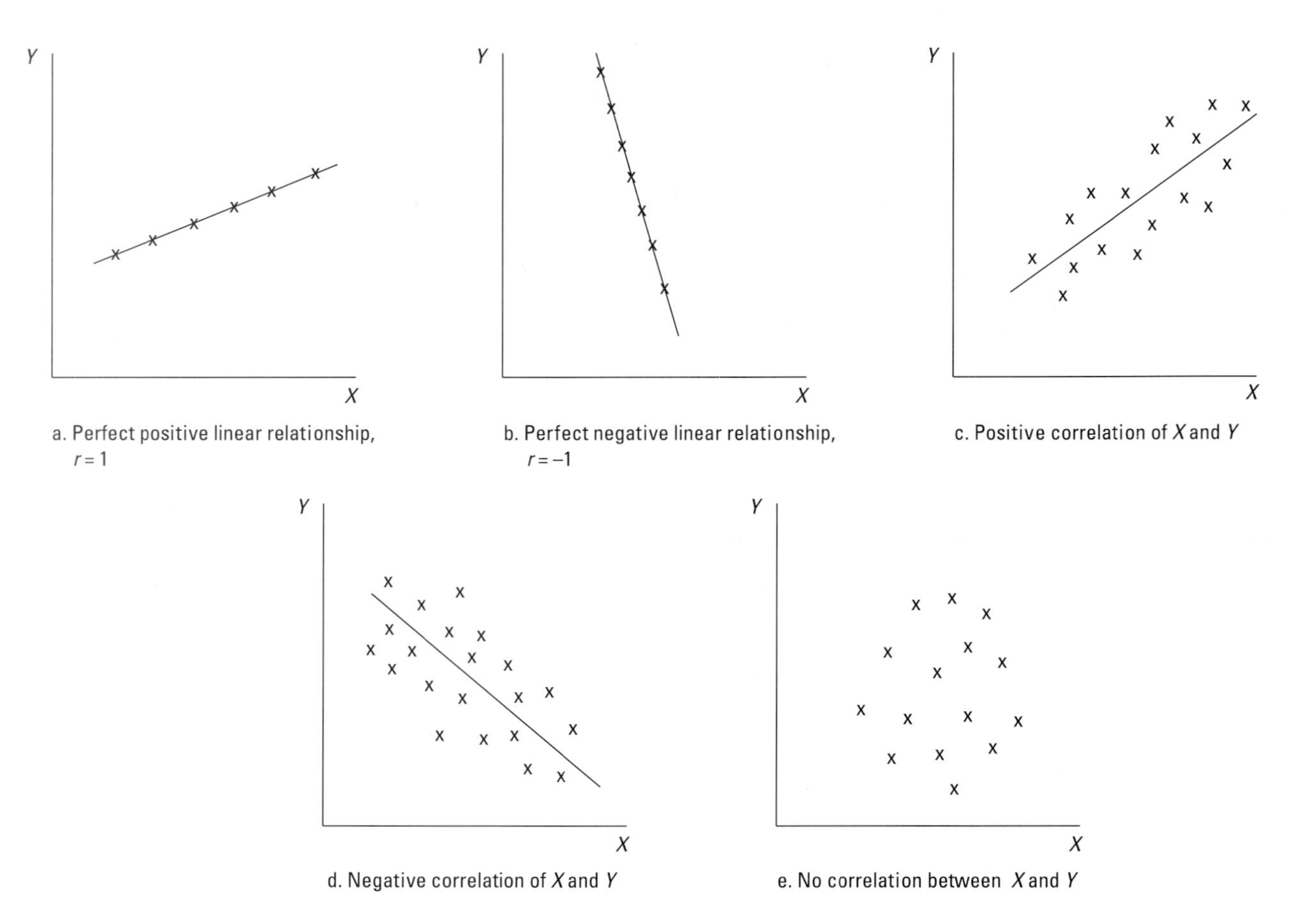

FIGURE 4-13 Scatter plots indicating different degrees of correlation.

TABLE 4-5 Milling Operation Data

Observation i	*Depth of Cut,* X_i	*Tool Wear,* Y_i	*Observation i*	*Depth of Cut,* X_i	*Tool Wear,* Y_i
1	2.1	0.035	21	5.6	0.073
2	4.2	0.041	22	4.7	0.064
3	1.5	0.031	23	1.9	0.030
4	1.8	0.027	24	2.4	0.029
5	2.3	0.033	25	3.2	0.039
6	3.8	0.045	26	3.4	0.038
7	2.6	0.038	27	3.8	0.040
8	4.3	0.047	28	2.2	0.031
9	3.4	0.040	29	2.0	0.033
10	4.5	0.058	30	2.9	0.035
11	2.6	0.039	31	3.0	0.032
12	5.2	0.056	32	3.6	0.038
13	4.1	0.048	33	1.9	0.032
14	3.0	0.037	34	5.1	0.052
15	2.2	0.028	35	4.7	0.050
16	4.6	0.057	36	5.2	0.058
17	4.8	0.060	37	4.1	0.048
18	5.3	0.068	38	4.3	0.049
19	3.9	0.048	39	3.8	0.042
20	3.5	0.036	40	3.6	0.045

Y as X increases or decreases. No general trend can be established from the plot, and X and Y are either uncorrelated or very weakly correlated. Statistical tests are available for testing the significance of the sample correlation coefficient and for determining if the population correlation coefficient is significantly different from zero (Neter et al. 1996).

Example 4-17: Consider the data shown in Table 4-5 on the depth of cut and tool wear in a milling operation. To find the strength of linear relationship between these two variables, we need to compute the correlation coefficient.

Let's use Microsoft Excel software to calculate the r-value, which provides a CORREL function for computing the correlation of the two variables entered. The data on depth of cut is entered in cells A2–A41, and that on tool wear is entered in cells B2–B41. In cell A42, type **Pearson r,** and in cell B42, type **=CORREL(A2:A41,B2:B41),** then press **[Tab].** The displayed correlation coefficient in cell B42 is .915. This value indicates a strong positive linear relationship between depth of cut and tool wear. That is, as the depth of cut increases, the tool wear increases.

Example 4-18: Now let's use the milling operation data in Table 4-5 to obtain summary descriptive statistics. Minitab (1996), a statistical software package for a Windows environment, does this task well. The data is entered in two columns for the variables depth of cut and tool wear. From the Windows menu, choose **Stat > Basic Statistics > Descriptive Statistics.** Now, select the variable you want to describe—say, depth of cut. Click on **Graphs** and then on **Graphical summary.**

A sample Minitab output of the descriptive statistics is shown in Figure 4-14. The mean and standard deviation of depth of cut are 3.527 and 1.138, indicating location and dispersion measures, respectively. The first and third quartiles are 2.45 and 4.45, respectively, yielding an interquartile range of 2.0, within which 50% of the observations are contained. Skewness and

FIGURE 4-14 Descriptive statistics on the variable depth of cut.

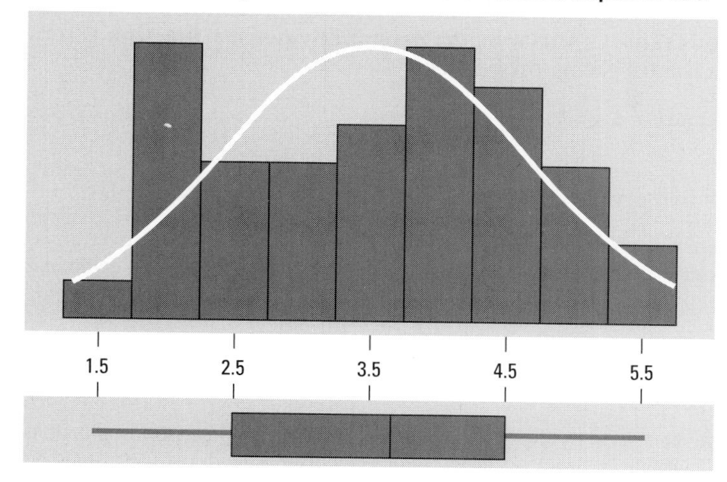

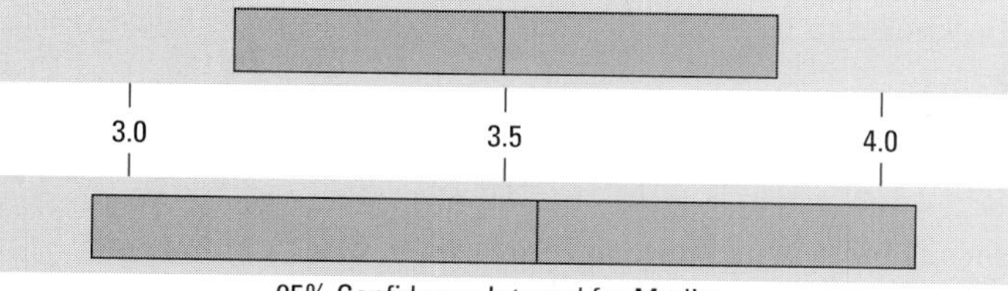

Variable: Depth of cut

Anderson-Darling Normality Test

A-Squared:	0.419
P-Value	0.312
Mean	3.52750
StDev	1.13815
Variance	1.29538
Skewness	−3.1E-02
Kurtosis	−1.19425
N	40
Minimum	1.50000
1st Quartile	2.45000
Median	3.60000
3rd Quartile	4.45000
Maximum	5.60000

95% Confidence Interval for Mu

3.16350 3.89150

95% Confidence Interval for Sigma

0.93233 1.46142

95% Confidence Interval for Median

3.00000 4.10000

kurtosis coefficients are also shown. The skewness coefficient is –0.031, indicating that the distribution is close to being symmetrical, though slightly negatively skewed. A kurtosis value of –1.194 indicates that the distribution is less peaked than the normal distribution, which would have a value of zero. The graphical summary yields four plots. The first is a frequency distribution with the normal curve superimposed on it. The confidence intervals shown are discussed in Section 4-7, as is the p-value, and the hypothesis testing associated with it. The box plot shown in Figure 4-14 is discussed in Chapter 5.

4-6 PROBABILITY DISTRIBUTIONS

Sample data can be described with frequency histograms or variations thereof (such as relative frequency or cumulative frequency, which we discuss later). Data values in a population are described by a probability distribution. As previously noted, random variables may be discrete or continuous. For discrete random variables, a probability distribution shows the values that the random variable can assume and their corresponding probabilities. Some examples of discrete random variables are the number of defects in an assembly, the number of customers served over a period of time, and the number of acceptable compressors.

Continuous random variables can take on an infinite number of values, so the probability distribution is usually expressed as a mathematical function of the random variable. This function can be used to find the probability that the random variable will be between certain bounds. Almost all variables for which numerical measurements can be obtained are continuous in nature—for example, the length of a pin, the diameter of a bolt, the tensile strength of a cable, or the specific gravity of a liquid.

For a discrete random variable X, which takes on the values x_1, x_2, and so on, a **probability distribution function** $p(x)$ has the following properties:

1. $p(x_i) \geq 0$ for all i, where $p(x_i) = P(X = x_i)$, $i = 1, 2, \ldots$
2. $\sum_{\text{all } i} p(x_i) = 1$

When X is a continuous random variable, the **probability density function** is represented by $f(x)$, which has the following properties:

1. $f(x) \geq 0$ for all x, where $P(a \leq x \leq b) = \int_a^b f(x)\,dx$
2. $\int_{-\infty}^{\infty} f(x)\,dx = 1$

Note the similarity of these two properties to those for discrete random variables.

Example 4-19: Let X denote a random variable that represents the number of defective transistors in an assembly. The probability distribution of the discrete random variable X may be given by

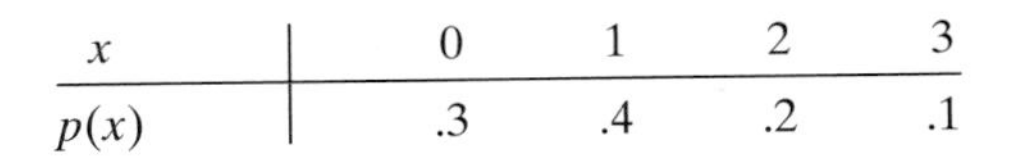

x	0	1	2	3
$p(x)$	.3	.4	.2	.1

This table gives the values taken on by the random variable and their corresponding probabilities. For instance, $P(X = 1) = .4$; that is, there is a 40% chance of finding one defective transistor. A graph of the probability distribution of this discrete random variable is shown in Figure 4-15.

Example 4-20: Consider a continuous random variable X representing the time taken to assemble a part. The variable X is known to be between 0 and 2 min, and its probability density function (pdf), $f(x)$, is given by

$$f(x) = \frac{x}{2}, \qquad 0 < x \leq 2$$

The graph of this probability density function is shown in Figure 4-16. Note that

$$\int_0^2 f(x)\,dx = \int_0^2 \frac{x}{2}\,dx = 1$$

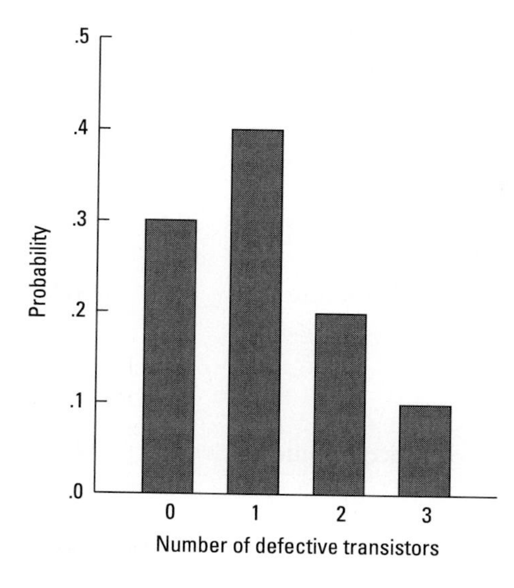

FIGURE 4-15 Probability distribution of a discrete random variable.

The probability that X is between 1 and 2 is

$$P(1 \le X \le 2) = \int_1^2 \frac{x}{2}\,dx = \frac{3}{4}$$

Cumulative Distribution Function

The **cumulative distribution function (cdf)** is usually denoted by $F(x)$ and represents the probability of the random variable X taking on a value less than or equal to x, that is

$$F(x) = P(X \le x)$$

For a discrete random variable,

$$F(x) = \sum_{\text{all } i} p(x_i) \qquad \text{for } x_i \le x \qquad \textbf{(4.21)}$$

If X is a continuous random variable,

$$F(x) = \int_{-\infty}^{x} f(t)\,dt \qquad \textbf{(4.22)}$$

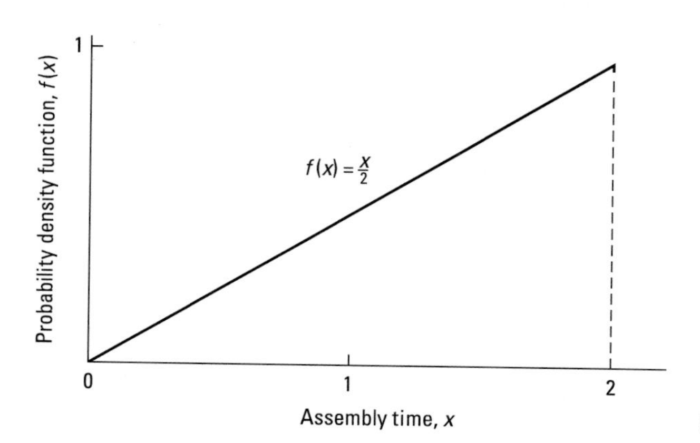

FIGURE 4-16 Probability density function, $0 < x \le 2$.

Note that $F(x)$ is a nondecreasing function of x such that

$$\lim_{x \to \infty} F(x) = 1 \qquad \text{and} \qquad \lim_{x \to -\infty} F(x) = 0$$

Expected Value

The **expected value** or mean of a distribution is given by

$$\mu = E(X) = \sum_{\text{all } i} x_i \, p(x_i) \qquad \text{if } X \text{ is discrete} \tag{4.23}$$

and

$$\mu = E(X) = \int_{-\infty}^{\infty} x \, f(x) \, dx \qquad \text{if } X \text{ is continuous} \tag{4.24}$$

The variance of a random variable X is given by

$$\begin{aligned} \text{Var}(X) &= E[(X - \mu)^2] \\ &= E(X^2) - [E(X)]^2 \end{aligned} \tag{4.25}$$

Example 4-21: For the probability distribution of Example 4-19, regarding the defective transistors, the mean μ or expected value $E(X)$ is given by

$$\begin{aligned} \mu = E(X) &= \sum_{\text{all } i} x_i \, p(x_i) \\ &= 0(.3) + 1(.4) + 2(.2) + 3(.1) \\ &= 1.1 \end{aligned}$$

The variance of X is

$$\sigma^2 = \text{Var}(X) = E(X^2) - [E(X)]^2$$

First, $E(X^2)$ is calculated as follows:

$$\begin{aligned} E(X^2) &= \sum_{\text{all } i} x_i^2 \, p(x_i) \\ &= (0)^2(.3) + (1)^2(.4) + (2)^2(.2) + (3)^2(.1) \\ &= 2.1 \end{aligned}$$

So

$$\text{Var}(X) = 2.1 - (1.1)^2 = 0.89$$

Hence, the standard deviation of X is $\sigma = \sqrt{0.89} = 0.943$.

Example 4-22: For the probability distribution function in Example 4-20, regarding a part's assembly time, the mean μ, or expected value $E(X)$, is given by

$$\begin{aligned} E(X) &= \int_{-\infty}^{\infty} x \, f(x) \, dx \\ &= \int_0^2 x \left(\frac{x}{2} \right) dx \\ &= \left. \frac{x^3}{6} \right|_0^2 \\ &= \frac{2^3}{6} = 1.333 \text{ min} \end{aligned}$$

Thus, the mean assembly time for this part is 1.333 min.

Discrete Distributions

The discrete class of probability distributions deals with those random variables that can take on a finite or countably infinite number of values. Several **discrete distributions** have applications in quality control, three of which are discussed in this section (Montgomery, 1996).

Hypergeometric Distribution

A **hypergeometric distribution** is useful in sampling from a finite population (or lot) without replacement (that is, without placing the sample elements back in the population) when the items or outcomes can be categorized into one of two groups (usually called success and failure). If we consider finding a nonconforming item a success, the probability distribution of the number of nonconforming items (x) in the sample is given by

$$p(x) = \frac{\binom{D}{x}\binom{N-D}{n-x}}{\binom{N}{n}}, \qquad x = 0, 1, 2, \ldots, \min(n, D) \tag{4.26}$$

where D = number of nonconforming items in the population
N = size of the population
n = size of the sample
x = number of nonconforming items in the sample
$\binom{D}{x}$ = combination of D items taken x at a time, $= \frac{D!}{x!(D-x)!}$

The factorial of a positive integer x is written as $x! = x(x-1)(x-2)\ldots 3 \cdot 2 \cdot 1$, and $0!$ is defined to be 1. The mean (or expected value) of a hypergeometric distribution is given by

$$\mu = E(X) = \frac{nD}{N} \tag{4.27}$$

The variance of a hypergeometric random variable is given by

$$\sigma^2 = \mathrm{Var}(X) = \frac{nD}{N}\left(1 - \frac{D}{N}\right)\left(\frac{N-n}{N-1}\right) \tag{4.28}$$

Example 4-23: A lot of 20 transistors contains 5 nonconforming ones. If an inspector randomly samples 4 items, find the probability of 3 nonconforming transistors.

Solution. In this problem, $N = 20$, $D = 5$, $n = 4$, and $x = 3$.

$$P(X = 3) = \frac{\binom{5}{3}\binom{15}{1}}{\binom{20}{4}} = .031$$

Binomial Distribution

Consider a series of independent trials where each trial results in one of two outcomes. These outcomes are labeled as either a success or a failure. The probability p of success on any trial is assumed to be constant. Let X denote the number of successes if n such trials are conducted. Then the probability of x successes is given by

$$p(x) = \binom{n}{x} p^x (1-p)^{n-x}, \qquad x = 0, 1, 2, \ldots, n \tag{4.29}$$

and X is said to have a binomial distribution. The mean of the binomial random variable is given by

$$\mu = E(X) = np \tag{4.30}$$

and the variance is expressed as

$$\sigma^2 = \text{Var}(X) = np(1-p) \tag{4.31}$$

A **binomial distribution** is a distribution using the two parameters n and p. If the values of these parameters are known, then all information associated with the binomial distribution can be determined. Such a distribution is applicable to sampling without replacement from a population (or lot) that is large compared to the sample, or to sampling with replacement from a finite population. It is also used for situations in which items are selected from an ongoing process (that is, the population size is very large). Tables of cumulative binomial probabilities are shown in Appendix A-1.

Example 4-24: A manufacturing process is estimated to produce 5% nonconforming items. If a random sample of five items is chosen, find the probability of getting two nonconforming items.

Solution. Here, $n = 5$, $p = .05$ (if success is defined as getting a nonconforming item), and $x = 2$.

$$P(X=2) = \binom{5}{2}(.05)^2(.95)^3 = .021$$

This probability may be checked using Appendix A-1.

$$\begin{aligned} P(X=2) &= P(X \le 2) - P(X \le 1) \\ &= .999 - .977 \\ &= .022 \end{aligned}$$

The discrepancy between the two values is due to rounding the values of Appendix A-1 to three decimal places. Using Appendix A-1, the complete probability distribution of X, the number of nonconforming items, may be obtained:

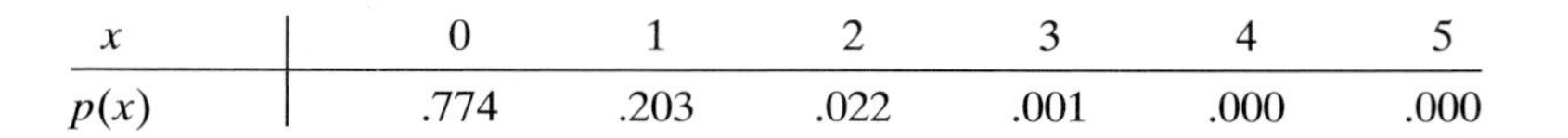

x	0	1	2	3	4	5
$p(x)$	.774	.203	.022	.001	.000	.000

Figure 4-17 is a graph of this probability distribution. The expected number of nonconforming items in the sample is

$$\mu = E(X) = 5(.05) = 0.25 \text{ item}$$

while the variance is

$$\sigma^2 = 5(.05)(.95) = 0.2375 \text{ item}^2$$

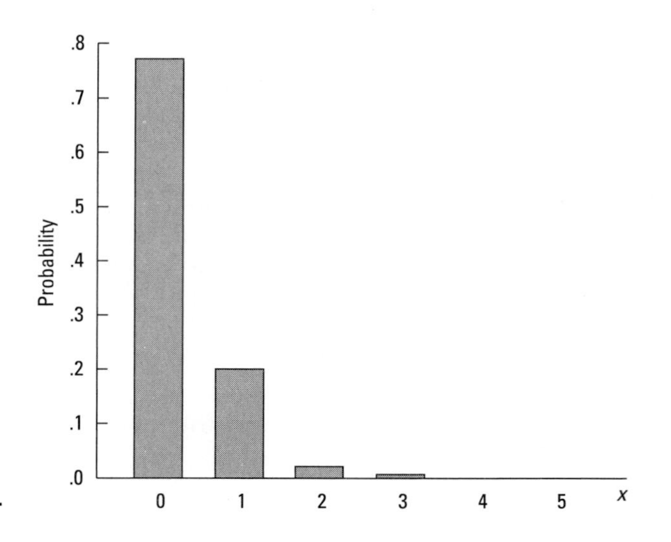

FIGURE 4-17 Binomial distribution with $n = 5$, $p = .05$.

The major differences between binomial and hypergeometric distributions are as follows: The trials are independent in a binomial distribution, whereas they are not in a hypergeometric one; the probability of success on any trial remains constant in a binomial distribution but not so in a hypergeometric one. A hypergeometric distribution approaches a binomial distribution as $N \to \infty$ and D/N remains constant.

The proportion of nonconforming items in a sample is frequently used in statistical quality control. This may be expressed as

$$\hat{p} = \frac{x}{n}$$

where X has a binomial distribution with parameters n and p, and x denotes an observed value of X. The probability distribution of $\hat{p}$ is obtained using

$$P(\hat{p} \le a) = P\left(\frac{x}{n} \le a\right) = P(x \le na)$$

$$= \sum_{x=0}^{[na]} \binom{n}{x} p^x (1-p)^{n-x} \qquad \textbf{(4.32)}$$

where $[na]$ is the largest integer less than or equal to na. It can be shown that the mean of $\hat{p}$ is p and that the variance of $\hat{p}$ is given by

$$\text{Var}(\hat{p}) = \frac{p(1-p)}{n}$$

Poisson Distribution

A **Poisson distribution** is used to model the number of events that happen within a product unit (number of defective rivets in an airplane wing), space or volume (blemishes per 200 square yards of fabric), or time period (machine breakdowns per month). It is assumed that the events happen randomly and independently.

The Poisson random variable is denoted by X. An observed value of X is represented by x. The probability distribution (or mass) function of the number of events (x) is given by

$$p(x) = \frac{e^{-\lambda} \lambda^x}{x!}, \qquad x = 0, 1, 2, \ldots \qquad \textbf{(4.33)}$$

where λ = mean or average number of events that happen over the specified product, volume, or time period

The symbol e represents the base of natural logarithms, which is equal to about 2.7183. The Poisson distribution has one parameter, λ. The mean and the variance of a Poisson distribution are equal and are given by

$$\mu = \sigma^2 = \lambda \qquad \textbf{(4.34)}$$

The Poisson distribution is sometimes used as an approximation to the binomial distribution when n is large ($n \to \infty$) and p is small ($p \to 0$), such that $np = \lambda$ is a constant. That is, a Poisson distribution can be used when all of the following hold:

1. The number of possible occurrences of defects or nonconformities per unit is large.
2. The probability or chance of a defect or nonconformity happening is small ($p \to 0$).
3. The average number of defects or nonconformities per unit is constant.

Appendix A-2 lists cumulative Poisson probabilities for various values of λ.

Example 4-25: It is estimated that the average number of surface defects in 20 m^2 of paper produced by a process is 3. What is the probability of finding no more than 2 defects in 40 m^2 of paper through random selection?

Solution. Here, one unit is 40 m^2 of paper. So, λ is 6 because the average number of surface defects per 40 m^2 is 6. The probability is

$$P(X \leq 2) = P(X=0) + P(X=1) + P(X=2)$$
$$= \frac{e^{-6}6^0}{0!} + \frac{e^{-6}6^1}{1!} + \frac{e^{-6}6^2}{2!}$$
$$= .062$$

Appendix A-2 also gives this probability as .062. The mean and variance of the distribution are both equal to 6. Using Appendix A-2, the probability distribution is as follows:

x	0	1	2	3	4	5	6	7	8
$p(x)$	.002	.015	.045	.089	.134	.161	.160	.138	.103

x	9	10	11	12	13	14	15	16
$p(x)$	.069	.041	.023	.011	.005	.003	.000	.000

A graph of this probability distribution is shown in Figure 4-18.

Continuous Distributions

Continuous random variables may assume an infinite number of values over a finite or infinite range. The probability distribution of a continuous random variable X is often called the probability density function $f(x)$. The total area under the probability density function is 1.

Normal Distribution

The most widely used distribution in the theory of statistical quality control is the **normal distribution.** The probability density function of a normal random variable is given by

$$f(x) = \frac{1}{\sqrt{2\pi}\sigma} \exp\left[\frac{-(x-\mu)^2}{2\sigma^2}\right] \qquad -\infty < x < \infty \qquad \textbf{(4.35)}$$

where μ = population mean
σ = population standard deviation

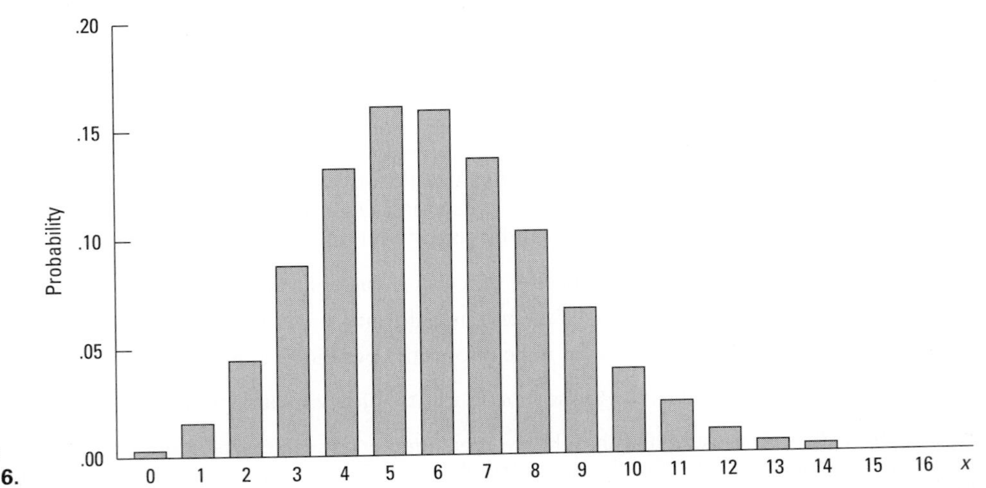

FIGURE 4-18 Poisson distribution with λ = 6.

The two parameters of a normal distribution are the mean and the variance (or standard deviation). Note that the variance σ^2 is the square of the standard deviation. Figure 4-19 shows normal probability density function. The effect of the parameters μ and σ^2 on the shape of the probability density function is shown in Figure 4-20a and b. A change in the mean μ causes a change in the location of the distribution. As the mean increases, the distribution shifts to the right, and as the mean decreases, the distribution shifts to the left. As the variance σ^2 (or standard deviation) increases, the spread about the mean increases. A normal distribution is symmetric about the mean; that is, the mean, median, and mode are equal.

The standard deviation is very important in a normal distribution. The proportion of population values that fall in range $\mu \pm \sigma$ is 68.26%. Similarly, 95.44% of the total area is within $\mu \pm 2\sigma$, and 99.74% of the area is between $\mu \pm 3\sigma$. This relationship is shown in Figure 4-21.

Finding the area under a normal curve requires integrating eq. (4.35) within the prescribed limits of the random variable, a fairly involved task. Fortunately, already constructed tables enable us to find this area. Note that because the shape of the density

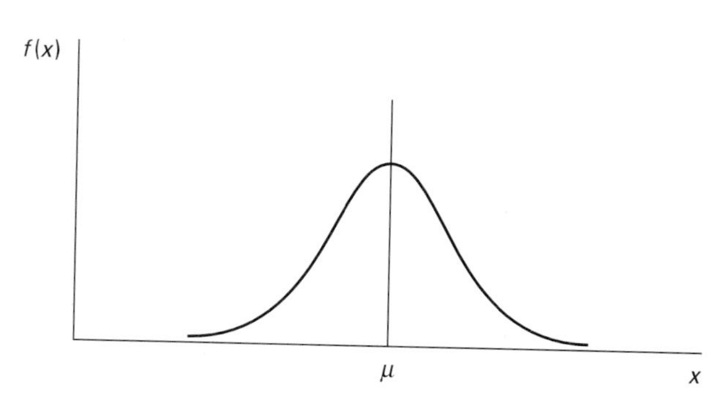

FIGURE 4-19 Normal distribution.

FIGURE 4-20 Effects of the parameters μ and σ^2 on the normal distribution.

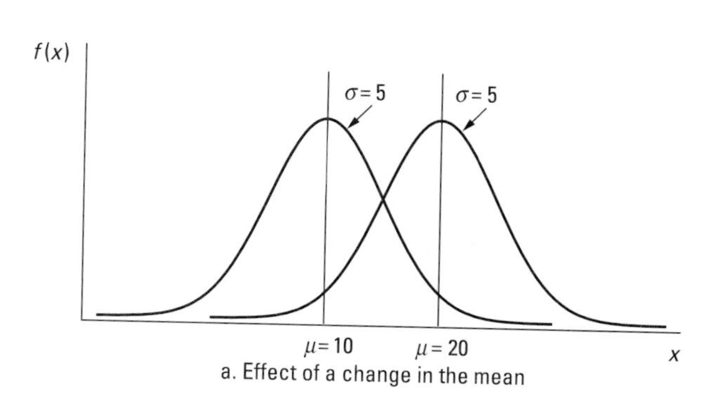

a. Effect of a change in the mean

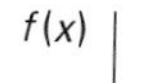
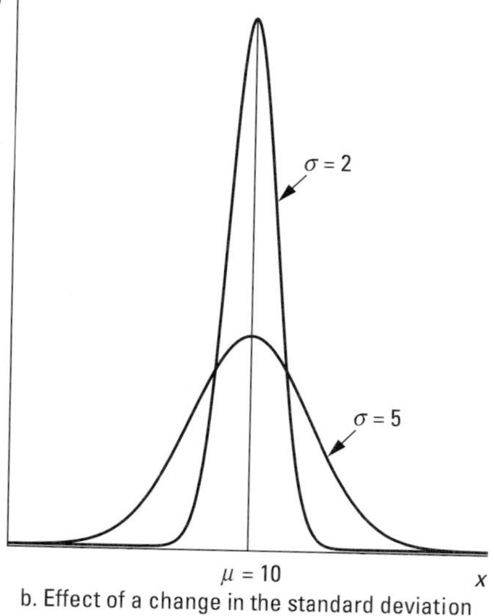

b. Effect of a change in the standard deviation

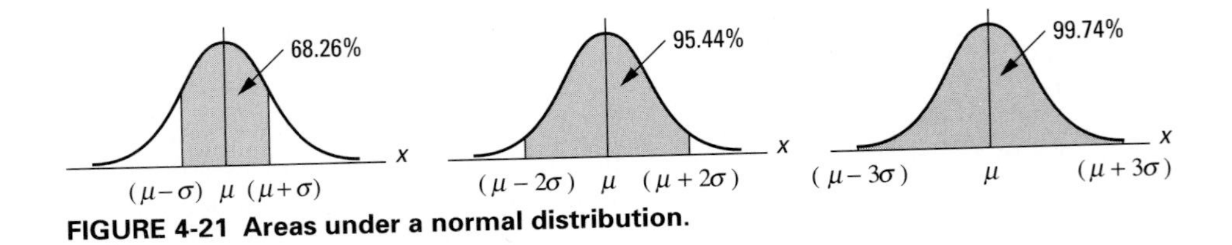

FIGURE 4-21 Areas under a normal distribution.

function changes with each possible combination of μ and σ^2, it is impossible to tabulate areas for each conceivable normal distribution. Nevertheless, the area within certain limits for any normal distribution can be found by looking up tabulated areas for a **standard normal distribution**. The standardized normal random variable Z is given by

$$Z = \frac{X - \mu}{\sigma} \tag{4.36}$$

The z-value, or standardized value, is the number of standard deviations a raw, or observed, value x is from the mean. The z-value can be positive or negative. If the z-value is positive, the raw value is to the right of the mean, whereas negative z-values indicate points to the left of the mean. At the mean, the z-value is 0. The distribution of the standardized normal random variable has a mean of 0 and a variance of 1. It is represented as an $N(0, 1)$ variable, where the first parameter represents the mean and the second the variance, and its density function is given by

$$f(z) = \frac{1}{\sqrt{2\pi}} e^{-z^2/2}, \qquad -\infty < z < \infty \tag{4.37}$$

The cumulative distribution function of Z is

$$\Phi(z) = F(z) = \int_{-\infty}^{z} f(t)\, dt \tag{4.38}$$

Figure 4-22a and b show the standard normal distribution and its relationship to the raw variable X.

Appendix A-3 gives values for the cumulative distribution function of Z. The normal distribution has the property that the area between certain limits a and b for a variable X is the same as the area between the standardized values for a and b under the standard normal distribution. Thus, we need only one set of tables—those for the standard normal distribution function—to calculate the area between certain limits for any normal distribution.

FIGURE 4-22 Normal distributions.

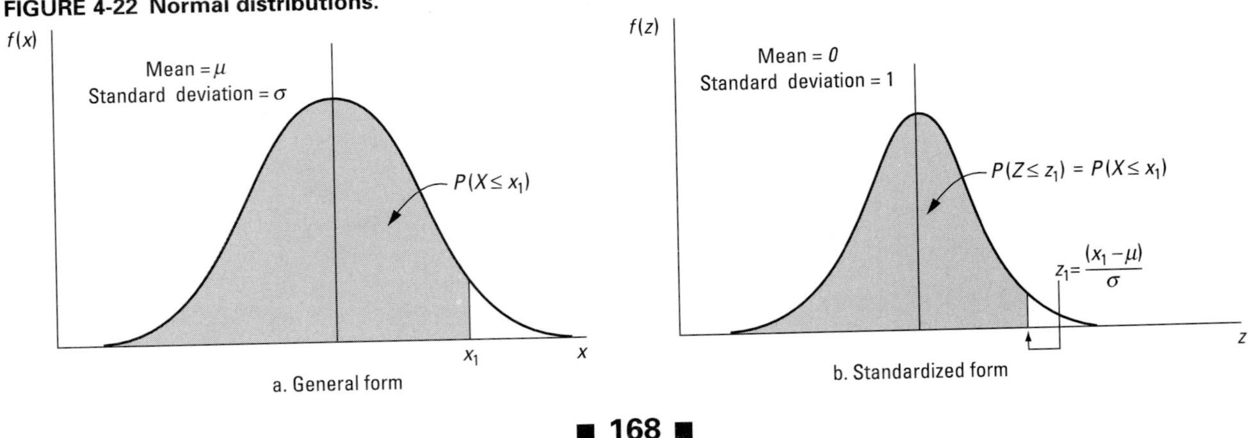

Example 4-26: The length of a machined part is known to have a normal distribution with a mean of 100 mm and a standard deviation of 2 mm.

a. What proportion of the parts will be above 103.3 mm?

Solution. Let X denote the length of the part. The parameter values for the normal distribution are $\mu = 100$ and $\sigma = 2$. The required probability is shown in Figure 4-23a. The standardized value of 103.3 corresponds to

$$z_1 = \frac{x_1 - \mu}{\sigma} = \frac{103.3 - 100}{2} = 1.65$$

Thus, $P(X > 103.3) = P(Z > 1.65)$. From Appendix A-3, $P(Z \leq 1.65) = .9505$, which also equals $P(X \leq 103.3)$. So,

$$\begin{aligned} P(Z > 1.65) &= 1 - P(Z \leq 1.65) \\ &= 1 - .9505 = .0495 \end{aligned}$$

The desired probability $P(X > 103.3)$ is .0495, or 4.95%.

b. What proportion of the output will be between 98.5 and 102.0 mm?

Solution. We wish to find $P(98.5 \leq X \leq 102.0)$, which is shown in Figure 4-23b. The standardized values are computed as

$$z_1 = \frac{102.0 - 100}{2} = 1.00$$

$$z_2 = \frac{98.5 - 100}{2} = -0.75$$

From Appendix A-3, we have $P(Z \leq 1.00) = .8413$ and $P(Z \leq -0.75) = .2266$. The required probability equals $.8413 - .2266 = .6147$. Thus, 61.47% of the output is expected to be between 98.5 and 102.0 mm.

c. What proportion of the parts will be shorter than 96.5 mm?

Solution. We want $P(X < 96.5)$, which is equivalent to $P(X \leq 96.5)$, since for a continuous random variable the probability that the variable equals a particular value is zero. The standardized value is

$$z_1 = \frac{96.5 - 100}{2} = -1.75$$

The required proportion is shown in Figure 4-23c. Using Appendix A-3, $P(Z \leq -1.75) = .0401$. Thus, 4.01% of the parts will have a length less than 96.5 mm.

d. It is important that not many of the parts exceed the desired length. If a manager stipulates that no more than 5% of the parts should be oversized, what specification limit should be recommended?

Solution. Let the specification limit be A. From the problem information, $P(X \geq A) = .05$. To find A, we first find the standardized value at the point where the raw value is A. Here, the approach will be the reverse of what was done for the previous three parts of this example. That is,

FIGURE 4-23 Calculation of normal probabilities.

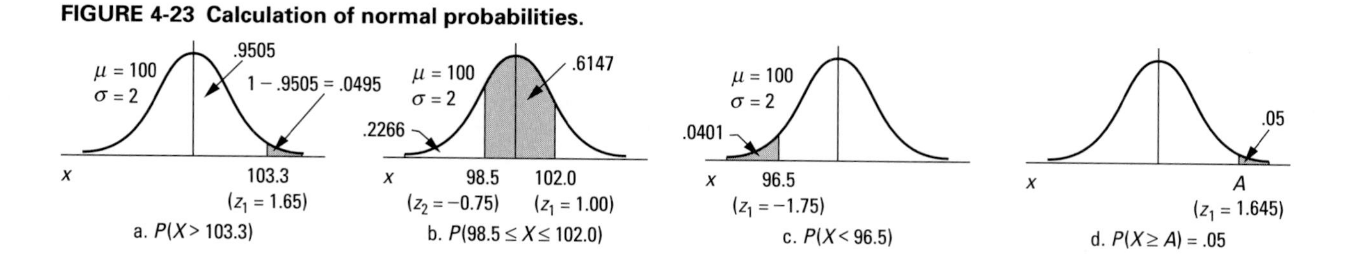

we are given an area, and we want to find the z-value. Here, $P(X \leq A) = 1 - .05 = .95$. We look for an area of .95 in Appendix A-3 and find that the linearly interpolated z-value is 1.645. Finally we unstandardize this value to determine the limit A.

$$1.645 = \frac{x_1 - 100}{2}$$

$$x_1 = 103.29 \text{ mm}$$

Thus, A should be set at 103.29 mm to achieve the desired stipulation.

Exponential Distribution

The **exponential distribution** is used in reliability analysis to describe the time to the failure of a component or system. Its probability density function is given by

$$f(x) = \lambda e^{-\lambda x}, \qquad x \geq 0 \tag{4.39}$$

where λ denotes the failure rate. Figure 4-24 shows the density function. An exponential distribution represents a constant failure rate and is used to model failures that happen randomly and independently. If we consider the typical life cycle of a product, its useful life occurs after the debugging phase and before the wearout phase. During its useful life, the failure rate is fairly constant, and failures happen randomly and independently. An exponential distribution, which has these properties, is therefore appropriate for modeling failures in the useful phase of a product. The mean and the variance of an exponential random variable are given by

$$\mu = \frac{1}{\lambda}, \qquad \sigma^2 = \frac{1}{\lambda^2} \tag{4.40}$$

Thus, the mean and the standard deviation are equal for an exponential random variable.

The exponential cumulative distribution function is obtained as follows:

$$\begin{aligned} F(x) &= P(X \leq x) \\ &= \int_0^x \lambda e^{-\lambda t}\, dt \\ &= 1 - e^{-\lambda x} \end{aligned} \tag{4.41}$$

This function is shown in Figure 4-25.

An exponential distribution has the property of being *memoryless*. This means that the probability of a component's life exceeding $(s + t)$ time units, given that it has lasted t time units, is the same as the probability of the life exceeding s time units. Mathematically, this property may be represented as

$$P(X > s + t \mid X > t) = P(X > s) \qquad \text{for all } s \text{ and } t \geq 0 \tag{4.42}$$

Example 4-27: It is known that a battery for a video game has an average life of 500 hours (h). The failures of batteries are known to be random and independent and may be described by an exponential distribution.

a. Find the probability that a battery will last at least 600 h.

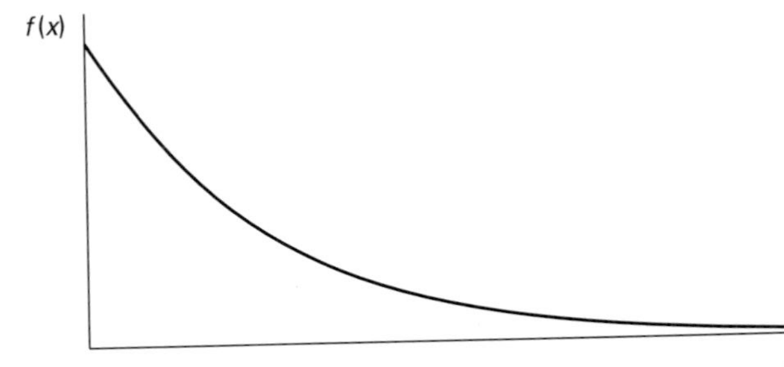

FIGURE 4-24 Exponential density function.

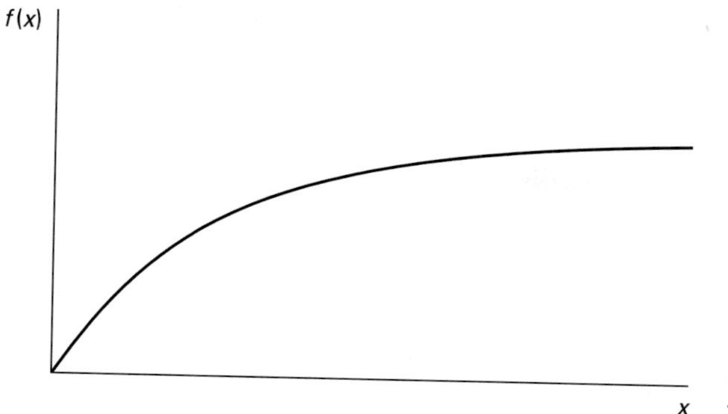

FIGURE 4-25 Exponential cumulative distribution function.

Solution. Since the average life of a battery, or the mean life, is given to be 500 h, the failure rate is

$$\lambda = \frac{1}{500}$$

If the life of a battery is denoted by X, we wish to find $P(X > 600)$.

$$\begin{aligned} P(X > 600) &= 1 - P(X \leq 600) \\ &= 1 - [1 - e^{-(1/500)600}] \\ &= e^{-1.2} = .301 \end{aligned}$$

b. Find the probability of a battery failing within 200 h.

Solution.

$$\begin{aligned} P(X \leq 200) &= 1 - e^{-(1/500)200} \\ &= 1 - e^{-0.4} = .330 \end{aligned}$$

c. Find the probability of a battery lasting between 300 and 600 h.

Solution.

$$\begin{aligned} P(300 \leq X \leq 600) &= F(600) - F(300) \\ &= e^{-(1/500)300} - e^{-(1/500)600} \\ &= e^{-0.6} - e^{-1.2} = .248 \end{aligned}$$

d. Find the standard deviation of the life of a battery.

Solution.

$$\sigma = \frac{1}{\lambda} = 500 \text{ h}$$

e. If it is known that a battery has lasted 300 h, what is the probability that it will last at least 500 h?

Solution.

$$\begin{aligned} P(X > 500 \mid X > 300) &= P(X > 200) = 1 - P(X \leq 200) \\ &= 1 - [1 - e^{-(1/500)200}] \\ &= e^{-0.4} = .670 \end{aligned}$$

Weibull Distribution

A Weibull random variable is typically used in reliability analysis to describe the time to failure of mechanical and electrical components. It is a three-parameter distribution (Banks, 1989; Henley and Kumamoto, 1981). A Weibull probability density function is given by

$$f(x) = \frac{\beta}{\alpha}\left(\frac{x-\gamma}{\alpha}\right)^{\beta-1} \exp\left[-\left(\frac{x-\gamma}{\alpha}\right)^{\beta}\right], \qquad x \geq \gamma \qquad \textbf{(4.43)}$$

The parameters are a location parameter $\gamma(-\infty < \gamma < \infty)$, a scale parameter α $(\alpha > 0)$, and a shape parameter β $(\beta > 0)$.

Figure 4-26 shows the probability density functions for $\gamma = 0$, $\alpha = 1$, and several values of β. The **Weibull distribution** as a general distribution is important because it can be used to model a variety of situations. The shape varies depending on the parameter values. For certain parameter combinations, it approaches a normal distribution. If $\gamma = 0$ and $\beta = 1$, a Weibull distribution reduces to an exponential distribution. The mean and the variance of the Weibull distribution are

$$\mu = E(X) = \gamma + \alpha\Gamma\left(\frac{1}{\beta} + 1\right) \tag{4.44}$$

$$\sigma^2 = \text{Var}(X) = \alpha^2\left\{\Gamma\left(\frac{2}{\beta} + 1\right) - \left[\Gamma\left(\frac{1}{\beta} + 1\right)\right]^2\right\} \tag{4.45}$$

where $\Gamma(t)$ represents the **gamma function** given by

$$\Gamma(t) = \int_0^\infty e^{-x}x^{t-1}\,dx$$

If u is an integer such that $u \geq 1$, then $\Gamma(u) = (u-1)!$. Note that $u! = u(u-1)(u-2)\ldots 1$ and $0! = 1$.

The cumulative distribution function of a Weibull random variable is given by

$$F(x) = 1 - \exp\left[-\left(\frac{x-\gamma}{\alpha}\right)^\beta\right], \qquad x \geq \gamma \tag{4.46}$$

Example 4-28: The time to failure for a cathode ray tube can be modeled using a Weibull distribution with parameters $\gamma = 0$, $\beta = \frac{1}{3}$, and $\alpha = 200$ h.

a. Find the mean time to failure and its standard deviation.

Solution. The mean time to failure is given by

$$\begin{aligned}\mu &= E(X)\\ &= 0 + 200\Gamma(3+1)\\ &= 200\Gamma(4) = 1200 \text{ h}\end{aligned}$$

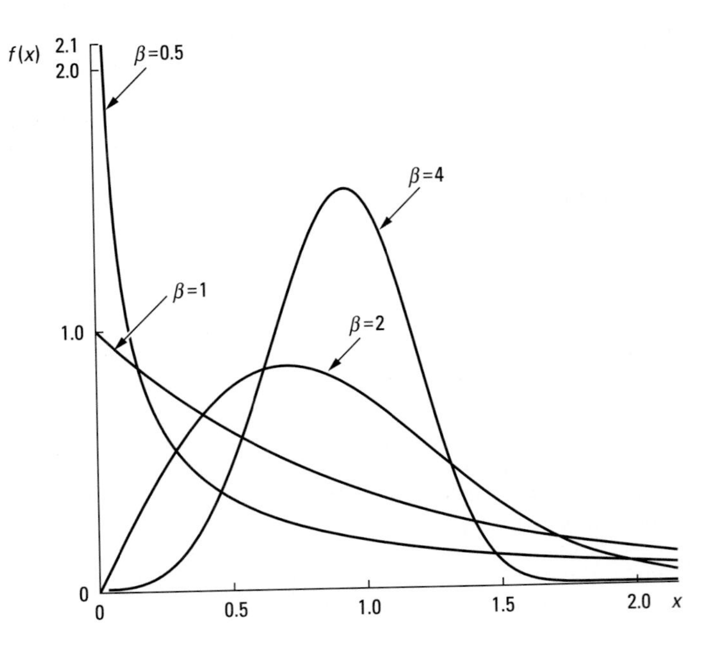

FIGURE 4-26 Weibull probability density functions ($\gamma = 0$, $\alpha = 1$, $\beta = 0.5, 1, 2, 4$).

The variance is given by

$$\sigma^2 = (200)^2\{\Gamma(6+1) - [\Gamma(3+1)]^2\}$$
$$= (200)^2\{\Gamma(7) - [\Gamma(4)]^2\} = 2736 \times 10^4$$

The standard deviation is $\sigma = 5230.679$ h.

b. What is the probability of a tube operating for at least 800 h?

Solution.

$$P(X > 800) = 1 - P(X \le 800)$$
$$= 1 - \{1 - \exp[-(800/200)^{1/3}]\}$$
$$= \exp[-(4)^{1/3}] = \exp[-1.587]$$
$$= .204$$

4-7 INFERENTIAL STATISTICS—DRAWING CONCLUSIONS ON PRODUCT AND PROCESS QUALITY

In this section we examine statistical procedures that are used to make inferences about a population (a process or product characteristic), on the basis of sample data. As mentioned previously, analysts use statistics to draw conclusions about a process based on limited information. The two main procedures of inferential statistics are estimation (point and interval) and hypothesis testing.

Usually, the parameters of a process, such as average furnace temperature, average component length, and average component diameter, are unknown, so these values must be estimated, or claims as to these parameter values must be tested for verification. For a more thorough treatment of estimation and hypothesis testing, see Duncan (1986) or Mendenhall, Reinmuth, and Beaver (1993).

Sampling Distributions

An estimator, or statistic (which is a characteristic of a sample), is used to make inferences on the corresponding parameter. For example, an estimator of sample mean is used to draw conclusions on the population mean. Similarly, a sample variance is an estimator of the population variance. Studying the behavior of these estimators through repeated sampling allows us to draw conclusions about the corresponding parameters. The behavior of an estimator in repeated sampling is known as the **sampling distribution** of the estimator, which is expressed as the probability distribution of the statistic. Sampling distributions will be discussed in greater detail in the section on interval estimation.

The sample mean is one of the most widely used estimators in quality control because analysts frequently need to estimate the population mean. It is therefore of interest to know the sampling distribution of the sample mean; this is described by the **Central Limit Theorem.**

Suppose we have a population with mean μ and standard deviation σ. If random samples of size n are selected from this population, the following holds if the sample size is large:

1. The sampling distribution of the sample mean will be approximately normal.
2. The mean of the sampling distribution of the sample mean ($\mu_{\bar{X}}$) will be equal to the population mean μ.
3. The standard deviation of the sample mean is given by $\sigma_{\bar{X}} = \sigma/\sqrt{n}$.

The degree to which a sampling distribution of a sample mean approximates a normal distribution becomes greater as the sample size n becomes larger. Figure 4-27 shows a sampling distribution of a sample mean. A sample size should be 30 or more to

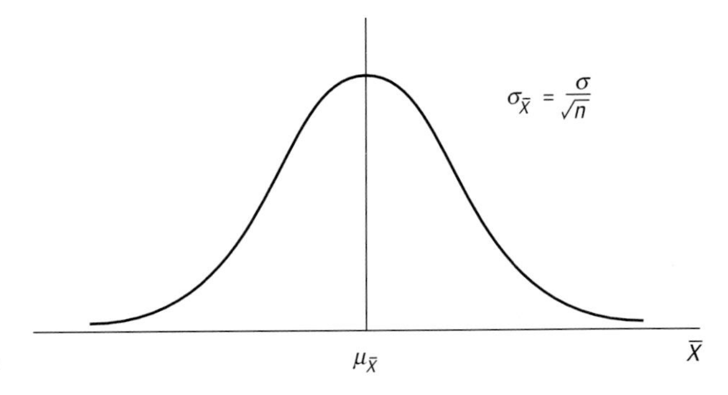

FIGURE 4-27 Sampling distribution of a sample mean.

allow a close approximation of a normal distribution. However, it has been shown that if a population distribution is symmetric and unimodal, then sample sizes as small as 4 or 5 yield sample means that are approximately normally distributed. In the case of a population distribution already being normal, samples of any size (even $n = 1$) will lead to sample means that are normally distributed. Note that the variability of the sample means, as measured by the standard deviation, decreases as the sample size increases.

Example 4-29: The tuft bind strength of a synthetic material used to make carpets is known to have a mean of 100 lb and a standard deviation of 20 lb. If a sample of size 40 is randomly selected, what is the probability that the sample mean will be less than 105 lb?

Solution. Using the Central Limit Theorem, the sampling distribution of the sample mean will be approximately normal with a mean $\mu_{\bar{X}}$ of 100 lb and a standard deviation of

$$\sigma_{\bar{X}} = \frac{20}{\sqrt{40}} = 3.162 \text{ lb}$$

We want to find $P(\bar{X} < 105)$, as shown in Figure 4-28. We first find the standardized value:

$$z_1 = \frac{\bar{X} - \mu_{\bar{X}}}{\sigma_{\bar{X}}} = \frac{105 - 100}{3.162} = 1.58$$

Then $P(\bar{X} < 105) = P(Z < 1.58) = .9429$ using Appendix A-3.

Estimation of Product and Process Parameters

One branch of statistical inference uses sample data to estimate unknown population parameters. There are two types of estimation: *point estimation* and *interval estimation.* In **point estimation,** a single numerical value is obtained as an estimate of the population parameter. In **interval estimation,** a range or interval is determined such that there is some desired level of probability that the true parameter value is contained within it. Interval estimates are also called confidence intervals.

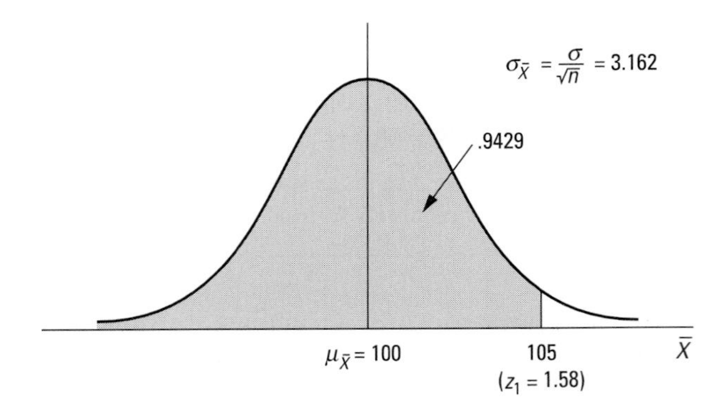

FIGURE 4-28 Determining probability of the sample mean being less than 105.

Point Estimation

A point estimate consists of a single numerical value that is used to make an inference about an unknown product or process parameter. Suppose we wish to estimate the mean diameter of all piston rings produced in a certain month. We randomly select 100 piston rings and compute the sample mean diameter, which is 50 mm. The value of 50 mm is thus a point estimate of the mean diameter of all piston rings produced that month. A common convention for denoting an estimator is to use "ˆ" above the corresponding parameter. For example, an estimator of the population mean μ is $\hat{\mu}$, which is the sample mean $\overline{X}$. An estimator of the population variance σ^2 is $\hat{\sigma}^2$, usually noted as the sample variance s^2.

Desirable Properties of Estimators Two desirable properties of estimators are worth noting here. A point estimator is said to be **unbiased** if the expected value, or mean, of its sampling distribution is equal to the parameter being estimated. A point estimator is said to have a *minimum variance* if its variance is smaller than that of any other point estimator for the parameter under consideration.

The point estimators $\overline{X}$ and s^2 are unbiased estimators of the parameters μ and σ^2, respectively. We know that $E(\overline{X}) = \mu$ and $E(s^2) = \sigma^2$. In fact, using a denominator of $(n - 1)$ in the computation of s^2 in eqs. (4.11) or (4.12) makes s^2 unbiased. The Central Limit Theorem supports the idea that the sample mean is unbiased. Also note from the Central Limit Theorem that the variance of the sample mean $\overline{X}$ is inversely proportional to the square root of the sample size.

Interval Estimation

Interval estimation consists of finding an interval defined by two end points—say, L and U—such that the probability of the parameter θ being contained in the interval is some value $(1 - \alpha)$. That is,

$$P(L \leq \theta \leq U) = 1 - \alpha \tag{4.47}$$

This expression represents a two-sided **confidence interval,** with L representing the lower confidence limit and U the upper confidence limit. If a large number of such confidence intervals were constructed from independent samples, then $100(1 - \alpha)\%$ of these intervals would be expected to contain the true parameter value of θ. (The methods for using sample data to construct such intervals are discussed in the next subsections.)

Suppose a 90% confidence interval for the mean piston ring diameter in millimeters is desired. One sample yields an interval of (48.5, 51.5)—that is, $L = 48.5$ mm and $U = 51.5$ mm. Then, if 100 such intervals were constructed (one each from 100 samples), we would expect 90 of them to contain the population mean piston ring diameter. Figure 4-29 shows this concept. The quantity $(1 - \alpha)$ is called the level of confidence or the confidence coefficient.

Confidence intervals can also be one-sided. An interval of the type

$$L \leq \theta, \qquad \text{such that } P(L \leq \theta) = 1 - \alpha$$

is a one-sided lower $100(1 - \alpha)\%$ confidence interval for θ. On the other hand, an interval of the type

$$\theta \leq U, \qquad \text{such that } P(\theta \leq U) = 1 - \alpha$$

is an upper $100(1 - \alpha)\%$ confidence interval for θ. The context of a situation will influence the type of confidence interval to be selected. For example, when the concern is breaking strength of steel cables, the customer may prefer a one-sided lower confidence interval. Since the exact expression for the confidence intervals is determined by the estimator, we discuss the estimation of several types of parameters next.

Confidence Interval About the Mean

1. *Variance known.* Suppose we want to estimate the mean μ of a product when the population variance σ^2 is known. A random sample of size n is chosen, and the sample mean $\overline{X}$ is calculated. From the Central Limit Theorem, we know that the sampling

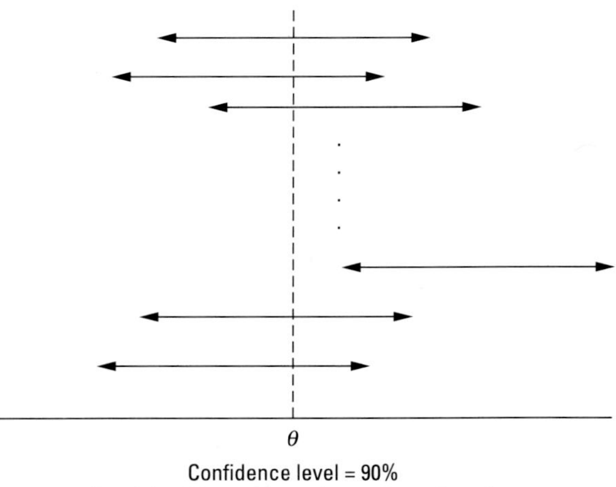

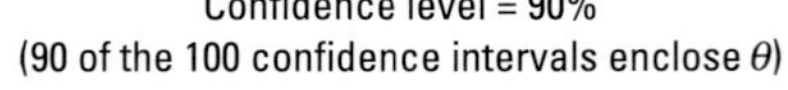

FIGURE 4-29 Interpreting confidence intervals.

distribution of the point estimator $\overline{X}$ is approximately normal with mean μ and variance σ^2/n. A $100(1-\alpha)\%$ two-sided confidence interval for μ is given by

$$\overline{X} - z_{\alpha/2}\frac{\sigma}{\sqrt{n}} \le \mu \le \overline{X} + z_{\alpha/2}\frac{\sigma}{\sqrt{n}} \tag{4.48}$$

The value of $z_{\alpha/2}$ is the standard normal variate, such that the tail area of the standardized normal distribution is $\alpha/2$. Figure 4-30 shows the location of $z_{\alpha/2}$, which can be found from the tables in Appendix A-3. Equation (4.48) represents an approximate $100(1-\alpha)\%$ confidence interval for any distribution of a random variable X. However, if X is normally distributed, then eq. (4.48) becomes an exact $100(1-\alpha)\%$ confidence interval.

Example 4-30: The output voltage of a power source is known to have a standard deviation of 10 V. Fifty readings are randomly selected, yielding an average of 118 V. Find a 95% confidence interval for the population mean voltage.

Solution. For this example, $n = 50$, $\sigma = 10$, $\overline{X} = 118$, and $1 - \alpha = .95$. From Appendix A-3, we have $z_{.025} = 1.96$. Hence a 95% confidence interval for the population mean voltage μ is

$$118 - \frac{(1.96)(10)}{\sqrt{50}} \le \mu \le 118 + \frac{(1.96)(10)}{\sqrt{50}}$$

or

$$115.228 \le \mu \le 120.772$$

Hence, there is a 95% chance that the population mean voltage falls within this range.

2. *Variance unknown.* Suppose we have a random variable X that is normally distributed with unknown mean μ and unknown variance σ^2. A random sample of size n is

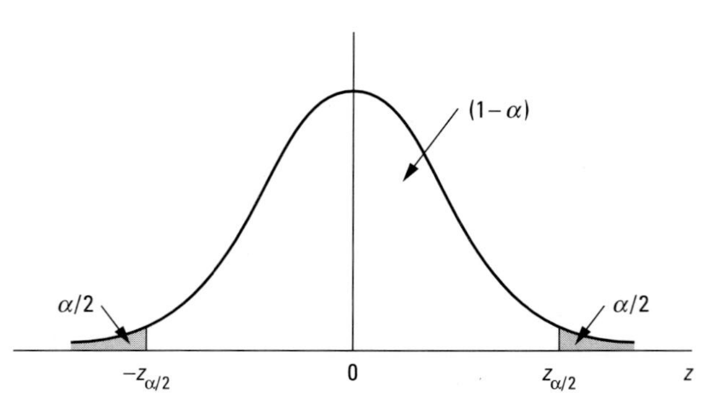

FIGURE 4-30 Finding $z_{\alpha/2}$ for confidence intervals.

selected, and the sample mean $\overline{X}$ and sample variance s^2 are computed. It is known that the sampling distribution of the quantity $(\overline{X} - \mu)/(s/\sqrt{n})$ is what is known as a ***t*-distribution** with $(n - 1)$ degrees of freedom, that is,

$$\frac{\overline{X} - \mu}{s/\sqrt{n}} \sim t_{n-1} \qquad \textbf{(4.49)}$$

where the symbol "~" stands for "is distributed as." The shape of a *t*-distribution is similar to that of the standard normal distribution and is shown in Figure 4-31. As the sample size n increases, the *t*-distribution approaches the standard normal distribution. The number of degrees of freedom of *t*, in this case $(n - 1)$, is the same as the denominator used to calculate s^2 in eqs. (4.11) or (4.12). The number of **degrees of freedom** represents the fact that if we are given the sample mean $\overline{X}$ of n observations, then $(n - 1)$ of the observations are free to be any value. Once these $(n - 1)$ values are found, there is only one value for the nth observation that will yield a sample mean of $\overline{X}$. Hence, one observation is "fixed," and $(n - 1)$ are "free."

The values of t corresponding to particular right-hand tail areas and numbers of degrees of freedom are given in Appendix A-4. For a right tail area of .025 and 10 degrees of freedom, the *t*-value is 2.228. As the number of degrees of freedom increases for a given right tail area, the *t*-value decreases. When the number of degrees of freedom is large (say, greater than 120), notice that the *t*-value given in Appendix A-4 is equal to the corresponding *z*-value given in Appendix A-3. A $100(1 - \alpha)\%$ two-sided confidence interval for the population mean μ is given by

$$\overline{X} - t_{\alpha/2,n-1}\frac{s}{\sqrt{n}} \leq \mu \leq \overline{X} + t_{\alpha/2,n-1}\frac{s}{\sqrt{n}} \qquad \textbf{(4.50)}$$

where $t_{\alpha/2,n-1}$ represents the axis point of the *t*-distribution where the right tail area is $\alpha/2$ and the number of degrees of freedom is $(n - 1)$.

Example 4-31: A new process has been developed that transforms ordinary iron into a kind of superiron called metallic glass. This new product is stronger than steel alloys and is much more corrosion-resistant than steel. However, it has a tendency to become brittle at high temperatures. It is desired to estimate the mean temperature at which it becomes brittle. A random sample of 20 pieces of metallic glass is selected. The temperature at which brittleness is first detected is recorded for each piece. The summary results give a sample mean $\overline{X}$ of 600°F and a sample standard deviation s of 15°F. Find a 90% confidence interval for the mean temperature at which metallic glass becomes brittle.

Solution. We have $n = 20$, $\overline{X} = 600$, and $s = 15$. Using the *t*-distribution tables in Appendix A-4, $t_{.05,19} = 1.729$. A 90% confidence interval for μ is

$$600 - (1.729)\frac{15}{\sqrt{20}} \leq \mu \leq 600 + (1.729)\frac{15}{\sqrt{20}}$$

or

$$594.201 \leq \mu \leq 605.799$$

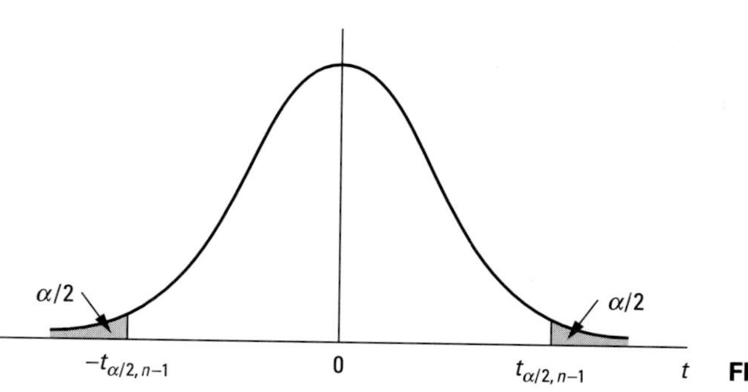

FIGURE 4-31 **A *t*-distribution.**

Example 4-32: Consider the milling operation data on depth of cut shown in Table 4-5. Figure 4-14 shows the Minitab output on descriptive statistics for this variable. Note the 95% confidence interval for the mean μ, which is

$$3.1635 \le \mu \le 3.8915$$

Confidence Interval for the Difference Between Two Means

1. *Variances known.* Suppose we have a random variable X_1 from a first population with mean μ_1 and variance σ_1^2; X_2 represents a random variable from a second population with mean μ_2 and variance σ_2^2. Assume that μ_1 and μ_2 are unknown, and σ_1^2 and σ_2^2 are known. Suppose a sample of size n_1 is selected from the first population and a sample of size n_2 is selected from the second population.

Let the sample means be denoted by $\overline{X}_1$ and $\overline{X}_2$. A 100(1 – α)% two-sided confidence interval for the difference between the two means is given by

$$(\overline{X}_1 - \overline{X}_2) - z_{\alpha/2}\sqrt{\frac{\sigma_1^2}{n_1} + \frac{\sigma_2^2}{n_2}} \le \mu_1 - \mu_2 \le (\overline{X}_1 - \overline{X}_2) + z_{\alpha/2}\sqrt{\frac{\sigma_1^2}{n_1} + \frac{\sigma_2^2}{n_2}} \quad \textbf{(4.51)}$$

2. *Variances unknown.* Let's consider two cases here. The first is the situation where the unknown variances are equal (or are assumed to be equal)—that is, $\sigma_1^2 = \sigma_2^2$. Suppose the random variable X_1 is from a normal distribution with mean μ_1 and variance σ_1^2—that is, $X_1 \sim N(\mu_1, \sigma_1^2)$, and the random variable X_2 is from $N(\mu_2, \sigma_2^2)$. Using the same notation as before, a 100(1 – α)% confidence interval for the difference in the population means $(\mu_1 - \mu_2)$ is

$$(\overline{X}_1 - \overline{X}_2) - t_{\alpha/2,n_1+n_2-2}s_p\sqrt{\frac{1}{n_1} + \frac{1}{n_2}} \le \mu_1 - \mu_2 \le (\overline{X}_1 - \overline{X}_2) + t_{\alpha/2,n_1+n_2-2}s_p\sqrt{\frac{1}{n_1} + \frac{1}{n_2}} \quad \textbf{(4.52)}$$

where a pooled estimate of the common variance, obtained by combining the information on the two sample variances, is given by

$$s_p^2 = \frac{(n_1 - 1)s_1^2 + (n_2 - 1)s_2^2}{n_1 + n_2 - 2} \quad \textbf{(4.53)}$$

The validity of assuming that the population variances are equal ($\sigma_1^2 = \sigma_2^2$) can be tested using a statistical test, which we discuss on page 190.

In the second case, the population variances are not equal—that is, $\sigma_1^2 \ne \sigma_2^2$ (a situation known as the Behrens–Fisher problem). A 100(1 – α)% two-sided confidence interval is

$$(\overline{X}_1 - \overline{X}_2) - t_{\alpha/2,v}\sqrt{\frac{s_1^2}{n_1} + \frac{s_2^2}{n_2}} \le \mu_1 - \mu_2 \le (\overline{X}_1 - \overline{X}_2) + t_{\alpha/2,v}\sqrt{\frac{s_1^2}{n_1} + \frac{s_2^2}{n_2}} \quad \textbf{(4.54)}$$

where the number of degrees of freedom of t is denoted by v, which is given by

$$v = \frac{\left(\dfrac{s_1^2}{n_1} + \dfrac{s_2^2}{n_2}\right)^2}{\dfrac{(s_1^2/n_1)^2}{n_1 - 1} + \dfrac{(s_2^2/n_2)^2}{n_2 - 1}} \quad \textbf{(4.55)}$$

Example 4-33: Two operators perform the same machining operation. Their supervisor wants to estimate the difference in the mean machining times between them. No assumption can be made as to whether the variabilities of machining time are the same for both operators. It can be assumed, however, that the distribution of machining times is normal for each operator. A random sample of 10 from the first operator gives an average machining time of 4.2 min with a standard deviation of 0.5 min. A random sample of 6 from the second operator yields an average machining time of 5.1 min with a standard deviation of 0.8 min. Find a 95% confidence interval for the difference in the mean machining times between the two operators.

Solution. We have $n_1 = 10$, $\overline{X}_1 = 4.2$, $s_1 = 0.5$, and $n_2 = 6$, $\overline{X}_2 = 5.1$, $s_2 = 0.8$. Since the assumption of equal variances cannot be made, eq. (4.54) must be used. From eq. (4.55), the number of degrees of freedom of t is

$$\nu = \frac{\left(\frac{0.25}{10} + \frac{0.64}{6}\right)^2}{\frac{(0.25/10)^2}{9} + \frac{(0.64/6)^2}{5}} = 7.393$$

As an approximation, using 7 degrees of freedom rather than the calculated value of 7.393, Appendix A-4 gives $t_{.025,7} = 2.365$. A 95% confidence interval for the difference in the mean machining times is

$$(4.2 - 5.1) - 2.365\sqrt{\frac{0.25}{10} + \frac{0.64}{6}} \le (\mu_1 - \mu_2) \le (4.2 - 5.1) + 2.365\sqrt{\frac{0.25}{10} + \frac{0.64}{6}}$$

or

$$-1.758 \le (\mu_1 - \mu_2) \le -0.042$$

Confidence Interval for a Proportion Now let's consider the parameter p, the proportion of successes in a binomial distribution. In statistical quality control, this parameter corresponds to the proportion of nonconforming items in a process or in a large lot. A point estimator of p is $\hat{p}$, the sample proportion of nonconforming items, which is found from $\hat{p} = x/n$, where x denotes the number of nonconforming items and n the number of trials or items sampled. When n is large, a $100(1-\alpha)\%$ two-sided confidence interval for p is given by

$$\hat{p} - z_{\alpha/2}\sqrt{\frac{\hat{p}(1-\hat{p})}{n}} \le p \le \hat{p} + z_{\alpha/2}\sqrt{\frac{\hat{p}(1-\hat{p})}{n}} \qquad \textbf{(4.56)}$$

For small n, the binomial tables should be used to determine the confidence limits for p. When n is large and p is small ($np < 5$), the Poisson approximation to the binomial can be used. If n is large and p is neither too small nor too large [$np \ge 5$, $n(1-p) \ge 5$], the normal distribution serves as a good approximation to the binomial.

Confidence Interval for the Difference Between Two Binomial Proportions Suppose a sample of size n_1 is selected from a binomial population with parameter p_1, while a sample of size n_2 is selected from another binomial population with parameter p_2. For large sample sizes of n_1 and n_2, a $100(1-\alpha)\%$ confidence interval for $(p_1 - p_2)$ is

$$(\hat{p}_1 - \hat{p}_2) - z_{\alpha/2}\sqrt{\frac{\hat{p}_1(1-\hat{p}_1)}{n_1} + \frac{\hat{p}_2(1-\hat{p}_2)}{n_2}} \le p_1 - p_2$$
$$\le (\hat{p}_1 - \hat{p}_2) + z_{\alpha/2}\sqrt{\frac{\hat{p}_1(1-\hat{p}_1)}{n_1} + \frac{\hat{p}_2(1-\hat{p}_2)}{n_2}} \qquad \textbf{(4.57)}$$

Example 4-34: Two operators perform the same operation of applying a plastic coating to Plexiglas. We want to estimate the difference in the proportion of nonconforming parts produced by the two operators. A random sample of 100 parts from the first operator shows that 6 are nonconforming. A random sample of 200 parts from the second operator shows that 8 are nonconforming. Find a 90% confidence interval for the difference in the proportion of nonconforming parts produced by the two operators.

Solution. We have $n_1 = 100$, x_1 (number of nonconforming parts produced by the first operator) $= 6$, $n_2 = 200$, $x_2 = 8$, $(1 - \alpha) = .90$. From Appendix A-3, using linear interpolation,

$z_{.05} = 1.645$ (for the right tail area of .05). So, $\hat{p}_1 = x_1/n_1 = 6/100 = .06$, and $\hat{p}_2 = x_2/n_2 = 8/200 = .04$. A 90% confidence interval for the difference in the proportion of nonconforming parts is

$$(.06 - .04) - 1.645\sqrt{\frac{(.06)(.94)}{100} + \frac{(.04)(.96)}{200}} \le p_1 - p_2 \le (.06 - .04) + 1.645\sqrt{\frac{(.06)(.94)}{100} + \frac{(.04)(.96)}{200}}$$

or

$$-.025 \le p_1 - p_2 \le .065$$

Confidence Interval for the Variance Consider a random variable X from a normal distribution with mean μ and variance σ^2 (both unknown). An estimator of σ^2 is the sample variance s^2. We know that the sampling distribution of $(n-1)s^2/\sigma^2$ is a **chi-squared (χ^2) distribution** with $(n-1)$ degrees of freedom. Notationally,

$$\frac{(n-1)s^2}{\sigma^2} = \chi^2_{n-1} \tag{4.58}$$

A chi-squared distribution is skewed to the right as shown in Figure 4-32. It is dependent on the number of degrees of freedom ν. Appendix A-5 shows the values of χ^2 corresponding to the right tail area α for various numbers of degrees of freedom ν. A $100(1-\alpha)\%$ two-sided confidence interval for the population variance σ^2 is given by

$$\frac{(n-1)s^2}{\chi^2_{\alpha/2,n-1}} \le \sigma^2 \le \frac{(n-1)s^2}{\chi^2_{1-\alpha/2,n-1}} \tag{4.59}$$

where $\chi^2_{\alpha/2,n-1}$ denotes the axis point of the chi-squared distribution with $(n-1)$ degrees of freedom and a right tail area of $\alpha/2$.

Example 4-35: The time to process customer orders is known to be normally distributed. A random sample of 20 orders is selected. The average processing time $\overline{X}$ is found to be 3.5 days with a standard deviation s of 0.5 day. Find a 90% confidence interval for the variance σ^2 of the order processing times.

Solution. We have $n = 20$, $\overline{X} = 3.5$, and $s = 0.5$. From Appendix A-5, $\chi^2_{.05,19} = 30.14$ and $\chi^2_{.95,19} = 10.12$. A 90% confidence interval for σ^2 is

$$\frac{19(0.5)^2}{30.14} \le \sigma^2 \le \frac{19(0.5)^2}{10.12}$$

or

$$0.158 \le \sigma^2 \le 0.469$$

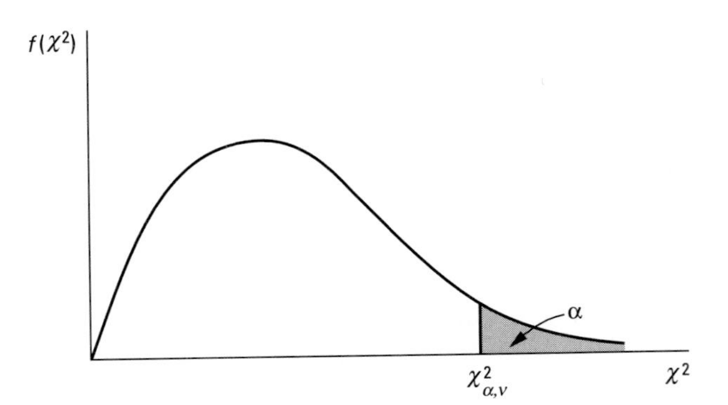

FIGURE 4-32 A chi-squared distribution.

Example 4-36: Consider the milling operation data on depth of cut shown in Table 4-5. Figure 4-14 shows the Minitab output on descriptive statistics for this variable with a 95% confidence interval for the standard deviation σ as

$$0.93233 \le \sigma \le 1.46142$$

So the 95% confidence interval for the variance σ^2 is

$$0.869 \le \sigma^2 \le 2.136$$

Confidence Interval for the Ratio of Two Variances Suppose we have a random variable X_1, from a normal distribution with mean μ_1 and variance σ_1^2, and a random variable X_2, from a normal distribution with mean μ_2 and variance σ_2^2. A random sample of size n_1 is chosen from the first population, yielding a sample variance s_1^2, and a random sample of size n_2 selected from the second population yields a sample variance s_2^2. We know that the ratio of these statistics, that is, the sample variances divided by the population variance, is an ***F*-distribution** with $(n_1 - 1)$ degrees of freedom in the numerator and $(n_2 - 1)$ in the denominator (Kendall and Stuart, 1967)—that is,

$$\frac{s_1^2/\sigma_1^2}{s_2^2/\sigma_2^2} \sim F_{(n_1-1),(n_2-1)} \tag{4.60}$$

An F-distribution is skewed to the right, as shown in Figure 4-33. It is dependent on both the numerator and denominator degrees of freedom. Appendix A-6 shows the axis points of the F-distribution corresponding to a specified right tail area α and various numbers of degrees of freedom of the numerator and denominator (ν_1 and ν_2), respectively. A $100(1 - \alpha)\%$ two-sided confidence interval for σ_1^2/σ_2^2 is given by

$$\frac{s_1^2}{s_2^2}\left(\frac{1}{F_{\alpha/2,\nu_1,\nu_2}}\right) \le \frac{\sigma_1^2}{\sigma_2^2} \le \frac{s_1^2}{s_2^2}\left(\frac{1}{F_{1-\alpha/2,\nu_1,\nu_2}}\right)$$

The lower-tail F-value, $F_{1-\alpha/2,\nu_1,\nu_2}$, can be obtained from the upper-tail F-value using the following relation:

$$F_{1-\alpha/2,\nu_1,\nu_2} = \frac{1}{F_{\alpha/2,\nu_2,\nu_1}} \tag{4.61}$$

Using eq. (4.61) yields a $100(1 - \alpha)\%$ two-sided confidence for σ_1^2/σ_2^2 of

$$\frac{s_1^2}{s_2^2}\left(\frac{1}{F_{\alpha/2,\nu_1,\nu_2}}\right) \le \frac{\sigma_1^2}{\sigma_2^2} \le \frac{s_1^2}{s_2^2} F_{\alpha/2,\nu_2,\nu_1} \tag{4.62}$$

Example 4-37: The chassis assembly time for a television set is observed for two operators. A random sample of 10 assemblies from the first operator gives an average assembly time of 22 min with a standard deviation of 3.5 min. A random sample of 8 assemblies from the second operator gives an average assembly time of 20.4 min with a standard deviation of 2.2 min. Find a 95% confidence interval for the ratio of the variances of the operators' assembly times.

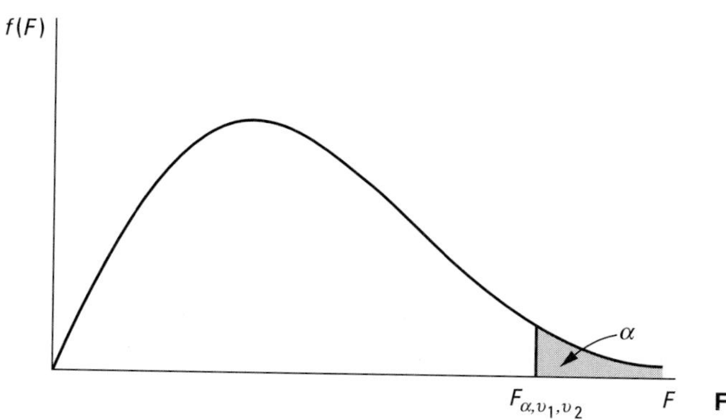

FIGURE 4-33 An *F*-distribution.

Solution. For this problem, $n_1 = 10$, $\overline{X}_1 = 22$, $s_1 = 3.5$, $n_2 = 8$, $\overline{X}_2 = 20.4$, and $s_2 = 2.2$. From Appendix A-6,

$$F_{.025,9,7} = 4.82 \qquad \text{and} \qquad F_{.025,7,9} = 4.20$$

Hence, a 95% confidence interval for the ratio of the variances of the assembly times is

$$\frac{(3.5)^2}{(2.2)^2}\left(\frac{1}{4.82}\right) \le \frac{\sigma_1^2}{\sigma_2^2} \le \frac{(3.5)^2}{(2.2)^2}(4.20)$$

or

$$0.525 \le \frac{\sigma_1^2}{\sigma_2^2} \le 10.630$$

Table 4-6 lists the formulas for the various confidence intervals and the assumptions required for each.

Hypothesis Testing

Concepts

Determining whether claims on product or process parameters are valid is the aim of **hypothesis testing.** Hypothesis tests are based on sample data. A sample statistic used to test hypotheses is known as a **test statistic.** For example, the sample mean

TABLE 4-6 Summary of Formulas for Confidence Intervals

Equation Number	*Parameter*	*Assumptions*	*Two-Sided Confidence Interval*
(4.48)	μ	σ^2 known n large	$\overline{X} \pm z_{\alpha/2}\dfrac{\sigma}{\sqrt{n}}$
(4.50)	μ	σ^2 unknown $X \sim N(\mu, \sigma^2)$	$\overline{X} \pm t_{\alpha/2,n-1}\dfrac{s}{\sqrt{n}}$
(4.51)	$\mu_1 - \mu_2$	σ_1^2, σ_2^2 known n_1, n_2 large	$(\overline{X}_1 - \overline{X}_2) \pm z_{\alpha/2}\sqrt{\dfrac{\sigma_1^2}{n_1} + \dfrac{\sigma_2^2}{n_2}}$
(4.52), (4.53)	$\mu_1 - \mu_2$	σ_1^2, σ_2^2 unknown $X_1 \sim N(\mu_1, \sigma_1^2)$ $X_2 \sim N(\mu_2, \sigma_2^2)$ $\sigma_1^2 = \sigma_2^2$	$(\overline{X}_1 - \overline{X}_2) \pm t_{\alpha/2,n_1+n_2-2}\, s_p\sqrt{\dfrac{1}{n_1} + \dfrac{1}{n_2}}$, where $s_p^2 = \dfrac{(n_1-1)s_1^2 + (n_2-1)s_2^2}{n_1+n_2-2}$
(4.54), (4.55)	$\mu_1 - \mu_2$	σ_1^2, σ_2^2 unknown $X_1 \sim N(\mu_1, \sigma_1^2)$ $X_2 \sim N(\mu_2, \sigma_2^2)$ $\sigma_1^2 \ne \sigma_2^2$	$(\overline{X}_1 - \overline{X}_2) \pm t_{\alpha/2,v}\sqrt{\dfrac{s_1^2}{n_1} + \dfrac{s_2^2}{n_2}}$, where $v = \dfrac{\left(\dfrac{s_1^2}{n_1} + \dfrac{s_2^2}{n_2}\right)^2}{\dfrac{(s_1^2/n_1)^2}{n_1-1} + \dfrac{(s_2^2/n_2)^2}{n_2-1}}$
(4.56)	p	$X \sim \text{binomial}(n, p)$ n large	$\hat{p} \pm z_{\alpha/2}\sqrt{\dfrac{\hat{p}(1-\hat{p})}{n}}$
(4.57)	$p_1 - p_2$	$X_1 \sim \text{binomial}(n_1, p_1)$ $X_2 \sim \text{binomial}(n_2, p_2)$ n_1, n_2 large	$(\hat{p}_1 - \hat{p}_2) \pm z_{\alpha/2}\sqrt{\dfrac{\hat{p}_1(1-\hat{p}_1)}{n_1} + \dfrac{\hat{p}_2(1-\hat{p}_2)}{n_2}}$
(4.59)	σ^2	$X \sim N(\mu, \sigma^2)$	$\left(\dfrac{(n-1)s^2}{\chi^2_{\alpha/2,n-1}}, \dfrac{(n-1)s^2}{\chi^2_{1-\alpha/2,n-1}}\right)$
(4.62)	$\dfrac{\sigma_1^2}{\sigma_2^2}$	$X_1 \sim N(\mu_1, \sigma_1^2)$ $X_2 \sim N(\mu_2, \sigma_2^2)$	$\left(\dfrac{s_1^2}{s_2^2}\dfrac{1}{F_{\alpha/2,v_1,v_2}}, \dfrac{s_1^2}{s_2^2}F_{\alpha/2,v_2,v_1}\right)$

length could be a test statistic. Usually, rather than using a point estimate (like the sample mean, which is an estimator of the population mean), a standardized quantity based on the point estimate is found and used as the test statistic. For instance, either the normalized or standardized value of the sample mean could be used as the test statistic, depending on whether or not the population standard deviation is known.

If the population standard deviation is known, the normalized value of the sample mean is the z-statistic, given by

$$z = \frac{\bar{x} - \mu}{\sigma/\sqrt{n}}$$

If the population standard deviation is unknown, the standardized value of the sample mean is the t-statistic, given by

$$t = \frac{\bar{x} - \mu}{s/\sqrt{n}}$$

Now, how do we test a hypothesis? Suppose the mean length of a part is expected to be 30 mm. We are interested in determining whether, for the month of March, the mean length differs from 30 mm. That is, we need to test this hypothesis. In any hypothesis-testing problem, there are two hypotheses: the **null hypothesis** H_0 and the **alternative hypothesis** H_a. The null hypothesis represents the status quo, or the circumstance being tested (which is not rejected unless proven incorrect). The alternative hypothesis represents what we wish to prove or establish. It is formulated to contradict the null hypothesis. For the situation we have just described, the hypotheses are

$$H_0: \quad \mu = 30$$
$$H_a: \quad \mu \neq 30$$

where μ represents the mean length of the part. This is a **two-tailed test;** that is, the alternative hypothesis is designed to detect departures of a parameter from a specified value in both directions. On the other hand, if we were interested in determining whether the average length *exceeds* 30 mm, the hypotheses would be

$$H_0: \quad \mu \leq 30$$
$$H_a: \quad \mu > 30$$

This is a **one-tailed test;** that is, the alternative hypothesis detects departures of a parameter from a specified value in only one direction. If our objective were to find whether the average part length is less than 30 mm, the two hypotheses would be

$$H_0: \quad \mu \geq 30$$
$$H_a: \quad \mu < 30$$

This is also a one-tailed test.

In hypothesis testing, the null hypothesis is assumed to be true unless proven otherwise. Hence, if we wish to establish the validity of a certain claim, that claim must be formulated as the alternative hypothesis. If there is statistically significant evidence contradictory to the null hypothesis, the null hypothesis is rejected; otherwise, it is not rejected. Defining what is statistically significant will, of course, depend on what the decision maker deems tolerable. Say we wish to prove that the mean length is less than 30—that is,

$$H_0: \quad \mu \geq 30$$
$$H_a: \quad \mu < 30$$

We'll assume that the population standard deviation σ is 2 mm. We take a sample of size 36 and find the sample mean length to be 25 mm. Is this difference statistically significant?

Detailed expressions for test statistics will be given later, in accordance with the parameters for which hypothesis tests are being performed.

Figure 4-34 shows the sampling distribution of the sample mean $\overline{X}$ under the assumption that $\mu = 30$. According to the Central Limit Theorem, the distribution of $\overline{X}$ will be approximately normal for large sample sizes. The important question for our scenario is whether the sample mean length of 25 mm is significantly less statistically than the specified value of 30 mm. Can we reject the null hypothesis?

To determine this, we need a cutoff point beyond which the null hypothesis will be rejected. That is, how small must the sample mean be for us to conclude that the mean length is less than 30 mm? There is a critical value, in this case on the left tail, such that if the sample mean (or test statistic) falls below it, we will reject the null hypothesis. This value defines the **rejection region** of the null hypothesis. If the test statistic does not fall in the rejection region, we do not have significant evidence to conclude that the population mean is less than 30, and so we will not reject the null hypothesis.

But how is the precise location of the critical value—and hence the rejection region—selected? How small must the sample mean be to be considered significantly less than 30? The answer to this question is influenced by the choice of the **level of significance** of the test. The rejection region is chosen such that if the null hypothesis is true, the probability of the test statistic falling in that region is small (say, .01 or .05); this probability is known as the level of significance and is denoted by α. Hence, the choice of α will dictate the rejection region.

Suppose that for a suitable choice of α (say, .05), the **critical value** is found to be $1.645\sigma_{\overline{X}}$ below the population mean of 30. (Details as to how to arrive at an expression for the critical value are given later.) The rejection region is then the shaded portion under the curve where the sample mean is at a distance more than $1.645\sigma_{\overline{X}}$ from the population mean, as shown in Figure 4-34. For our scenario then,

$$\sigma_{\overline{X}} = \frac{\sigma}{\sqrt{n}} = \frac{2}{\sqrt{36}} = 0.333 \text{ mm}$$

So the critical value is 0.548 [1.645(0.333)] units below 30, and the rejection region is $\overline{X} < 29.452$ mm. If a smaller value of α were chosen, the rejection region would shift further to the left.

For a given α, once the rejection region is selected, a framework for decision making in hypothesis testing is defined. Only if the test statistic falls in the rejection region will the null hypothesis be rejected. In our example, the rejection region was found to be $\overline{X} < 29.452$ mm, which is equivalent to $Z < 1.645$. The observed sample mean is 25 mm. The appropriate decision then is to reject the null hypothesis; that is, 25 mm is significantly less.

Errors in Hypothesis Testing

There are two types of errors in hypothesis testing: Type I and Type II. In a **Type I error,** the null hypothesis is rejected when it is actually true. The probability of a Type I error is indicated by α, the level of significance of the test. Thus, $\alpha = P(\text{Type I error}) = P(\text{Rejecting } H_0 \mid H_0 \text{ is true})$. For example, in testing (H_0: $\mu \geq 30$) against (H_a: $\mu < 30$),

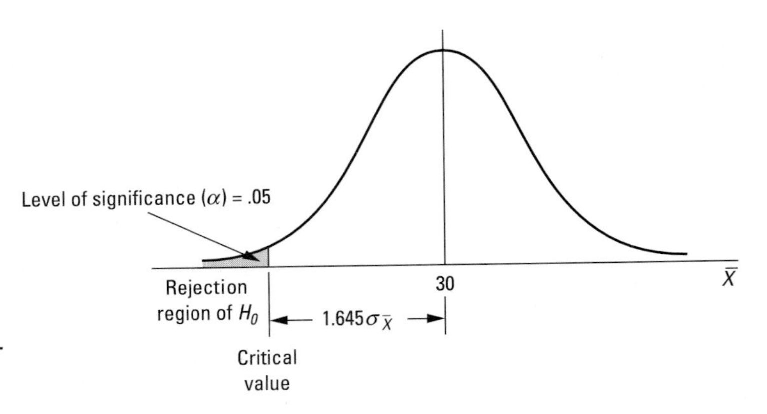

FIGURE 4-34 Sampling distribution of $\overline{X}$ assuming $\mu = 30$.

suppose a random sample of 36 parts yields a sample average length of 28 mm when the true mean length of all parts is really 30 mm. If our rejection region is $\overline{X} < 29.542$, then we must reject the null hypothesis. The magnitude of such an error can be controlled by selecting an acceptable level of α.

In a **Type II error,** the null hypothesis is not rejected even though it is false. The probability of a Type II error is denoted by β. Thus, $\beta = P(\text{Type II error}) = P(\text{Not rejecting } H_0 \mid H_0 \text{ is false})$. For example, let's test (H_0: $\mu \geq 30$) against (H_a: $\mu < 30$) with a rejection region of $\overline{X} < 29.452$. Now, suppose the true population mean length of all parts is 28 mm and a sample of 36 parts yields a sample mean of 29.8 mm. In this case, we do not reject the null hypothesis (because 29.8 does not lie in the region $\overline{X} < 29.452$). This is a Type II error.

Calculating the probability of a Type II error requires information about the population parameter (or at least an assumption about it). In such instances, we predict the probability of a Type II error based on the actual or assumed parameter value; this prediction serves as a measure of the goodness of the testing procedure and the acceptability of the chosen rejection region. The values of α and β are inversely related. If all other problem parameters remain the same, β will decrease as α increases, and vice versa. Increasing the sample size can reduce both α and β.

The **power** of a test is the complement of β and is defined as

$$\text{Power} = 1 - \beta = P(\text{Rejecting } H_0 \mid H_0 \text{ is false})$$

The power is the probability of correctly rejecting a null hypothesis that is false. Obviously, tests with high powers are the most desirable.

Steps in Hypothesis Testing

In hypothesis testing, different formulas are used with different parameters (such as the population mean or difference between two population means). For each situation, the appropriate test statistic is based on an estimator of the population parameter, and the rejection region is found accordingly. The following steps summarize the hypothesis testing procedure:

1. Formulate the null and alternative hypotheses.
2. Determine the test statistic.
3. Determine the rejection region of the null hypothesis based on a chosen level of significance α.
4. Make the decision. If the test statistic lies in the rejection region, reject the null hypothesis. Otherwise, do not reject the null hypothesis.

In an alternative procedure, the rejection region is not specifically found, though a chosen level of significance α is given. Upon determining the test statistic, the probability of obtaining that value (or an even more extreme value) for the test statistic, assuming that the null hypothesis is true, is computed. This is known as the probability value or the ***p*-value** associated with the test statistic. This p-value, also known as the observed level of significance, is then compared to α, the chosen level of significance. If the p-value is smaller than α, the null hypothesis is rejected.

Let's reconsider the mean part length example (see Figure 4-34). Suppose the observed sample mean ($\overline{X}$) is 25, for a sample of size 36. The standard deviation of the sample mean, $\sigma_{\overline{X}}$, is $2/\sqrt{36} = 0.333$ mm (assuming a population standard deviation of 2). The observed sample mean of 25 is 5 less than the population mean of 30, which corresponds to $(25 - 30)/0.333 = -15.015$ standard deviations away from the population mean. The probability of observing a sample mean of 25 or less represents the p-value and is found using the standard normal table (Appendix A-3):

$$\begin{aligned} p\text{-value} &= P(\overline{X} \leq 25) \\ &= P(Z \leq -15.015) \simeq .0000 \end{aligned}$$

Therefore, if the chosen level of significance α is .05, the p-value is essentially zero (which is less than α), so we reject H_0. This means that if the null hypothesis is true, the chance of observing an average of 25 or something even more extreme is highly unlikely. Therefore, since we observed a sample mean of 25, we would be inclined to conclude that the null hypothesis must not be true, and we would therefore reject it.

Hypothesis Testing of the Mean

1. *Variance known.* For this situation, we assume that the sample size is large (allowing the Central Limit Theorem to hold) or that the population distribution is normal. The appropriate test statistics and rejection regions are as follows:

Hypotheses	H_0: $\mu = \mu_0$ H_a: $\mu \neq \mu_0$	H_0: $\mu \leq \mu_0$ H_a: $\mu > \mu_0$	H_0: $\mu \geq \mu_0$ H_a: $\mu < \mu_0$
Rejection Region	$\lvert z_0 \rvert > z_{\alpha/2}$	$z_0 > z_\alpha$	$z_0 < -z_\alpha$

$$\textbf{Test Statistic} \quad z_0 = \frac{\overline{X} - \mu_0}{\sigma/\sqrt{n}} \qquad \textbf{(4.63)}$$

The steps for testing the hypothesis are the same four steps as described previously. Let α denote the chosen level of significance and z_α denote the axis point of the standard normal distribution such that the right tail area is α.

2. *Variance unknown.* In this situation, we assume that the population distribution is normal. If the sample size is large ($n \geq 30$), slight departures from normality do not strongly influence the test. The notation refers to t-distributions as described in the section on interval estimation.

Hypotheses	H_0: $\mu = \mu_0$ H_a: $\mu \neq \mu_0$	H_0: $\mu \leq \mu_0$ H_a: $\mu > \mu_0$	H_0: $\mu \geq \mu_0$ H_a: $\mu < \mu_0$
Rejection Region	$\lvert t_0 \rvert > t_{\alpha/2,n-1}$	$t_0 > t_{\alpha,n-1}$	$t_0 < -t_{\alpha,n-1}$

$$\textbf{Test Statistic} \quad t_0 = \frac{\overline{X} - \mu_0}{s/\sqrt{n}} \qquad \textbf{(4.64)}$$

Example 4-38: In Example 4-31, the mean temperature at which metallic glass becomes brittle was of interest. Now suppose we would like to determine whether this mean temperature exceeds 595°F. A random sample of 20 is taken, yielding a sample mean $\overline{X}$ of 600°F and a sample standard deviation s of 15°F. Use a level of significance α of .05.

Solution. The hypotheses are

$$H_0: \mu \leq 595$$
$$H_a: \mu > 595$$

The test statistic is

$$t_0 = \frac{600 - 595}{15/\sqrt{20}} = 1.491$$

From Appendix A-4, $t_{.05,19} = 1.729$. The rejection region is therefore $t_0 > 1.729$. Since the test statistic t_0 does not lie in the rejection region, we do not reject the null hypothesis. Thus, even though the sample mean is 600°F, the 5% level of significance does not allow us to conclude that there is statistically significant evidence that the mean temperature exceeds 595°F.

Hypothesis Testing for the Difference Between Two Means

1. *Variances known.* In this situation, we assume that the sample sizes are large enough for the Central Limit Theorem to hold. However, if the population distribution is normal, then the test statistic as shown will be valid for any sample size.

Hypotheses	H_0: $\mu_1 - \mu_2 = \mu_0$	H_0: $\mu_1 - \mu_2 \leq \mu_0$	H_0: $\mu_1 - \mu_2 \geq \mu_0$
	H_a: $\mu_1 - \mu_2 \neq \mu_0$	H_a: $\mu_1 - \mu_2 > \mu_0$	H_a: $\mu_1 - \mu_2 < \mu_0$
Rejection Region	$\lvert z_0 \rvert > z_{\alpha/2}$	$z_0 > z_\alpha$	$z_0 < -z_\alpha$

Test Statistic
$$z_0 = \frac{(\overline{X}_1 - \overline{X}_2) - \mu_0}{\sqrt{\sigma_1^2/n_1 + \sigma_2^2/n_2}} \qquad \textbf{(4.65)}$$

Example 4-39: The owner of a local logging operation wants to examine the average unloading time of logs. Two methods are used for unloading. A random sample of size 40 for the first method gives an average unloading time $\overline{X}_1$ of 20.5 min. A random sample of size 50 for the second method yields an average unloading time $\overline{X}_2$ of 17.6 min. We know that the variance of the unloading times using the first method is 3, while that for the second method is 4. At a significance level α of .05, can we conclude that there is a difference in the mean unloading times for the two methods?

Solution. The hypotheses are

$$H_0: \mu_1 - \mu_2 = 0$$
$$H_a: \mu_1 - \mu_2 \neq 0$$

The test statistic is

$$z_0 = \frac{(20.5 - 17.6) - 0}{\sqrt{\frac{3}{40} + \frac{4}{50}}} = 7.366$$

From Appendix A-3, $z_{.025} = 1.96$. The critical values are ± 1.96, and the rejection region is $\lvert z_0 \rvert > 1.96$. Since the test statistic z_0 lies in the rejection region, we reject the null hypothesis and conclude that there is a difference in the mean unloading times for the two methods.

2. *Variances unknown.* Here we assume that each population is normally distributed. If we assume that the population variances, though unknown, are equal (that is, $\sigma_1^2 = \sigma_2^2$), then we get the following:

Hypotheses	H_0: $\mu_1 - \mu_2 = \mu_0$	H_0: $\mu_1 - \mu_2 \leq \mu_0$	H_0: $\mu_1 - \mu_2 \geq \mu_0$
	H_a: $\mu_1 - \mu_2 \neq \mu_0$	H_a: $\mu_1 - \mu_2 > \mu_0$	H_a: $\mu_1 - \mu_2 < \mu_0$
Rejection Region	$\lvert t_0 \rvert > t_{\alpha/2, n_1+n_2-2}$	$t_0 > t_{\alpha, n_1+n_2-2}$	$t_0 < -t_{\alpha, n_1+n_2-2}$

Test Statistic
$$t_0 = \frac{(\overline{X}_1 - \overline{X}_2) - \mu_0}{s_p\sqrt{(1/n_1) + (1/n_2)}} \qquad \textbf{(4.66)}$$

Note: s_p^2 is given by eq. (4.53).

If the population variances cannot be assumed to be equal ($\sigma_1^2 \neq \sigma_2^2$), we have the following:

Hypotheses	H_0: $\mu_1 - \mu_2 = \mu_0$	H_0: $\mu_1 - \mu_2 \leq \mu_0$	H_0: $\mu_1 - \mu_2 \geq \mu_0$
	H_a: $\mu_1 - \mu_2 \neq \mu_0$	H_a: $\mu_1 - \mu_2 > \mu_0$	H_a: $\mu_1 - \mu_2 < \mu_0$
Rejection Region	$\lvert t_0 \rvert > t_{\alpha/2, v}$	$t_0 > t_{\alpha, v}$	$t_0 < -t_{\alpha, v}$

Test Statistic
$$t_0 = \frac{(\overline{X}_1 - \overline{X}_2) - \mu_0}{\sqrt{s_1^2/n_1 + s_2^2/n_2}} \qquad \textbf{(4.67)}$$

Note: v is given by eq. (4.55).

Example 4-40: A large corporation is interested in determining whether the average days of sick leave taken annually is more for night-shift employees than for day-shift employees. It is assumed that the distribution of the days of sick leave is normal for both shifts and that the variances of sick leave taken are equal for both shifts. A random sample of 12 employees from the night shift yields an average sick leave $\overline{X}_1$ of 16.4 days with a standard deviation s_1 of 2.2 days. A random sample of 15 employees from the day shift yields an average sick leave $\overline{X}_2$ of

12.3 days with a standard deviation s_2 of 3.5 days. At a level of significance α of .05, can we conclude that the average sick leave for the night shift exceeds that in the day shift?

Solution. The hypotheses are

$$H_0: \quad \mu_1 - \mu_2 \le 0$$
$$H_a: \quad \mu_1 - \mu_2 > 0$$

The pooled estimate of the variance, s_p^2, from eq. (4.53) is

$$s_p^2 = \frac{11(2.2)^2 + 14(3.5)^2}{25} = 8.990$$

So, $s_p = \sqrt{8.990} = 2.998$. The test statistic is

$$t_0 = \frac{(16.4 - 12.3) - 0}{2.998\sqrt{\frac{1}{12} + \frac{1}{15}}} = 3.531$$

From Appendix A-4, $t_{.05,25} = 1.708$. Since the test statistic t_0 exceeds 1.708 and falls in the rejection region, we reject the null hypothesis and conclude that the average sick leave for the night shift exceeds that for the day shift.

Hypothesis Testing for a Proportion

The assumption here is that the number of trials n in a binomial experiment is large, that $np \ge 5$, and that $n(1 - p) \ge 5$. This allows the distribution of the sample proportion of successes ($\hat{p}$) to approximate a normal distribution.

Hypotheses	$H_0: \ p = p_0$	$H_0: \ p \le p_0$	$H_0: \ p \ge p_0$
	$H_a: \ p \ne p_0$	$H_a: \ p > p_0$	$H_a: \ p < p_0$
Rejection Region	$\lvert z_0 \rvert > z_{\alpha/2}$	$z_0 > z_\alpha$	$z_0 < -z_\alpha$

Test Statistic
$$z_0 = \frac{\hat{p} - p_0}{\sqrt{p_0(1 - p_0)/n}} \qquad \textbf{(4.68)}$$

Example 4-41: The timeliness with which due dates are met is an important factor in maintaining customer satisfaction. A medium-sized organization wants to test whether the proportion of times that it does not meet due dates is less than 6%. Based on a random sample of 100 customer orders, they found that they missed the due date five times. What is your conclusion? Test at a level of significance α of .05.

Solution. The hypotheses are

$$H_0: \quad p \ge .06$$
$$H_a: \quad p < .06$$

The test statistic is

$$z_0 = \frac{.05 - .06}{\sqrt{(.06)(.94)/100}} = -0.421$$

From Appendix A-3, $z_{.05} = 1.645$. Since the test statistic z_0 is not less than -1.645, it does not lie in the rejection region. Hence, we do not reject the null hypothesis. At the 5% level of significance, we cannot conclude that the proportion of due dates missed is less than 6%.

Hypothesis Testing for the Difference Between Two Binomial Proportions

Here we assume that the sample sizes are large enough to allow a normal distribution for the difference between the sample proportions. Also, we consider the case for the null hypothesis, where the difference between the two proportions is zero. For a treatment of other cases, such as the hypothesized difference between two proportions being 3%, where the null hypothesis is given by H_0: $p_1 - p_2 = .03$, consult Mendenhall, Reinmuth, and Beaver (1993) and Duncan (1986).

Hypotheses	H_0: $p_1 - p_2 = 0$ H_a: $p_1 - p_2 \neq 0$	H_0: $p_1 - p_2 \leq 0$ H_a: $p_1 - p_2 > 0$	H_0: $p_1 - p_2 \geq 0$ H_a: $p_1 - p_2 < 0$
Rejection Region	$\lvert z_0 \rvert > z_{\alpha/2}$	$z_0 > z_\alpha$	$z_0 < -z_\alpha$

Test Statistic
$$z_0 = \frac{\hat{p}_1 - \hat{p}_2}{\sqrt{\hat{p}(1-\hat{p})(1/n_1 + 1/n_2)}} \quad \textbf{(4.69)}$$

Note: $\hat{p} = \dfrac{n_1\hat{p}_1 + n_2\hat{p}_2}{n_1 + n_2}$ is the pooled estimate of the proportion of nonconforming items.

Example 4-42: A company is interested in determining whether the proportion of nonconforming items is different for two of its vendors. A random sample of 100 items from the first vendor revealed 4 nonconforming items. A random sample of 200 items from the second vendor showed 10 nonconforming items. What is your conclusion? Test at a level of significance α of .05.

Solution. The hypotheses are

$$H_0: \ p_1 - p_2 = 0$$
$$H_a: \ p_1 - p_2 \neq 0$$

The pooled estimate of the proportion of nonconforming items is

$$\hat{p} = \frac{100(.04) + 200(.05)}{300} = 0.047$$

The test statistic is

$$z_0 = \frac{.04 - .05}{\sqrt{(.047)(.953)(\frac{1}{100} + \frac{1}{200})}} = -0.386$$

From Appendix A-3, $z_{.025} = 1.96$. Since the test statistic z_0 does not lie in the rejection region, we do not reject the null hypothesis. We cannot conclude that the proportion of nonconforming items between the two vendors differs.

Hypothesis Testing for the Variance

Assume here that the population distribution is normal.

Hypotheses	H_0: $\sigma^2 = \sigma_0^2$ H_a: $\sigma^2 \neq \sigma_0^2$	H_0: $\sigma^2 \leq \sigma_0^2$ H_a: $\sigma^2 > \sigma_0^2$	H_0: $\sigma^2 \geq \sigma_0^2$ H_a: $\sigma^2 < \sigma_0^2$
Rejection Region	$\chi_0^2 > \chi^2_{\alpha/2,n-1}$ or $\chi_0^2 < \chi^2_{1-\alpha/2,n-1}$	$\chi_0^2 > \chi^2_{\alpha,n-1}$	$\chi_0^2 < \chi^2_{1-\alpha,n-1}$

Test Statistic
$$\chi_0^2 = \frac{(n-1)s^2}{\sigma_0^2} \quad \textbf{(4.70)}$$

Example 4-43: The variability of the downtime of equipment in a job shop is of concern to the owner. A random sample of 15 machines that had to be repaired shows a mean downtime $\overline{X}$ of 2.2 h with a standard deviation s of 0.2 h. Can we conclude that the variance of downtimes is less than 0.06? Use a level of significance α of .01.

Solution. The hypotheses are

$$H_0: \ \sigma^2 \geq 0.06$$
$$H_a: \ \sigma^2 < 0.06$$

The test statistic is

$$\chi_0^2 = \frac{(14)(0.2)^2}{0.06} = 9.333$$

From Appendix A-5, $\chi^2_{.99,14} = 4.66$. The test statistic value of 9.333 is not less than 4.66 and so does not lie in the rejection region. Hence, we do not reject the null hypothesis. At the 1% level of significance, we cannot conclude that the variance of downtimes is less than 0.06 h.

Hypothesis Testing for the Ratio of Two Variances

We assume that both populations are normally distributed.

Hypotheses	H_0: $\sigma_1^2 = \sigma_2^2$ H_a: $\sigma_1^2 \neq \sigma_2^2$	H_0: $\sigma_1^2 \leq \sigma_2^2$ H_a: $\sigma_1^2 > \sigma_2^2$	H_0: $\sigma_1^2 \geq \sigma_2^2$ H_a: $\sigma_1^2 < \sigma_2^2$
Rejection Region	$F_0 > F_{\alpha/2,v_1,v_2}$ or $F_0 < F_{1-\alpha/2,v_1,v_2}$	$F_0 > F_{\alpha,v_1,v_2}$	$F_0 < F_{1-\alpha,v_1,v_2}$

$$\textbf{Test Statistic} \quad F_0 = \frac{s_1^2}{s_2^2} \qquad \textbf{(4.71)}$$

Example 4-44: The variabilities of the service times of two bank tellers are of interest. Their supervisor wants to determine whether the variance of service time for the first teller is greater than that for the second. A random sample of 8 observations from the first teller yields a sample average $\overline{X}_1$ of 3.4 min with a standard deviation s_1 of 1.8 min. A random sample of 10 observations from the second teller yields a sample average $\overline{X}_2$ of 2.5 min with a standard deviation of 0.9 min. Can we conclude that the variance of the service time is greater for the first teller than for the second? Use a level of significance α of .05.

The hypotheses are

$$H_0: \sigma_1^2 \leq \sigma_2^2$$
$$H_a: \sigma_1^2 > \sigma_2^2$$

The test statistic is

$$F = \frac{s_1^2}{s_2^2} = \frac{(1.8)^2}{(0.9)^2} = 4.00$$

From Appendix A-6, $F_{.05,7,9} = 3.29$. The test statistic lies in the rejection region, and so we reject the null hypothesis.

4-8 CONCEPTS IN SAMPLING

Introduction

In quality control, it is often not feasible to obtain data regarding a certain quality characteristic for each item in a population due to a lack of time and resources. Furthermore, sample data provides adequate information about a product or process characteristic at a fraction of the cost. It is therefore important to know how samples are selected and the properties of various sampling procedures.

A **sampling design** is a description of the procedure by which the observations in a sample are to be chosen. It does not necessarily deal with the measuring instrument to be used. For example, a sampling design might specify choosing every tenth item produced.

In the context of sampling, an **element** is an object (or group of objects) for which data or information is gathered. A **sampling unit** is an individual element or a collection of nonoverlapping elements from the population. A **sampling frame** is a list of all sampling units. For example, if our interest is confined to a set of parts produced in the month of July (sampling element), the sampling unit could be an individual part, while the sampling frame would be a list of the part numbers of all of the items produced.

Sampling Designs and Schemes

A major objective of any sampling design or scheme is to select the sample in such a way as to accurately portray the population from which it is drawn. After all, a sample is supposed to be representative of the population.

Sampling in general has certain advantages. If the measurement requires destroying the item being measured (destructive testing), we cannot afford to obtain data from each item in the population. Also, in measurements involving manual methods or high production rates, inspector fatigue may result, which would yield inaccurate data.

Errors in Sampling

There are three sources of errors in sample surveys. The first source is **random variation.** The inherent nature of sampling variability sometimes causes such errors to occur. The more sophisticated the measuring instrument, the lower the random variation.

Misspecification of the population is a second source of error. This type of error occurs in public opinion polling, in obtaining responses regarding consumer satisfaction with the product, in listing a sampling frame incorrectly, and so on.

The third source of error deals with **nonresponses** (usually in sample surveys). This category also includes situations where a measurement is not feasible due to an inoperative measuring instrument, a shortage of people responsible for taking the measurement, or other such reasons.

Simple Random Sample

One of the most widely used sampling designs in quality control is the simple random sample. Suppose we have a finite population of N items from which a sample of n items is to be selected. If samples are chosen such that each possible sample of size n has an equal chance of being selected, the sampling process is said to be random, and the sample obtained is known as a **simple random sample.**

Random-number tables (or computer-generated random numbers) can be used to draw a simple random sample. For example, if there are 1000 elements in the population, the three-digit numbers 000–999 are used to identify each element. To start, a random number is selected; one element corresponds to this number. This selection continues until the desired sample of size n is chosen. If a random number that has already been used comes up, it is ignored, and another one is chosen.

Stratified Random Sample

Sometimes the population from which samples are selected is heterogeneous. For instance, consider the output from two operators who are known to differ greatly in their performance. Rather than randomly selecting a sample from their combined output, a random sample is selected from the output of each operator. This way both operators are fairly represented so we can determine whether there are significant differences between them. Thus, a **stratified random sample** is obtained by separating the elements of the population into nonoverlapping distinct groups (called strata) and then selecting a simple random sample from each stratum.

Cluster Sample

In the event that a sampling frame is not available, or if obtaining samples from all segments of the population is not feasible for geographic reasons, a **cluster sample** can be used. Here, the population is first divided into groups of elements, or clusters. Clusters are randomly selected, and a census of data is obtained (that is, all of the elements within the chosen clusters are examined). If a company has plants throughout the southeastern United States, it may not be feasible to sample from each plant. Clusters are then defined (say, one for each plant), and some of the clusters are then randomly chosen (say, three of the five clusters). A census of data is then obtained for each selected cluster.

Sample Size Determination

The size of a sample has a direct impact on the reliability of the information provided by the data. The larger the sample size, the more valuable the data. It is usually of interest to know the minimum sample size that can be used to estimate an unknown product or process parameter accurately.

Estimating Population Mean

Suppose the mean of a product or process characteristic is of interest (for example, the mean width of a component used in an assembly). There is a $(1 - \alpha)$ probability that the difference between the estimated value and the actual value will be no greater than some number B. Figure 4-35 shows the principle behind selecting an appropriate sample size for estimating the population mean. The quantity B is sometimes referred to as the tolerable error bound and is given by

$$B = z_{\alpha/2}\sigma_{\bar{X}} = z_{\alpha/2}\frac{\sigma}{\sqrt{n}} \tag{4.72}$$

or

$$n = \frac{z_{\alpha/2}^2\sigma^2}{B^2} \tag{4.73}$$

Example 4-45: An analyst wishes to estimate the average bore size of a large casting. Based on historical data, it is estimated that the standard deviation of the bore size is 4.2 mm. If it is desired to estimate with a probability of .95 the average bore size to within 0.8 mm, find the appropriate sample size.

Solution. We have $\sigma = 4.2$, $B = 0.8$, $z_{.025} = 1.96$

$$\text{Sample size } n = \frac{(1.96)^2(4.2)^2}{(0.8)^2} = 105.88 \simeq 106$$

Estimating Population Proportion

Consider a binomial population where the objective is to estimate the proportion of "successes" (p). Examples might include estimating the proportion of nonconforming stamped parts in a press shop or the proportion of unsatisfied customers in a restaurant. Here again, we must select a tolerable error bound B such that that estimate will have a probability of $(1 - \alpha)$ of being within B units of the parameter value. Figure 4-36 shows the sampling distribution of the sample proportion of successes ($\hat{p}$), which is approximately normal for large sample sizes. The equation for determining the sample size n is given by

$$B = z_{\alpha/2}\sigma_{\hat{p}} = z_{\alpha/2}\sqrt{\frac{p(1-p)}{n}} \tag{4.74}$$

or

$$n = \frac{z_{\alpha/2}^2 p(1-p)}{B^2} \tag{4.75}$$

Since the true parameter value p is not known, there are a couple of ways in which eq. (4.75) can be modified. First, if a historical estimate of p is available (say, $\hat{p}$), it can be used in place of p in eq. (4.75). Second, if no prior information is available, a conservative estimate of n can be calculated from eq. (4.75) by using $p = .5$. Using $p = .5$ maximizes the value of $p(1-p)$ and hence produces a conservative estimate.

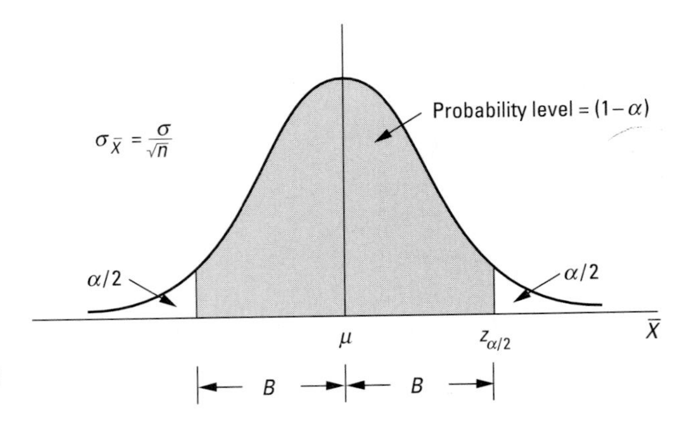

FIGURE 4-35 Sample size determination for estimating population mean μ.

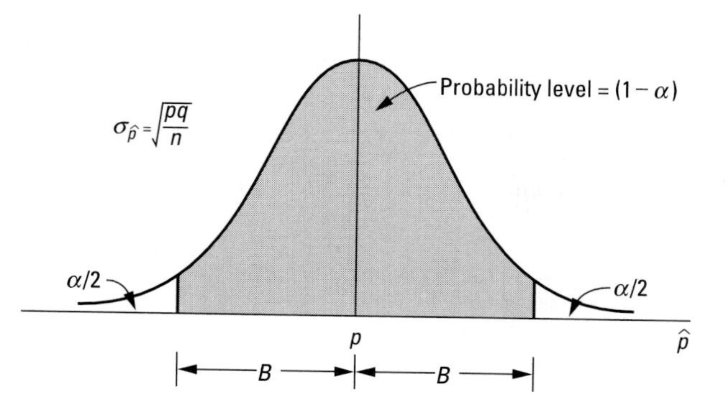

FIGURE 4-36 Sampling distribution of $\hat{p}$.

Example 4-46: In the production of rubber tubes, the tube stock has to be first cut into a piece of a specified length. This piece is then formed into a circular shape and joined using pressure and the right temperature. The operator training and such process parameters as temperature, pressure, and die size influence the production of conforming tubes. We want to estimate with a probability of .90 the proportion of nonconforming tubes to within 4%. How large a sample should be chosen if no prior information is available on the process?

Solution. Using the preceding notation, $B = 0.04$. From Appendix A-3, $z_{.05} = 1.645$. Since no information on p, the proportion of nonconforming tubes, is available, use $p = .5$. Hence

$$n = \frac{(1.645)^2(.5)(.5)}{(0.04)^2} = 422.8 \simeq 423$$

If a prior estimate of p had been available, it would have reduced the required sample size.

4-9 SUMMARY

This chapter has presented the statistical foundations necessary for quality control and improvement. The procedures for summarizing data that describe product or process characteristics have been discussed. A review of common discrete and continuous probability distributions with applications in quality control has been included. The chapter has also presented inferential statistics that can be used for drawing conclusions as to product and process quality. In particular, the topics of estimation and hypothesis testing have been emphasized. Finally, some fundamental concepts of sampling designs have been presented. The variety of statistical procedures presented in this chapter are meant to serve as an overview. Technical details have been purposely limited. For further discussion of any of the topics, you should consult corresponding references on the subject matter of interest.

APPENDIX: APPROXIMATIONS TO SOME PROBABILITY DISTRIBUTIONS

In some situations, if tables for the needed probability distributions are not available for the parameter values in question or if the calculations using the formula become tedious and prone to error due to roundoffs, approximations for the probability distribution under consideration can be considered.

Binomial Approximation of the Hypergeometric

When the ratio of sample size to population size is small—that is, n/N is small (≤ 0.1, as a rule of thumb)—the binomial distribution serves as a good approximation to the hypergeometric distribution. The parameter values to be used for the binomial distribution would be the same value of n as in the hypergeometric distribution, and $p = D/N$.

Example A-1: Consider a lot of 100 parts, of which 6 are nonconforming. If a sample of 4 parts is selected, what is the probability of obtaining 2 nonconforming items? If a binomial approximation is used, what is the required probability?

Solution. Using the hypergeometric distribution, $N = 100$, $D = 6$, $n = 4$, and $x = 2$, we have

$$P(X=2) = \frac{\binom{6}{2}\binom{94}{2}}{\binom{100}{4}} = .017$$

Note that $n/N = .04$, which is less than .1. Using the binomial distribution as an approximation to the hypergeometric, with $p = 6/100 = .06$, yields

$$P(X=2) = \binom{4}{2}(.06)^2(.94)^2 = .019$$

Poisson Approximation to the Binomial

In a binomial distribution, if n is large and p is small ($p < .1$) such that np is constant, the Poisson distribution serves as a good approximation to the binomial. The parameter λ in the Poisson distribution is used as np. The larger the value of n and smaller the value of p, the better the approximation. As a rule of thumb, when $np < 5$, the approximation is acceptable.

Example A-2: A process is known to have a nonconformance rate of .02. If a random sample of 100 items is selected, what is the probability of finding 3 nonconforming items?

Solution. Using the binomial distribution, $n = 100$, $p = .02$, and $x = 3$, we have

$$P(X=3) = \binom{100}{3}(.02)^3(.98)^{97} = .182$$

Next, we use the Poisson distribution as an approximation to the binomial. Using $\lambda = (100)(.02) = 2$ yields

$$P(X=3) = \frac{e^{-2}2^3}{3!} = .180$$

Normal Approximation to the Binomial

In a binomial distribution, if n is large and p is close to .5, the normal distribution may be used to approximate the binomial. Usually, if p is neither too large nor too small ($.1 \le p \le .9$), the normal approximation is acceptable when $np \ge 5$. A continuity correction factor is used by finding an appropriate z-value because the binomial distribution is discrete, whereas the normal distribution is continuous.

Example A-3: A process is known to produce about 6% nonconforming items. If a random sample of 200 items is chosen, what is the probability of finding between 6 and 8 nonconforming items?

Solution. If we decide to use the binomial distribution with $n = 200$, $p = .06$, we have

$$\begin{aligned} P(6 \le X \le 8) &= P(X=6) + P(X=7) + P(X=8) \\ &= \binom{200}{6}(.06)^6(.94)^{194} \\ &\quad + \binom{200}{7}(.06)^7(.94)^{193} \\ &\quad + \binom{200}{8}(.06)^8(.94)^{192} \\ &= .0235 + .0416 + .0641 = .1292 \end{aligned}$$

Using the normal distribution to approximate a binomial probability, the mean is given by $np = 200(.06) = 12$, and the variance is given by $np(1 - p) = 200(.06)(1 - .06) = 11.28$. The required probability is

$$P(6 \le X \le 8) \simeq P(5.5 \le X \le 8.5)$$

The .5 adjustment is often known as the continuity correction factor. The binomial random variable is discrete. When using a continuous random variable such as the normal to approximate it, this adjustment makes the approximation better. In other words, $P(X \le 8)$ is written as $P(X \le 8.5)$, while $P(X \ge 6)$ is written as $P(X \ge 5.5)$.

$$\begin{aligned}\text{Required probability} &= P\left(\frac{5.5 - 12}{\sqrt{11.28}} \le z \le \frac{8.5 - 12}{\sqrt{11.28}}\right)\\ &= P(-1.94 \le z \le -1.04)\\ &= .1492 - .0262 = .1230\end{aligned}$$

Normal Approximation to the Poisson

If the mean λ of a Poisson distribution is large ($\lambda \ge 10$), a normal distribution may be used to approximate it. The parameters of this normal distribution are the mean μ, which is set equal to λ and the variance σ^2, which is also set equal to λ. As in Example A-3, the continuity correction factor of .5 may be used since a discrete random variable is approximated by a continuous one.

Example A-4: The number of small businesses that fail each year is known to have a Poisson distribution with a mean of 16. Find the probability that in a given year there will be no more than 18 small business failures.

Solution. Using the Poisson distribution ($\lambda = 16$), the required probability is $P(X \le 18)$. From the table in Appendix A-2, this probability is .742. When using the normal distribution as an approximation, the mean and variance are set equal to λ. The .5 continuity correction factor is used in a similar manner as in Example A-3.

$$\begin{aligned}P(X \le 18) &\simeq P(X \le 18.5)\\ &= P\left(z \le \frac{18.5 - 16}{\sqrt{16}}\right)\\ &= P(z \le .625 = .7340 \text{ (using linear interpolation)}\end{aligned}$$

Figure 4A-1 summarizes the conditions under which the different approximations to these common distributions may be used.

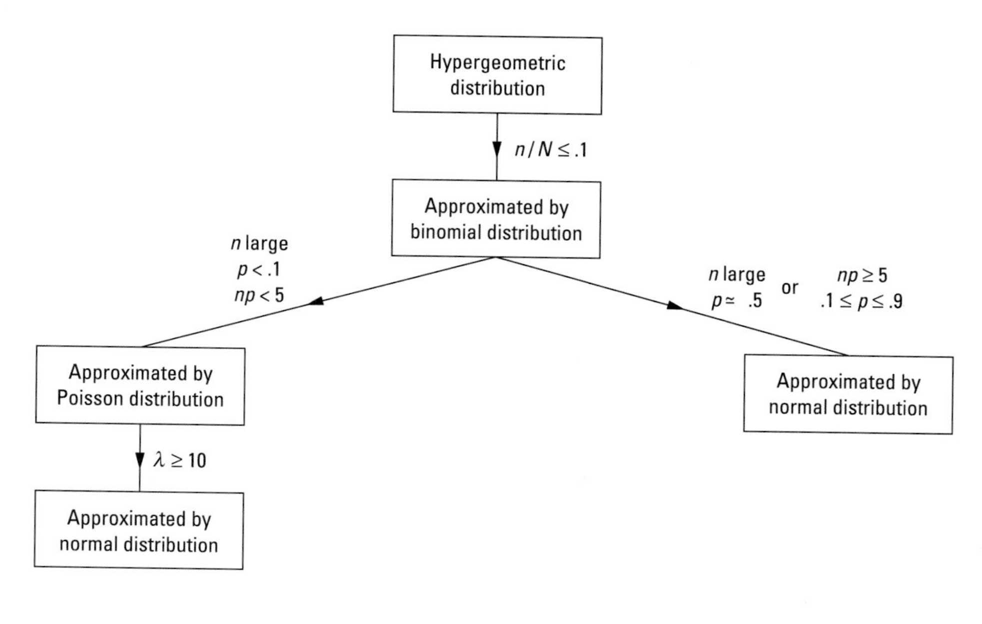

FIGURE 4A-1 Necessary conditions for using approximations to distributions.

Key Terms

- accuracy
- additive law
- alternative hypothesis
- association, measures of
- binomial distribution
- calibration
- central tendency, measures of
- Central Limit Theorem
- chi-squared distribution
- confidence interval
- continuous variable
- correlation coefficient
- critical value
- cumulative distribution function (cdf)
- data collection
- degrees of freedom
- descriptive statistics
- discrete variable
- dispersion, measures of
- distributions
  - continuous
  - discrete
- estimation
  - point
  - interval
- events
  - complementary
  - compound
  - independent
  - mutually exclusive
  - simple
- expected value
- exponential distribution
- *F*-distribution
- gamma function
- hypergeometric distribution
- hypothesis testing
  - one-tailed
  - two-tailed
- inferential statistics
- interquartile range
- interval estimation
- kurtosis coefficient
- level of significance
- mean
  - population
  - sample
- measurement scales
  - interval
  - nominal
  - ordinal
  - ratio
- median
- misspecification
- mode
- multiplicative law
- nonresponse
- normal distribution
  - standard normal
- null hypothesis
- outlier
- *p*-value
- parameter
- point estimation
- Poisson distribution
- population
- power
- precision
- probability
- probability density function
- probability distribution function
- random variable
- random variation
- range
- rejection region
- sample
  - cluster
  - simple random
  - stratified
- sample size, determination of
- sample space
- sampling
  - design
  - element
  - errors in
  - frame
  - unit
- sampling distribution
- skewness coefficient
- standard deviation
- statistic
- statistics
  - descriptive
  - inferential
- *t*-distribution
- test statistic
- trimmed mean
- Type I error
- Type II error
- unbiased
- variance
- Weibull distribution

Exercises

Discussion Questions

1. Explain the difference between a parameter and a statistic. Give examples of each.
2. Explain the concept of independence of events. How does this relate to events being mutually exclusive?
3. Find an expression for the probability of the union of three events that are mutually independent of each other.
4. Are complementary events independent? Are they mutually exclusive?
5. How do procedures in descriptive statistics help in obtaining information about a product or a process?
6. Explain the importance of inferential statistics in decision making regarding a product or a process.
7. Explain the difference between accuracy and precision of measurements.
8. Explain the different types of measurement scales, and give examples of each type.

9. Distinguish between the usage of the mean, median, and mode in quality control applications. When would you prefer to use the trimmed mean?
10. What are the advantages of the standard deviation over the range as a measure of dispersion? Explain and interpret the interquartile deviation.
11. Why are measures of central tendency insufficient to describe the characteristics of a product or a process?
12. Explain the skewness and kurtosis coefficients. How does information about these two measures aid in quality control and improvement?
13. What does the correlation coefficient measure? How may this information contribute to quality improvement?
14. Describe the properties that must be satisfied by the probability distribution function of a discrete random variable.
15. Distinguish between a hypergeometric and binomial random variable.
16. State and explain the Central Limit Theorem. Explain its role in quality control.
17. What are the desirable properties of estimators?
18. A 95% confidence interval for the mean thickness of a part in millimeters is (10.2, 12.9). Interpret this interval.
19. Distinguish between a stratified sample and a cluster sample. Give some examples of where these two might be used.
20. What guidelines pertain to selecting a sample size for the purpose of estimation?
21. Explain the difference between a null hypothesis and alternative hypothesis.
22. Discuss a Type I error and Type II error in hypothesis testing. Explain these in the context of a quality control setting.

Problems

23. To satisfy certain environmental standards set by the EPA, a utility company conducts some tests of its own. One sample is chosen from each of the company's 6 plants. It is suspected that 2 of the plants are in violation of the standards. If an inspector randomly chooses samples for inspection, answer the following:
    a. What is the probability that at least 1 of the suspected plants is represented in the sample?
    b. What is the probability that none of the suspected plants are represented in the sample?
24. Based on historical data, it is estimated that 12% of new products will obtain a profitable market share. However, if two products are newly introduced in the same year, there is only a 5% chance of both products becoming profitable. A company is planning to market two new products, 1 and 2, this coming year. What is the probability of
    a. only product 1 becoming profitable?
    b. only product 2 becoming profitable?
    c. at least one of the products becoming profitable?
    d. neither product becoming profitable?
    e. either product 1 or product 2 (but not both) becoming profitable?
    f. product 2 becoming profitable, if product 1 is found to be profitable?
25. Two types of defects are observed in the production of integrated circuit boards. It is estimated that 6% of the boards have solder defects, while 3% of them have some surface-finish defects. The occurrences of the two types of defects are assumed to be independent of each other. If a circuit board is randomly selected from the output, find the probabilities for the following situations:
    a. Either a solder defect or a surface-finish defect or both is found.
    b. Only a solder defect is found.
    c. Both types of defects are found.
    d. The board is free of defects.
    e. If a surface-finish defect is found, what are the chances of also finding a solder defect?
26. The settling of unwanted material in a mold is causing some defects in the output. Based on recent data, it is estimated that 5% of the output has one or more defects. In spot checking some parts, an inspector randomly selects 2 parts. Find the probabilities that
    a. the first part is defect-free.
    b. the second part is defect-free.
    c. both parts are defect-free.
    d. one of the parts is acceptable.
    e. at least one part is acceptable.
27. The following times (in minutes) to process hot-rolled steel are observed for a random sample of size 10:

5.4	6.2	7.9	4.8	7.5
6.2	5.5	4.5	7.2	6.2

    a. Find the mean, median, and mode of the processing times. Interpret the differences between them.
    b. Compute the range, variance, and standard deviation of the processing times, and interpret them.

28. A random sample of 50 observations on the mileage per gallon of a particular brand of gasoline is shown:

```
33.2 29.4 36.5 38.1 30.0
29.1 32.2 29.5 36.0 31.5
34.5 33.6 27.4 30.4 28.4
32.6 30.4 31.8 29.8 34.6
30.7 31.9 32.3 28.2 27.5
34.9 32.8 27.7 28.4 28.8
30.2 26.8 27.8 30.5 28.5
31.8 29.2 28.6 27.5 28.5
30.8 31.8 29.1 26.9 34.2
33.5 27.4 28.5 34.8 30.5
```

a. Find the mean, median, and standard deviation, and interpret them.
b. Find the interquartile range, skewness, and kurtosis coefficient, and interpret their values.

29. An insurance company is interested in determining whether life insurance coverage is influenced by disposable income. A randomly chosen sample of size 20 produced the data shown. Calculate the correlation coefficient between disposable income and the amount of life insurance coverage, and interpret it. How can the company use this information in its decision making?

Disposable Income (in $1000)	*Life Insurance Coverage (in $1000)*	*Disposable Income (in $1000)*	*Life Insurance Coverage (in $1000)*
45	60	65	80
40	58	60	90
65	100	45	50
50	50	40	50
70	120	55	70
80	100	55	60
70	80	60	80
40	50	75	100
50	70	45	50
45	60	65	70

30. A distribution company has ordered 12 computers. Unknown to the company, three of the computers have defects. If the purchasing division of the company selects four computers from the shipment, what is the probability that none of them will have defects? What is the probability that no more than two will have defects?

31. A sampling plan currently in use in a software company producing diskettes involves selecting 5 items from a lot of 35 diskettes. If no more than 1 nonconforming diskette is found, the lot is accepted. If a lot actually has 3 nonconforming diskettes, what is the probability of accepting such a lot under this plan?

32. A pharmaceutical company making antibiotics has to abide by certain standards set by the Food and Drug Administration. The company performs some testing on the strength of the antibiotic. In a case of 25 bottles, 4 bottles are selected for testing. If the case actually contains 5 understrength bottles, what is the probability that the chosen sample will contain no understrength bottles? Exactly 1 understrength bottle? How many understrength bottles would be expected in the sample? What is the standard deviation of understrength bottles in the sample?

33. A company involved in making solar panels estimates that 3% of its product is nonconforming. If a random sample of 5 items is selected from the production output, what is the probability that none are nonconforming? That 2 are nonconforming? The cost of rectifying a nonconforming panel is estimated to be $5. For a shipment of 1000 panels, what is the expected cost of rectification?

34. A university has purchased a service contract for its computers and pays $20 annually for each computer. Maintenance records show that 8% of the computers require some sort of servicing during the year. Furthermore, it is estimated that the average expenses for each repair, had the university not been covered by the service contract, would be about $200. If the university currently has 20 computers, would you advise buying the service contract? Based on expected costs, for what annual premium per computer will the university be indifferent to purchasing the service contract? What is the probability of the university spending no more than $500 annually on repairs if it does not buy the service contract?

35. The probability of an electronic sensor malfunctioning is known to be .10. A random sample of 12 sensors is chosen. Find the probability that
a. at least 3 will malfunction.
b. no more than 5 will malfunction.
c. at least 1 but no more than 5 will malfunction.
d. What is the expected number of sensors that will malfunction?
e. What is the standard deviation of the number of sensors that will malfunction?

36. A process is known to produce 5% nonconforming items. A sample of 40 items is selected from the process.
a. What is the distribution of the nonconforming items in the sample?
b. Find the probability of obtaining no more than 3 nonconforming items in the sample.
c. Using the Poisson distribution as an approximation to the binomial, calculate the probability of the event in part b.

d. Compare the answers to parts b and c. What are your observations?

37. The guidance system design of a satellite places several components in parallel. The system will function as long as at least 1 of the components is operational. In a particular satellite, 4 such components are placed in parallel. If the probability of a component operating successfully is .9, what is the probability of the system functioning? What is the probability of the system failing? Assume that the components operate independently of each other.

38. In a lot of 200 electrical fuses, 20 are known to be nonconforming. A sample of 10 fuses is selected.
    a. What is the probability distribution of the number of nonconforming fuses in the sample? What are its mean and standard deviation?
    b. Using the binomial distribution as an approximation to the hypergeometric, find the probability of getting 2 nonconforming fuses. What is the probability of getting at most 2 nonconforming fuses?

39. A local hospital estimates that the number of patients admitted daily to the emergency room has a Poisson probability distribution with a mean of 4.0. What is the probability that on a given day
    a. only 2 patients will be admitted?
    b. at most 6 patients will be admitted?
    c. no one will be admitted?
    d. What is the standard deviation of the number of patients admitted?
    e. For each patient admitted, the expected daily operational expenses to the hospital are $800. If the hospital wants to be 94.9% sure of meeting daily expenses, how much money should it retain for operational expenses daily?

40. In an auto body shop, it is known that the average number of paint blemishes per car is 3. If 2 cars are randomly chosen for inspection, what is the probability that
    a. the first car has no more than 2 blemishes?
    b. each of the cars has no more than 2 blemishes?
    c. the total number of blemishes in both of the cars combined is no more than 2?

41. The number of bank failures per year among those insured by the Federal Deposit Insurance Company has a mean of 7.0. The failures occur independently. What is the probability that
    a. there will be at least 4 failures in the coming year?
    b. there will be between 2 and 8 failures, inclusive, in the coming year?
    c. during the next two years there will be at most 8 failures?

42. A legal consulting firm is examining the prospects of expansion. Based on information over the last five years, it is found that the average number of suits filed in the local county court is 4.0 per month. The suits are independent of each other. Find the probability that
    a. no more than 6 suits will be filed in a given month.
    b. at least 5 suits will be filed in a given month.
    c. at least 6 suits will be filed in a span of two months.

43. The outside diameter of a part used in a gear assembly is known to be normally distributed with a mean of 40 mm and a standard deviation of 2.5 mm. The specifications on the diameter are (36, 45), which means that part diameters between 36 mm and 45 mm are considered acceptable. The unit cost of rework is $0.20, while the unit cost of scrap is $0.50. If the daily production rate is 2000, what is the total daily cost of rework and scrap?

44. The breaking strength of a cable is known to be normally distributed with a mean of 4000 kg and a standard deviation of 25 kg. The manufacturer prefers that at least 95% of its product meet a strength requirement of 4050 kg. Is this requirement being met? If not, by changing the process parameter, what should the process mean target value be?

45. The specifications for the thickness of nonferrous washers are 1.0 ± 0.04 mm. From process data, the distribution of the washer thickness is estimated to be normal with a mean of 0.98 mm and a standard deviation of 0.02 mm. The unit cost of rework is $0.10, and the unit cost of scrap is $0.15. For a daily production of 10,000 items:
    a. What proportion of the washers are conforming? What is the total daily cost of rework and scrap?
    b. In its study of constant improvement, the manufacturer changes the mean setting of the machine to 1.0 mm. If the standard deviation is the same as before, what is the total daily cost of rework and scrap?
    c. The manufacturer is trying to further improve on the process and reduces its standard deviation to 0.015 mm. If the process mean is maintained at 1.0 mm, what is the percent decrease in the total daily cost of rework and scrap compared to that of part a?

46. A company has been able to restrict the use of electrical power through energy conservation measures. The monthly use is known to be normal with a mean of 60,000 kWh (kilowatt-hour) and a standard deviation of 400 kWh.
    a. What is the probability that the monthly consumption will be less than 59,100 kWh?

b. What is the probability that the monthly consumption will be between 59,000 and 60,300 kWh?

c. The capacity of the utility that supplies this company is 61,000 kWh. What is the probability that demand will not exceed supply by more than 100 kWh?

47. A component is known to have an exponential time-to-failure distribution with a mean life of 10,000 h.

a. What is the probability of the component lasting at least 8000 h?

b. If the component is in operation at 9000 h, what is the probability that it will last another 6000 h?

c. Two such components are put in parallel, so that the system will be in operation if at least one of the components is operational. What is the probability of the system being operational for 12,000 h? Assume that the components operate independently.

48. The time to repair an equipment is known to be exponentially distributed with a mean of 45 min.

a. What is the probability of the machine being repaired within half an hour?

b. If the machine breaks down at 3 P.M. and a repairman is available immediately, what is the probability of the machine being available for production by the start of the next day? Assume that the repairman is available until 5 P.M.

c. What is the standard deviation of the repair time?

49. A limousine service catering to a large metropolitan area has found that the time for a trip (from dispatch to return) is exponentially distributed with a mean of 30 min.

a. What is the probability that a trip will take more than an hour?

b. If a limousine has already been gone for 45 min, what is the probability that it will return within the next 20 min?

c. If two limousines have just been dispatched, what is the probability that both will not return within the next 45 min? Assume that the trips are independent of each other.

50. The time to failure of an electronic component can be described by a Weibull distribution with $\gamma = 0$, $\beta = 0.25$, and $\alpha = 800$ h.

a. Find the mean time to failure.

b. Find the standard deviation of the time to failure.

c. What is the probability of the component lasting at least 1500 h?

51. The time to failure of a mechanical component under friction may be modeled by a Weibull distribution with $\gamma = 20$ days, $\beta = 0.2$, and $\alpha = 35$ days.

a. What proportion of these components will fail within 30 days?

b. What is the expected life of the component?

c. What is the probability of a component lasting between 40 and 50 days?

52. The diameter of bearings is known to have a mean of 35 mm with a standard deviation of 0.5 mm. A random sample of 36 bearings is selected. What is the probability that the average diameter of these selected bearings will be between 34.95 and 35.18 mm?

53. Refer to Exercise 52. Suppose the machine is considered to be out of statistical control if the average diameter of a sample of 36 bearings is less than 34.75 mm or greater than 35.25 mm.

a. If the true mean diameter of all bearings produced is 35 mm, what is the probability of the test indicating that the machine is out of control?

b. Suppose the setting of the machine is accidentally changed such that the mean diameter of all bearings produced is 35.05 mm. What is the probability of the test indicating that the machine is in statistical control?

54. Vendor quality control is an integral part of a total quality system. A soft drink bottling company requires its vendors to produce bottles with an internal pressure strength of at least 300 kg/cm^2. A vendor claims that its bottles have a mean strength of 310 kg/cm^2 with a standard deviation of 5 kg/cm^2. As part of a vendor surveillance program, the bottling company samples 50 bottles from the production and finds the average strength to be 308.6 kg/cm^2.

a. What are the chances of getting that sample average that was observed, or even less, if the assertion by the vendor is correct?

b. If the standard deviation of the strength of the vendor's bottles is 8 kg/cm^2, with the mean (as claimed) of 310 kg/cm^2, what are the chances of seeing what the bottling company observed (or an even smaller sample average)?

55. The mean time to assemble a product as found from a sample of size 40 is 10.4 min. The standard deviation of the assembly times is known to be 1.2 min.

a. Find a two-sided 90% confidence interval for the mean assembly time, and interpret it.

b. Find a two-sided 99% confidence interval for the mean assembly time, and interpret it.

c. What assumptions are needed to answer parts a and b?
d. The manager in charge of the assembly line believes that the mean assembly time is less than 10.8 min. Can he make this conclusion at a significance level α of .05?

56. A company that dumps its industrial waste into a river has to meet certain restrictions. One particular constraint involves the minimum amount of dissolved oxygen that is needed to support aquatic life. A random sample of 10 specimens taken from a given location gives the following results of dissolved oxygen (in parts per million, ppm):

9.0 8.6 9.2 8.4 8.1
9.5 9.3 8.5 9.0 9.4

a. Find a two-sided 95% confidence interval for the mean dissolved oxygen, and interpret it.
b. What assumptions do you need to make to answer part a?
c. Suppose the environmental standards stipulate a minimum of 9.5 ppm of average dissolved oxygen. Is the company violating the standard? Test at a level of significance α of .05.

57. An automobile company is working on changes in a fuel injection system to improve gasoline mileage. A random sample of 15 test runs gives the following mileage (in miles per gallon):

38 42 40 39 44 37 39 45
40 42 38 39 44 41 42

a. Find a two-sided 90% confidence interval for the mean gasoline mileage. State the assumptions necessary to find the confidence interval.
b. The mean miles-per-gallon rating when using the previous fuel injection system was 35. Can we conclude that the new system has improved gasoline mileage? Use a level of significance of .01.

58. A beverage bottling company fills bottles that are labeled as 12 oz. The variability of the machine that dispenses the beverage has a standard deviation of 0.10 oz. A random sample of size 60 yielded a mean content weight of 12.05 oz.
a. Find a 98% confidence interval for the mean beverage content weight.
b. Can we conclude that the mean beverage content weight exceeds the advertised weight? Use a level of significance of .05.

59. The Occupational Safety and Health Administration (OSHA) mandates certain regulations that have to be adopted by corporations. Prior to the implementation of the OSHA program, a company found that for a sample of 40 randomly selected months, the mean employee time lost due to job-related accidents was 45 h. After the implementation of the OSHA program, for a random sample of 45 months, the mean employee time lost due to job-related accidents was 39 h. It can be assumed that the variability of time lost due to accidents is about the same before and after implementation of the OSHA program (with a standard deviation being 3.5 h):
a. Find a 90% confidence interval for the difference in the mean time lost due to accidents.
b. Test the hypothesis that implementation of the OSHA program has reduced the mean employee lost time. Use a level of significance of .10.

60. Refer to Exercise 59. Suppose the standard deviations of the values of lost time due to accidents before and after usage of the OSHA program are unknown but are assumed to be equal. The first sample of size 40 gave a mean of 45 h with a standard deviation of 3.8 h. Similarly, the second sample of size 45, taken after the implementation of the OSHA program, yielded a mean of 39 h with a standard deviation of 3.5 h.
a. Find a 95% confidence interval for the difference in the mean time lost due to job-related accidents.
b. What assumptions are needed to answer part a?
c. Can you conclude that the mean employee time lost due to accidents has decreased due to the OSHA program? Use a level of significance of .05.
d. How would you test the assumption of equality of the standard deviations of time lost due to job-related accidents before and after implementation of the OSHA program? Use a level of significance of .05.

61. A company is experimenting with synthetic fibers as a substitute for natural fibers. The quality characteristic of interest is the breaking strength. A random sample of 8 natural fibers yields an average breaking strength of 540 kg with a standard deviation of 55 kg. A random sample of 10 synthetic fibers gives a mean breaking strength of 610 kg with a standard deviation of 22 kg.
a. Can you conclude that the variances of the breaking strengths of natural and synthetic fibers are different? Use a level of significance α of .05. What assumptions are necessary to perform this test?
b. Based on the conclusions in part a, test to determine if the mean breaking strength for synthetic fibers exceeds that for natural fibers. Use a significance level α of .10.

c. Find a two-sided 95% confidence interval for the ratio of the variances of the breaking strengths of natural and synthetic fibers.
d. Find a two-sided 90% confidence interval for the difference in the mean breaking strength for synthetic and natural fibers.

62. Consider the data in Exercise 27 on the time (in minutes) to process hot-rolled steel for a sample of size 10.
 a. Find a 98% confidence interval for the mean time to process hot-rolled steel. What assumptions do you have to make to solve this problem?
 b. Find a 95% confidence interval for the variance.
 c. Test the hypothesis that the process variability, as measured by the variance, exceeds 0.80. Use $\alpha = .05$.

63. The machine shop in an aircraft manufacturing company deals with a variety of jobs. The floor supervisor wants to estimate the proportion of nonconforming parts that are output from this shop since rework is affecting the product cost. A random sample of 200 parts shows that 6 are nonconforming.
 a. Find a 90% confidence interval for the true proportion of nonconforming parts.
 b. Can we conclude that the percent nonconforming of all parts output from the shop is less than 4%? Use a level of significance of .05.

64. Price deregulation in the airline industry has promoted competition and a variety of fare structures. Prior to deciding on a price change, a particular airline is interested in obtaining an estimate of the proportion of the market that it presently captures for a certain city. A random sample of 300 passengers indicates that 80 used that airline.
 a. Find a point estimate of the proportion of the market that uses this particular airline.
 b. Find a 95% confidence interval for the proportion that uses this airline.
 c. Can the airline conclude that its market share is more than 25%? Use a level of significance of .01.

65. An advertising agency is judged by the increase in the proportion of people who buy a particular product after the advertising campaign is conducted. In a random sample of 200 people prior to the campaign, 40 said they prefer the product in question. After the advertising campaign, out of a random sample of 300, 80 say they prefer the product.
 a. Find a 90% confidence interval for the difference in the proportion of people who prefer the stipulated product before and after the advertising campaign.
 b. Can you conclude that the advertising campaign has been successful in increasing the proportion of people who prefer the product? Use a level of significance of .10.

66. A state university is seeking to increase its enrollment by providing more financial assistance, by making tuition competitive, and by improving the academic rigor. Prior to the institution of these measures, the university found that out of a random sample of 100 prospective students, 30 were interested in attending. After implementation of the proposed measures, a random sample of 200 shows that 90 are interested in attending.
 a. Find a 90% confidence interval for the difference in the proportion of students who are interested in attending the university before and after the proposed measures.
 b. Were the proposed measures effective in increasing the interest of students to attend the university? Use a significance level of .05.

67. Two machines used in the same operation are to be compared. A random sample of 80 parts from the first machine yields 6 nonconforming ones. A random sample of 120 parts from the second machine shows 14 nonconforming ones.
 a. Can we conclude that there is a difference in the output of the machines? Use a level of significance of .10.
 b. Find a 95% confidence interval for the difference in the proportion of nonconforming parts between the two machines.

68. Budget deficits have reduced the funding of several federal programs. A random sample of 100 people shows that 20 receive either full or partial federal aid that they had applied for.
 a. Find a 98% confidence interval for the proportion of people who apply for and receive some federal aid.
 b. Can we conclude that the proportion of people receiving federal aid has dropped below 22%? Use a level of significance of .10.

69. The precision of equipment and instruments is measured by the variability of their operation under repeated conditions. The output from an automatic lathe producing the diameter (in millimeters) of a part gave the following readings for a random sample of size 10:

10.3	9.7	9.6	9.9	9.5
10.2	9.8	10.1	9.8	10.2

 a. Find a 90% confidence interval for the variance of the diameters.
 b. Find a 90% confidence interval for the standard deviation of the diameters.

c. Test the null hypothesis that variance of the diameters does not exceed 0.05 mm. Use a significance level of .10.
d. Does the mean setting for the machine need adjustment? Is it significantly different from 9.5 mm? Test at a significance level of .10.

70. A company is investigating two potential vendors on the timeliness of their deliveries. A random sample of size 10 from the first vendor produced an average delay time of 4.5 days with a standard deviation of 2.3 days. A random sample of size 12 from the second vendor yielded an average delay time of 3.4 days with a standard deviation of 6.2 days.
    a. Find a 90% confidence interval for the ratio of the variances of the delay times for the two vendors.
    b. What assumptions are needed to solve part a?
    c. Can we conclude that the first vendor has a smaller variability regarding delay times than the second? Use a significance level of .05.
    d. Which vendor would you select and why?
71. A company's quality manager wants to estimate with a probability of .90 the copper content in a mixture to within 4%. How many samples must be selected if no prior information is available on the proportion of copper in the mixture?
72. The oven temperature in a steel smelting process is an important process parameter. We would like to estimate with a probability of .95 the average oven temperature to within 40°C. Based on previous information, the standard deviation of the oven temperature is estimated to be 150°C. How many samples must be selected?
73. The production of nonconforming items is of critical concern because it increases costs and reduces productivity. Identifying the causes behind the production of unacceptable items and taking remedial action are steps in the right direction. To begin, we decide to estimate the proportion of nonconforming items in a process to within 3% with a probability of .98.
    a. Previous information suggests that the percentage nonconforming is approximately 4%. How large a sample must be chosen?
    b. If no prior information is available on the proportion nonconforming, how large a sample must be chosen?
74. Management is exploring the possibility of adding more agents at the check-in counter of an airline to reduce the waiting time of customers. Based on available information, the standard deviation of the waiting times is approximately 5.2 min. If management wants to estimate the mean waiting time to within 2 min, with a probability of .95, how many samples must be selected?

References

Banks, J. (1989). *Principles of Quality Control.* New York: John Wiley.

Duncan, A. J. (1986). *Quality Control and Industrial Statistics,* 5th ed. Homewood, Ill.: Richard D. Irwin.

Henley, E. J., and H. Kumamoto (1981). *Reliability Engineering and Risk Assessment.* Upper Saddle River, N.J.: Prentice Hall.

Kendall, M. G., and A. Stuart (1967). *The Advanced Theory of Statistics,* 2nd ed., vol. 1. New York: Hafner.

Mendenhall, W., J. E. Reinmuth, and R. Beaver (1993). *Statistics for Management and Economics,* 7th ed. Belmont, Calif.: Duxbury Press.

Microsoft Excel User's Guide Version 5.0, (1994). Redmond, Wash.: Microsoft.

Minitab Reference Manual (1996), Release 11. State College, Penn.: Minitab.

Montgomery, D. C. (1996). *Introduction to Statistical Quality Control,* 3rd ed. New York: John Wiley.

Neter, J., M. H. Kutner, C. J. Nachtsheim, and W. Wasserman (1996). *Applied Linear Statistical Models,* 4th ed. Homewood, Ill.: Richard D. Irwin.

CHAPTER

5

Graphical Methods of Data Presentation and Quality Improvement

Chapter Outline

Symbols	
$\overline{X}$	Sample average
s	Sample standard deviation
n	Sample size
X_i	ith observation in a sample
s_m	Standard deviation of the sample median
$F(x)$	Cumulative distribution function
M	Median
Q_1	First quartile
Q_3	Third quartile
IQR	Interquartile range

5-1 INTRODUCTION

In this chapter we look at methods for describing data sets graphically. These methods help us learn about the characteristics of a process, its operating state of affairs, and the kind of output we may expect from it. Several methods are used for quality improvement. As we discussed in previous chapters, quality control involves taking action to remove the identifiable factors that cause the output of a process to be unstable or off-target. Once the process is in a state of statistical control, if the output does not meet desirable norms, then fundamental changes are needed in the process, the product, or both. Such changes are the focus of quality improvement.

Because graphical methods are easy to understand and provide comprehensive information, they are a viable tool for the analysis of product and process data. The information they provide on existing product or process characteristics helps us determine whether these characteristics are close to the desired norm. Several of the graphical techniques described here help us identify where to focus quality improvement efforts. Given the emphasis on neverending quality improvement, such tools are beneficial to any company, regardless of its current level of quality achievement.

5-2 FREQUENCY DISTRIBUTIONS AND HISTOGRAMS

Seldom do we get an idea of process characteristics just by looking at the individual data values gathered from the process. Such data is often voluminous. Frequency distributions and histograms summarize such information and present it in a format that allows us to draw conclusions regarding the process condition.

A **frequency distribution** is a rearrangement of raw data in ascending or descending order of magnitude, such that the quality characteristic is subdivided into classes and the number of occurrences in each class is presented.

Table 5-1 shows the inside diameter (in millimeters) of metal sleeves produced in a machine shop for 100 randomly selected parts. Twenty samples, each of size 5, were taken. Simply looking at the data in Table 5-1 provides little insight about the process. Even though we know there is variability in the sleeve diameters, we can hardly identify a pattern in the data (what is the degree of variability?) or comment about the central tendency of the process (about which value are most of the observations concentrated?).

Using the data in Table 5-1, we can construct a frequency distribution like the one shown in Table 5-2. Here, the diameters are categorized into classes (49.89 to 49.91, 49.91 to 49.93, and so on), and the number, or frequency, in each group is reported. Care must be taken to set the class, or cell, boundaries such that there is no overlap between them. Double-counting of observations should be avoided. Classes are usually of equal width. However, when there are outliers (values that are too large or too small compared to the majority of the observations), the two end classes may be kept open-ended.

TABLE 5-1 Inside Diameter of Metal Sleeves (in mm)

Sample	*Observations X (Five per Sample)*				
1	50.04	50.03	50.02	50.00	49.94
2	49.96	49.99	50.03	50.01	49.98
3	50.01	50.01	50.01	50.00	49.92
4	49.95	49.97	50.02	50.10	50.02
5	50.00	50.01	50.00	50.00	50.09
6	50.02	50.05	49.97	50.02	50.09
7	50.01	49.99	49.96	49.99	50.00
8	50.02	50.00	50.04	50.02	50.00
9	50.06	49.93	49.99	49.99	49.95
10	49.96	49.93	50.08	49.92	50.03
11	50.01	49.96	49.98	50.00	50.02
12	50.04	49.94	50.00	50.03	49.92
13	49.97	49.90	49.98	50.01	49.95
14	50.00	50.01	49.95	49.97	49.94
15	49.97	49.98	50.03	50.08	49.96
16	49.98	50.00	49.97	49.96	49.97
17	50.03	50.04	50.03	50.01	50.01
18	49.98	49.98	49.99	50.05	50.00
19	50.07	50.00	50.02	49.99	49.93
20	49.99	50.06	49.95	49.99	50.02

Table 5-2 shows the frequency in each class (for example, 1 observation between 49.89 and 49.91, 3 observations between 49.91 and 49.93, and so on). It also depicts the *relative frequency* in each cell, which is found by dividing the frequency in each cell by the total number of observations. For example, class $49.93 \leq X < 49.95$, contains 6% of all observations.

Table 5-2 also shows the *cumulative frequency* for each cell. The cumulative frequency for a given class is the number of observations in that class and in all classes preceding it. Note that the cumulative frequency for class $49.93 \leq X < 49.95$ is 10, which means that of the 100 metal sleeves, 10 have inside diameters that are less than 49.95 mm. The *cumulative relative frequency* of a class is simply the cumulative frequency for that class divided by the total number of observations. For class $49.93 \leq X < 49.95$, the cumulative relative frequency is 0.10, meaning that 10% of the sleeves have a diameter less than 49.95 mm.

TABLE 5-2 Frequency Distribution of Metal Sleeves

Classes for Sleeve Diameter X (in mm)	*Tally*	*Frequency*	*Relative Frequency*	*Cumulative Frequency*	*Cumulative Relative Frequency*
$49.89 \leq X < 49.91$	I	1	0.01	1	0.01
$49.91 \leq X < 49.93$	III	3	0.03	4	0.04
$49.93 \leq X < 49.95$	~~IIII~~ I	6	0.06	10	0.10
$49.95 \leq X < 49.97$	~~IIII~~ ~~IIII~~ I	11	0.11	21	0.21
$49.97 \leq X < 49.99$	~~IIII~~ ~~IIII~~ IIII	14	0.14	35	0.35
$49.99 \leq X < 50.01$	~~IIII~~ ~~IIII~~ ~~IIII~~ ~~IIII~~ III	23	0.23	58	0.58
$50.01 \leq X < 50.03$	~~IIII~~ ~~IIII~~ ~~IIII~~ ~~IIII~~ I	21	0.21	79	0.79
$50.03 \leq X < 50.05$	~~IIII~~ ~~IIII~~ I	11	0.11	90	0.90
$50.05 \leq X < 50.07$	IIII	4	0.04	94	0.94
$50.07 \leq X < 50.09$	III	3	0.03	97	0.97
$50.09 \leq X < 50.11$	III	3	0.03	100	1.00
	Total	100	1.00		

A *histogram* is a graphical display of data such that the characteristic is subdivided into classes, or cells. In a **frequency histogram,** the vertical axis usually represents the number of observations in each class. The following steps are used to construct a frequency histogram:

Step 1: Find the range of the observations—the difference between the largest and smallest value.

Step 2: Choose the number of classes, or cells. Usually, 5–20 classes should be selected. If too few classes are chosen, specific details of the data are lost. On the other hand, if too many classes are selected, a summary of how the data is distributed (which is the objective of a histogram) will not be achieved. As a rule of thumb, if n represents the number of data points, the number of classes should be approximately $\sqrt{n}$.

Step 3: Determine the width of the classes. Usually, all classes are of equal width except for the first or last class. If outliers are present, the first or last class can be kept open-ended to include them. The class width is found by dividing the range by the number of classes.

Step 4: Determine the class boundaries. Find the number of observations in each class. Make sure the classes are not overlapping.

Step 5: Draw the frequency histogram. Construct rectangles above the classes such that the heights of the rectangles correspond to the frequencies.

Example 5-1: For the data in Table 5-1, the range is $50.10 - 49.90 = 0.2$. The number of data points is 100, corresponding to $\sqrt{100} = 10$ classes. The class width is $0.2/10 = 0.02$. Note that if we put the midpoint of the first class at 49.90, then we end up with 11 classes in order to account for the value 50.10, and this is acceptable. There is no rule saying that the number of classes must be exactly 10. If an initially calculated class width does not turn out to be a round number, it may be modified.

The classes should be designated such that an observation will fall in one class only. The midpoint of our first class is chosen as 49.90. Thus, the first class includes values from 49.89 to 49.91. The upper boundary of each class will be noninclusive; if we have a data point with a value of 49.91, it will be included in the second class. Such explicit labeling avoids ambiguity as to where to place observations, and it avoids double-counting. The class boundaries, their midpoints, and the frequencies of the classes are shown in Table 5-3.

Using the class boundaries and midpoints shown in Table 5-3, we can construct a frequency histogram using Minitab. The data is first entered in a column, representing the variable of inside diameter. Choose **Graph > Character > Histogram.** Type the name of the variable—in this case, Diameter—in **X.** Figure 5-1 shows the resulting frequency histogram. This histogram gives us a sense of where the observations cluster and the degree of their variability. Such histograms also tell us whether there are outliers and provide information regarding the uniformity of a process. Furthermore, they enable us to determine the conformance of a process with respect to established specification limits. Using the data in Table 5-3 and the same classes as in the frequency histogram in Figure 5-1, we can construct a cumulative frequency histogram, as shown in Figure 5-2.

TABLE 5-3 Class Boundaries, Midpoints, and Frequencies for Example 5-1

Class Boundaries	*Midpoint*	*Frequency*	*Cumulative Frequency*
$49.89 \le X < 49.91$	49.90	1	1
$49.91 \le X < 49.93$	49.92	3	4
$49.93 \le X < 49.95$	49.94	6	10
$49.95 \le X < 49.97$	49.96	11	21
$49.97 \le X < 49.99$	49.98	14	35
$49.99 \le X < 50.01$	50.00	23	58
$50.01 \le X < 50.03$	50.02	21	79
$50.03 \le X < 50.05$	50.04	11	90
$50.05 \le X < 50.07$	50.06	4	94
$50.07 \le X < 50.09$	50.08	3	97
$50.09 \le X < 50.11$	50.10	3	100

FIGURE 5-1 Frequency histogram of sleeve diameters using Minitab.

Histogram of Diameter $N = 100$

Midpoint	Count	
49.90	1	*
49.92	3	***
49.94	6	******
49.96	11	***********
49.98	14	**************
50.00	23	***********************
50.02	21	*********************
50.04	11	***********
50.06	4	****
50.08	3	***
50.10	3	***

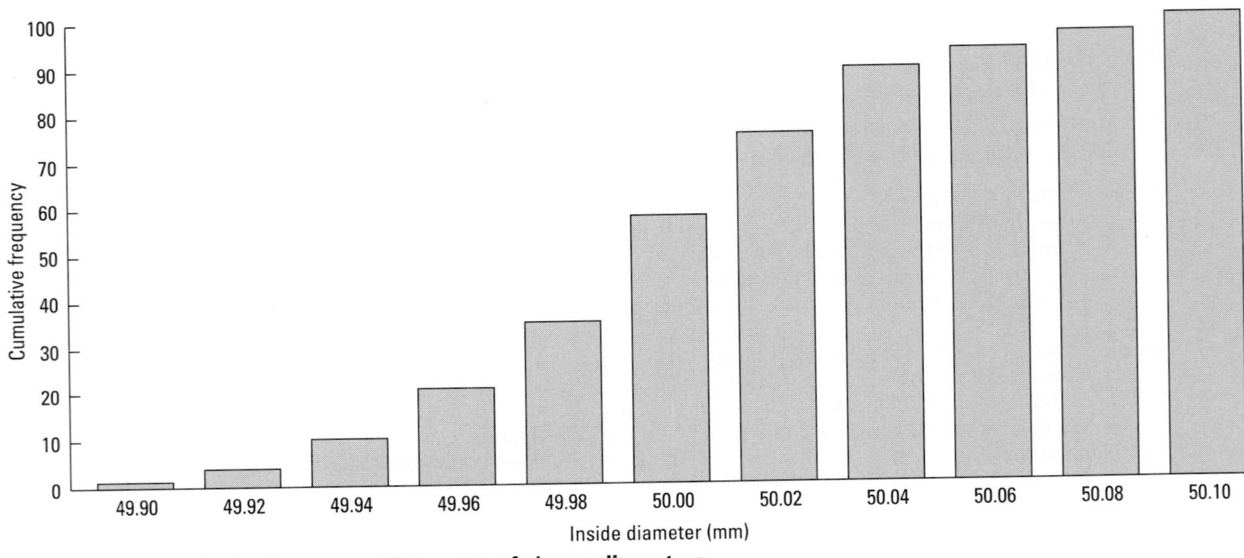

FIGURE 5-2 Cumulative frequency histogram of sleeve diameters.

Example 5-2: Refer to the milling operation data in Table 4-5 (page 158) on the depth of cut and also to Example 4-18. Figure 4-14 (page 159) gives the Minitab output of descriptive statistics for this variable, including a frequency histogram.

5-3 RUN CHARTS

A **run chart** is a plot of a quality characteristic as a function of time. Such a chart usually plots data point-by-point in the order in which it is obtained. Though run charts do not summarize any information, they do provide some idea of the general trend and degree of variability present in the process.

Example 5-3: In a chemical process, the acidity of a compound used to dye fabrics is of interest. Twenty random observations are selected from the process, and their pH values measured. The data values are shown in Table 5-4, and a run chart is shown in Figure 5-3. From the data in Table 5-4, not much of an opinion can be formed concerning the process. The run chart, on the other hand, immediately gives us some notion as to the process operating conditions. We know instantly, for example, there is a trend in increasing pH between observations 8 and 15. Such information gives us some insight into the process.

TABLE 5-4 pH Values of Dye Compound

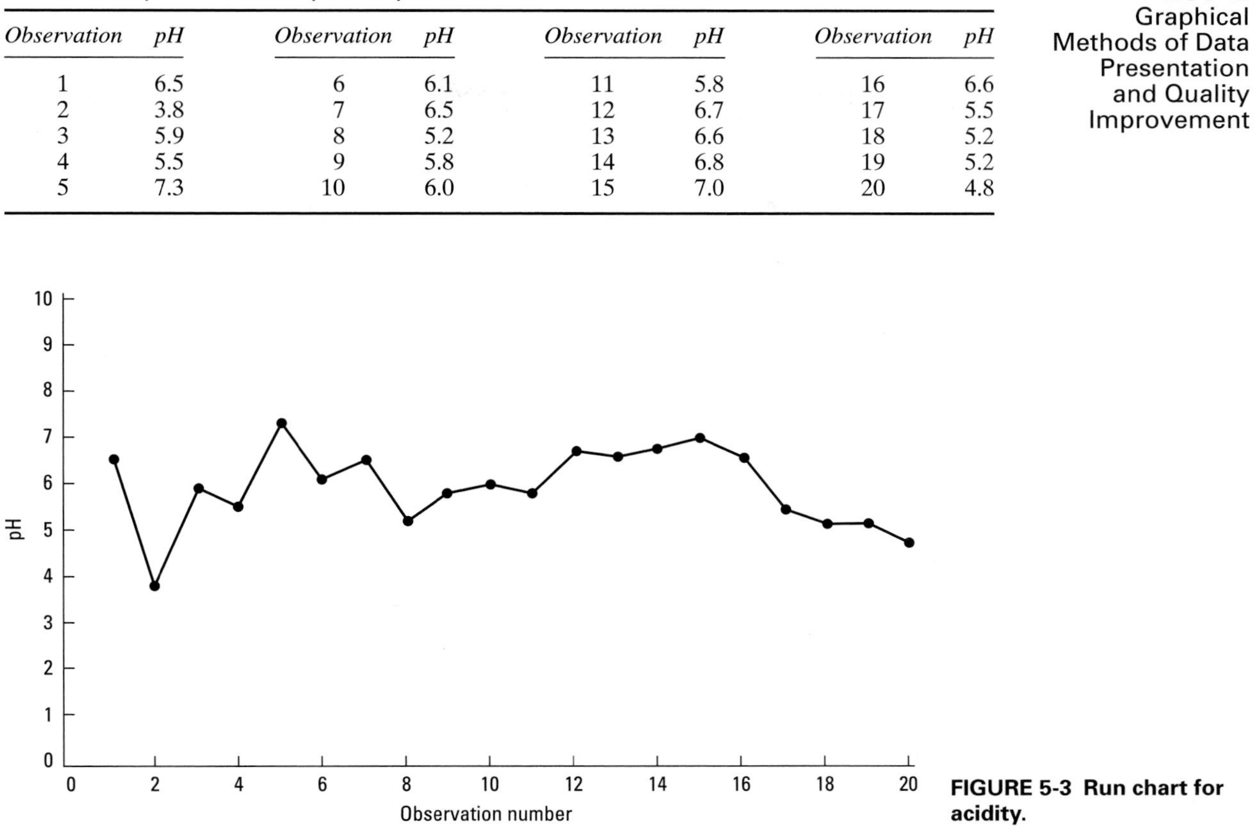

Observation	*pH*	*Observation*	*pH*	*Observation*	*pH*	*Observation*	*pH*
1	6.5	6	6.1	11	5.8	16	6.6
2	3.8	7	6.5	12	6.7	17	5.5
3	5.9	8	5.2	13	6.6	18	5.2
4	5.5	9	5.8	14	6.8	19	5.2
5	7.3	10	6.0	15	7.0	20	4.8

FIGURE 5-3 Run chart for acidity.

5-4 STEM-AND-LEAF PLOTS

Stem-and-leaf plots are another graphical approach to plotting observations and interpreting process characteristics. With frequency histograms, the identities of the individual observations are lost in the process of plotting. In the stem-and-leaf plot, however, individual numerical values are retained. When we plot histograms, we have to decide on the number of classes and the class width; this is not required in a stem-and-leaf plot. Let's construct a stem-and-leaf plot using the metal sleeves data from Table 5-1.

Example 5-4: Each data value is split into two parts, the "stem" and the "leaf." For example, 50.04 is displayed as follows:

Data value	Stem	Leaf
50.04	500	4

↑ implicit decimal point

A decimal point is implicit to the left of the rightmost digit in the stem. The digit in the leaf portion represents hundredths.

The choice of the stem and leaf positions is influenced by the magnitude and variability of the data values. For the metal sleeve data, the values are found to have either 499 or 500 in the stem, so the following modification is used; in the stem, use

- an asterisk (*) for leaves 0 and 1
- a "t" for leaves 2 and 3 (first letter of *t*wo and *t*hree)

```
499* | 1
499t | 2 2 2 3 3 3
499f | 4 4 4 5 5 5 5 5
499s | 6 6 6 6 6 6 7 7 7 7 7 7 7
499· | 8 8 8 8 8 8 8 9 9 9 9 9 9 9 9 9
500* | 0 0 0 0 0 0 0 0 0 0 0 0 0 0 1 1 1 1 1 1 1 1 1 1
500t | 2 2 2 2 2 2 2 2 2 2 3 3 3 3 3 3 3
500f | 4 4 4 4 5 5
500s | 6 6 7
500· | 8 8 9 9
501* | 0
```

FIGURE 5-4 Stem-and-leaf plot for inside diameter of metal sleeves.

- an "f" for leaves 4 and 5 (*f*our and *f*ive)
- an "s" for leaves 6 and 7 (*s*ix and *s*even)
- a dot (·) for leaves 8 and 9

Thus, in this modified scheme the data value 50.04 is represented as follows:

Data value	Stem	Leaf
50.04	500f	4

When we repeat this procedure for other data values, we obtain the complete stem-and-leaf plot shown in Figure 5-4.

5-5 PARETO DIAGRAMS

Pareto diagrams are important tools in the quality improvement process. Alfredo Pareto, an Italian economist in the 1848–1923 era, found that wealth is concentrated in a few people. This observation led him to formulate the Pareto principle—that the majority of wealth is held by a disproportionately small segment of the population. Joseph Juran, who we met in Chapter 2, realized that this principle applies to quality improvement as well. In manufacturing or service organizations, for example, problem areas or defect types follow a similar distribution. Of all the problems that occur, only a few are quite frequent. The others seldom occur. Thus, in grouping these problem areas into two categories, they are labeled as the *vital few* and the *trivial many.* The Pareto principle also lends support to the 80/20 rule, which states that 80% of the problems (nonconformities or defects) are created by 20% of the causes.

Pareto diagrams help management quickly identify the critical areas (those causing most of the problems) that deserve immediate attention. They identify those problems whose resolution will lead to substantial improvements in quality. They also guide management in allocating limited resources to problem-solving activities. Pareto diagrams arrange problems in order of importance. "Importance" may, for example, refer to the financial impact of a problem or the relative number of occurrences of the problem.

The steps for constructing Pareto diagrams are as follows:

Step 1: Decide on the data categorization system—say, by problem type, type of nonconformity (critical, major, minor), or whatever else seems appropriate.

Step 2: Determine how relative importance is to be judged—that is, whether it should be based on dollar values or the frequency of occurrence.

Step 3: Rank the categories from most important to least important.

Step 4: Compute the cumulative frequency of the data categories in their chosen order.

Step 5: Plot a bar graph, showing the relative importance of each problem area in descending order. Identify the vital few that deserve immediate attention.

Example 5-5: Data is gathered regarding the types of nonconformance in a textile mill. Table 5-5 lists the problem areas, along with the percentage of occurrence of each and the associated annual cost. Management wants to focus on the problems that represent at least 80% of the occurrences. The four most significant problem areas are a subpar quality of cotton, an improper tension setting, inadequate operator training, and bale storage problems; they are responsible for 84% of the nonconformities. Together, they are costing the company $31,000 annually; this implies that eliminating them would result in a savings of the same amount. Table 5-5 also shows the cumulative percentages of occurrence for the different problem areas.

A Pareto diagram, constructed with the data from Table 5-5, is shown in Figure 5-5. Management should definitely address the issue of subpar quality of cotton (problem type A) first. Eliminating this problem will result in a savings of about $20,000, more than for any other problem listed. Pareto diagrams can help crystallize thinking on priorities—that is, eliminating which problem will give the most benefit (in this case, cost savings).

Pareto diagrams are also used to compare process conditions (in terms of either the problem causes or the nonconformities) before and after action is taken to improve the process. For instance, after a thorough investigation of the problem areas of Example 5-5, management would move to take remedial actions. They might thereafter wish to determine whether there really is a reduction in nonconforming product.

TABLE 5-5 Types of Nonconformities in a Textile Mill

Problem Type	*Description*	*Percentage of Occurrence*	*Annual Cost (in $1000)*	*Cumulative Percentage of Occurrence*
A	Subpar quality of cotton	40	20	40
B	Improper tension setting	20	6	60
C	Inadequate operator training	14	3	74
D	Bale storage problems	10	2	84
E	Drop in hydraulic pressure in presses	8	2	92
F	Cutter not sharp	5	1.5	97
G	Dye for use in color not adequate	3	1.8	100

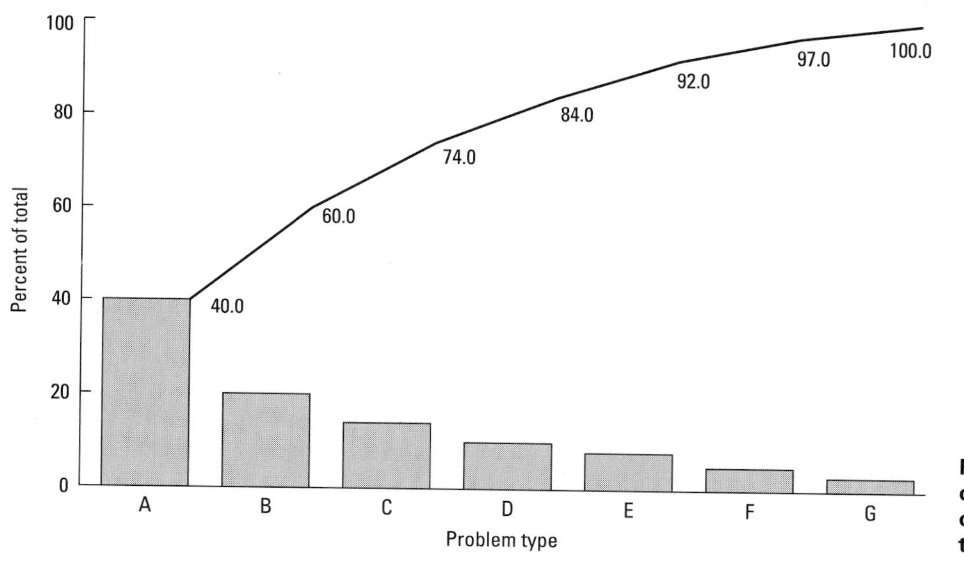

FIGURE 5-5 Pareto diagram for problem areas of nonconformance in a textile mill.

They might also decide what impact the remaining problems are having on nonconformance, now that four of them have been eliminated. "Before and after" Pareto diagrams can be compared to determine the impact of the remedial actions.

5-6 CAUSE-AND-EFFECT DIAGRAMS

Cause-and-effect diagrams were developed by Kaoru Ishikawa in 1943 and thus are often called **Ishikawa diagrams.** They are also known as **fishbone diagrams** because of their appearance (in the plotted form). Basically, **cause-and-effect diagrams** are used to identify and systematically list the different causes that can be attributed to a problem (or an effect) (Ishikawa 1976). These diagrams thus help determine which of several causes has the greatest effect. A cause-and-effect diagram can aid in identifying the reasons why a process goes out of control. Alternatively, if a process is stable, these diagrams can help management decide which causes to investigate for process improvement. There are three main applications of cause-and-effect diagrams: cause enumeration, dispersion analysis, and process analysis.

Cause Enumeration

Cause enumeration is one of the most widely used graphical techniques for quality control and improvement. It usually develops through a brainstorming session in which all possible types of causes (however remote they may be) are listed to show their influence on the problems (or effect) in question.

The procedure consists of first defining the problem or quality characteristic selected for study so that everyone knows what is being tackled. Next, the major causes influencing the characteristic are noted. In a manufacturing process, for example, the major causes for a nonconformance (say, a length not meeting specifications) could be equipment, operator, methods, environment, and so forth. After this step, subcauses within each of the major causes are listed. Before evaluating each cause, more thought is given to defining and identifying them clearly and also to evaluating appropriate methods of measurement. Next, one cause is singled out and analyzed. This, of course, is done systematically so that the predominant cause is analyzed first.

One advantage of using cause-and-effect diagrams is that the process of their construction creates a better understanding of the components of the process and their relationships, and thus a better understanding of the process itself.

Example 5-6: One of the quality characteristics of interest in automobile tires is the bore size, which should be within certain specifications. In a cause-and-effect diagram, the final bore size is the effect. Some of the main causes that influence the bore size are the incoming material, mixing process, tubing operation, splicing, press operation, operator, and measuring equipment. For each main cause, subcauses are identified and listed. For the raw material category, the incoming quality is affected by such subcauses as vendor selection process (for instance, is the vendor certified?), the content of scrap tire in the raw material, the density, and the ash content. Figure 5-6 shows the completed cause-and-effect diagram.

Dispersion Analysis

In **dispersion analysis,** each major cause is thoroughly analyzed by investigating the subcauses and their impact on the quality characteristic (or effect) in question. This process is repeated for each major cause in a prioritized order. The cause-and-effect diagram helps us analyze the reasons for any variability, or dispersion. Although this approach is similar to cause enumeration, there are some differences. In cause enumeration, smaller causes that are considered insignificant are still listed because that method tries to list all causes. However, in dispersion analysis, causes that don't fit the selected categories are not listed. Hence, it is possible that the root cause will not be identified in dispersion analysis. Cause enumeration facilitates the identification of root causes because all conceivable causes are listed.

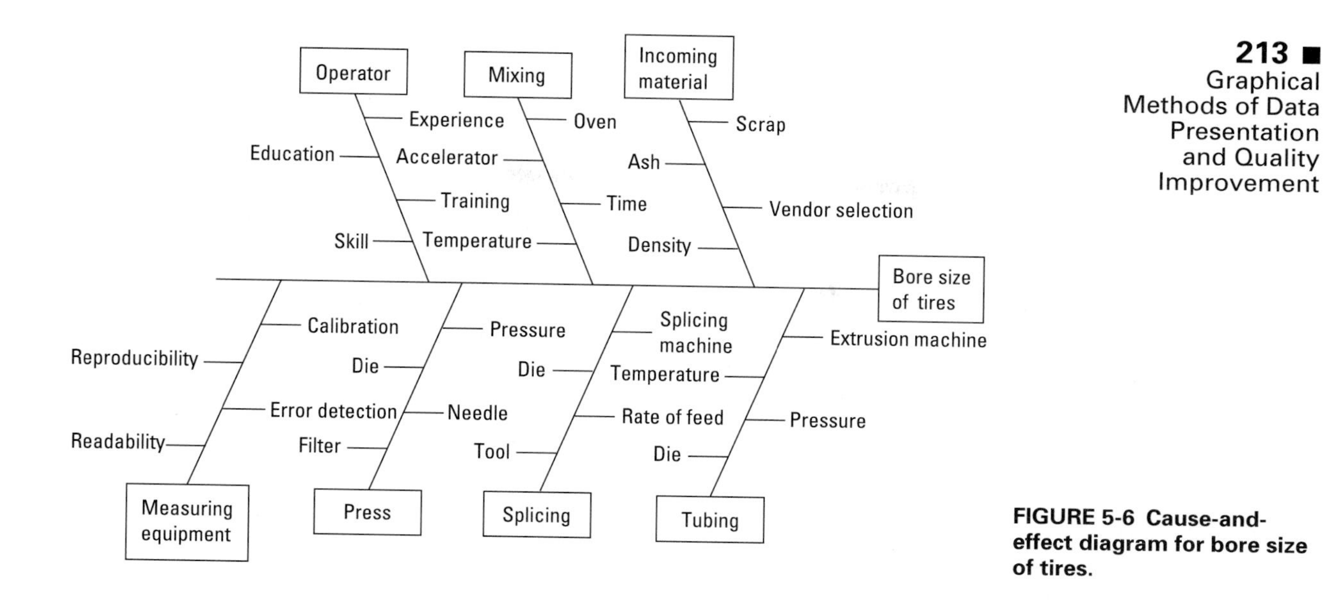

FIGURE 5-6 Cause-and-effect diagram for bore size of tires.

Example 5-7: In the production of plastic reels for computer tapes, the width of the reel (the effect) is an important product characteristic. The main causes can be grouped into the following categories: materials, methods, machines, measurements, and people. In a dispersion analysis, each of these is analyzed individually in a prioritized sequence.

Figure 5-7 shows a partially completed cause-and-effect diagram for the reel width characteristic. Here, one of the main causes, material, has been broken down into subcauses. The subcauses include composition of the material, availability of materials, the choice of vendor, and the type of incoming tests performed on the material. These subcauses are further broken down. For instance, under composition, the chemical properties of the compound are considered. These properties influence the flow of the compound in the molding process, which in turn influences the width of the reel. The uniformity of the raw material mixture (that is, its degree of consistency within a given shipment) influences the quality of the plastic produced. This too has an impact on the reel width. The other subcauses are similarly subdivided and further analyzed. The eventual structure of the cause-and-effect diagram would be similar to one found using cause enumeration.

FIGURE 5-7 Dispersion analysis using cause-and-effect diagram for reel width of computer tapes.

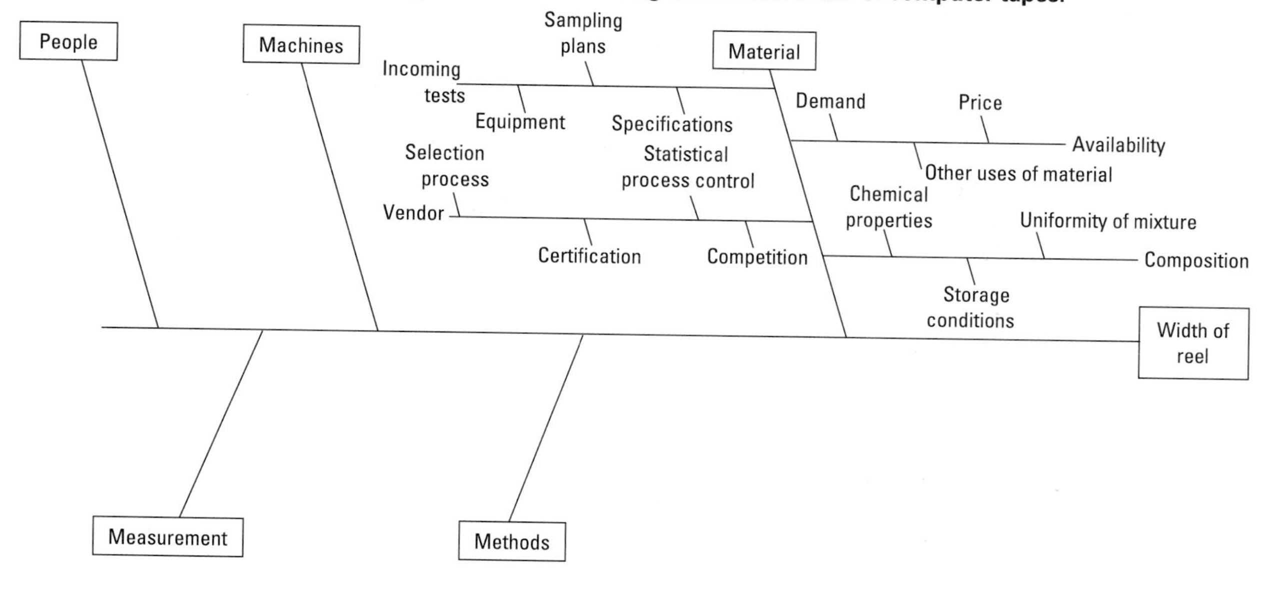

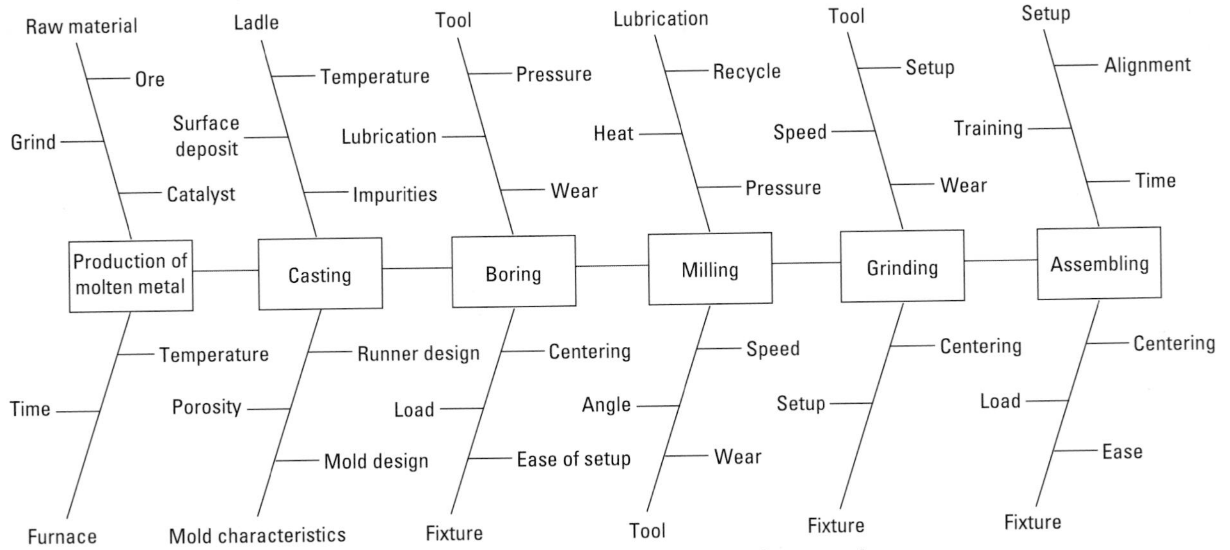

FIGURE 5-8 Process analysis using cause-and-effect diagram for aluminum casings.

Process Analysis

When cause-and-effect diagrams are constructed for **process analysis,** the emphasis is on listing the causes in the sequence in which the operations are actually conducted. This process is similar to creating a flow diagram, except that a cause-and-effect diagram lists in detail the causes that influence the quality characteristic of interest at each step of the production process.

Example 5-8: A company's quality analysts are looking at the aluminum casings produced for gear boxes. An increase in the production of unacceptable casings has warranted an analysis of the causes responsible, and the analysts have decided to use process analysis. Figure 5-8 shows their process analysis.

The steps in the production process are shown as the major causes. Each production step is then broken down into subcauses, and a similar decomposition continues until the root level of causes is reached. The main operations are the production of molten metal, casting, boring, milling, grinding, and assembly, and the breakdown of each of these operations is listed. This diagram tracks the product from one step to the next in a sequential manner. It is easy to construct. Note, however, that a given cause can be repeated several times in the diagram if similar concerns exist in two or more operations. For example, in Figure 5-8, "tool" appears three times, as does "fixture." The advantage of conducting process analysis through cause-and-effect diagrams is that it systematically develops the sequence of causes—analysis of which will lead to remedies.

5-7 BOX PLOTS

Box plots graphically depict data and also display summary measures (Chambers 1977; Chambers et al. 1983). A **box plot** shows the central tendency and dispersion of a data set and indicates the **skewness** (deviation from symmetry) and **kurtosis** (measure of tail length). The plot also shows outliers. The steps involved in the construction of a box plot are as follows:

Step 1: Determine the first **quartile** Q_1 (the 25th percentile). This value determines the lower edge of the box.

Step 2: Determine the third quartile Q_3 (the 75th percentile). This value determines the upper edge of the box. The length of the box is the difference between Q_3 and Q_1. This is known as the **interquartile range** (IQR).

Step 3: Find the median of the data set. A line is drawn at the median to divide the box.

Step 4: Two lines, known as **whiskers,** are drawn outward from the box. One line extends from the top edge of the box at Q_3 to either the maximum data value or $Q_3 + 1.5(\text{IQR})$, whichever is lower. Similarly, a line from the bottom edge of the box at Q_1 extends downward to a value that is either the minimum data value or $Q_1 - 1.5(\text{IQR})$, whichever is greater. The end points of the whiskers are known as the upper and lower adjacent values.

Step 5: Values that fall outside the adjacent values are candidates for consideration as **outliers.** They are plotted as asterisks (*).

Example 5-9: The Rockwell hardness values of metal fasteners are found for a randomly chosen sample of 20 parts. The observed values are as follows:

31.5 36.2 30.1 44.6 35.8
30.2 34.3 34.5 49.2 35.4
37.2 38.2 34.6 33.0 36.1
34.8 36.4 34.8 30.1 37.0

To construct a box plot, the data is first ranked:

30.1 30.1 30.2 31.5 33.0
34.3 34.5 34.6 34.8 34.8
35.4 35.8 36.1 36.2 36.4
37.0 37.2 38.2 44.6 49.2

The first quartile Q_1 is found to be 33.0, and the third quartile Q_3 is found to be 37.0. The median is 35.1. The interquartile range is $Q_3 - Q_1 = 4.0$, so $1.5(\text{IQR}) = 6.0$. Further calculations yield $Q_3 + 1.5(\text{IQR}) = 43.0$ and $Q_1 - 1.5(\text{IQR}) = 27.0$. The length of the top whisker is the lesser of 49.2 and 43.0, while the length of the bottom whisker is the greater of 30.1 and 27.0. The box plot is shown in Figure 5-9. Observe that two data values, 44.6 and 49.2, are beyond the top whisker and are plotted as asterisks.

Several insights can be gained from the box plot. The box itself, extending from 33 to 37, contains 50% of the observations. The relative location of the median, which has a value of 35.1, indicates the degree of skewness or asymmetry in the distribution of the data.

Half of the values are less than or equal to the median, which is closer to the top edge of the box Q_3. If the distribution were symmetric, the median would be located midway between the edges of the box—that is, at a value of $(Q_1 + Q_3)/2$. Therefore, since the median is closer to Q_3, the data is negatively skewed. Conversely, if the median had been closer to Q_1, the distribution would be positively skewed.

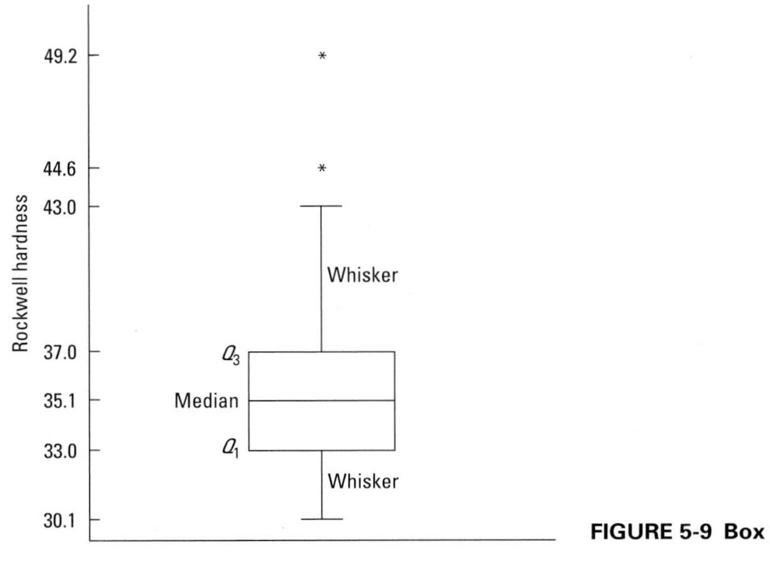

FIGURE 5-9 Box plot of Rockwell hardness data.

The lengths of the whiskers indicate the **tail lengths.** Since the top whisker is much longer than the bottom one, the distribution has a longer right tail. Furthermore, two values (plotted as asterisks) are beyond the top whisker, and there are no values below the bottom whisker. These outliers should be further investigated to determine why they are so large. Doing so may provide management with a course of action.

Example 5-10: Refer to Example 4-18, which uses the milling operation from Table 4-5. A box plot is shown in Figure 4-14 as part of the Minitab graphics summary. Note that the first and third quartiles are 2.45 and 4.45, respectively, with an interquartile range of 2.0. This is also the width of the box. The whiskers, in this case, extend to 1.5 and 5.6, which represent the minimum and maximum values, respectively.

Variations of the Basic Box Plot

One variation of the basic form of the box plot is the **notched box plot.** A notch, the width of which corresponds to the length of the confidence interval for the median, is constructed on the box around the median. Assuming the data values are normally distributed, the standard deviation of the median is given by Kendall and Stuart (1967) as

$$s_m = \frac{1.25(\text{IQR})}{1.35\sqrt{n}} \quad \textbf{(5.1)}$$

The notch around the median M should start at values of

$$M \pm C s_m \quad \textbf{(5.2)}$$

where C is a constant representing the level of confidence. For a level of confidence of 95%, $C = 1.96$ can be used. For further details, consult McGill, Tukey, and Larsen (1978).

Example 5-11: Refer to the Rockwell hardness data from Example 5-9. The sample size n is 20. The sample median M is found to be 35.1, with an interquartile range of 4.0. Using eq. (5.1), the standard deviation of the median is found to be

$$s_m = \frac{1.25(4.0)}{1.35\sqrt{20}} = 0.828$$

A 95% confidence interval for the median, found using eq. (5.2), is

$$\begin{aligned} 35.1 \pm (1.96)(0.828) &= 35.1 \pm 1.623 \\ &= (33.477, 36.723) \end{aligned}$$

This notched box plot is shown in Figure 5-10.

Notched box plots are used to determine whether there are significant differences between the medians of the quality characteristic of two plots. If there is no overlap between the notches of two box plots, we can conclude that there is a significant difference between the two medians. This may indicate that the actions taken have significantly changed the process parameter conditions.

Another variation of the basic box plot is the **variable-width box plot.** If we are comparing two or more data sets that have different sample sizes, then the widths of the boxes are set such that they are proportional to the sample sizes. That is, we will construct the box so that its width is proportional to the natural logarithm of the sample size.

Example 5-12: After some process changes for the situation described in Example 5-9, a sample of 50 observations of the Rockwell hardness is taken. The new sample mean $\overline{X}$ is 34.2, the sample standard deviation s is 3.550, the sample median is 34.0, the first quartile Q_1 is 31.6, and the third quartile Q_3 is 36.3. The observations range from 27.5 to 43.0.

The interquartile range is $Q_3 - Q_1 = 4.7$, so $1.5(\text{IQR}) = 7.05$. Next, $Q_1 - 1.5(\text{IQR}) = 24.55$, and $Q_3 + 1.5(\text{IQR}) = 43.35$. Hence, the length of the top whisker is the lesser of 43.35 and 43.0. The length of the bottom whisker is the greater of 24.55 and 27.5.

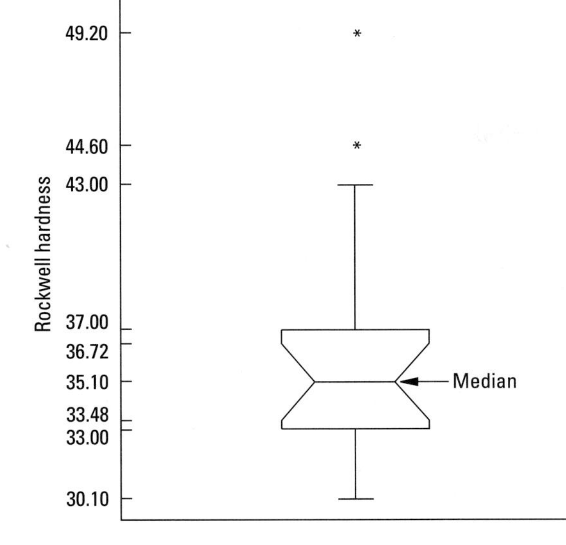

FIGURE 5-10 Notched box plot of Rockwell hardness data.

Using a 95% level of confidence, the confidence interval for the median [using eqs. (5.1) and (5.2)] is

$$34.0 \pm 1.96\left[\frac{1.25(4.7)}{1.35\sqrt{50}}\right] = 34.0 \pm 1.206$$

$$= (32.794, 35.206)$$

The width of this box plot should be proportional to $\ln(n) = \ln(50) = 3.912$. The width of the box plot in Example 5-9 (with sample size 20) is proportional to $\ln(20) = 2.996$. Figure 5-11 shows the variable-width box plots before and after the process changes. Note that since the notches of the two plots overlap, we can conclude that at a 95% level of confidence there is no significant difference between the medians.

FIGURE 5-11 Variable-width box plots before and after process changes.

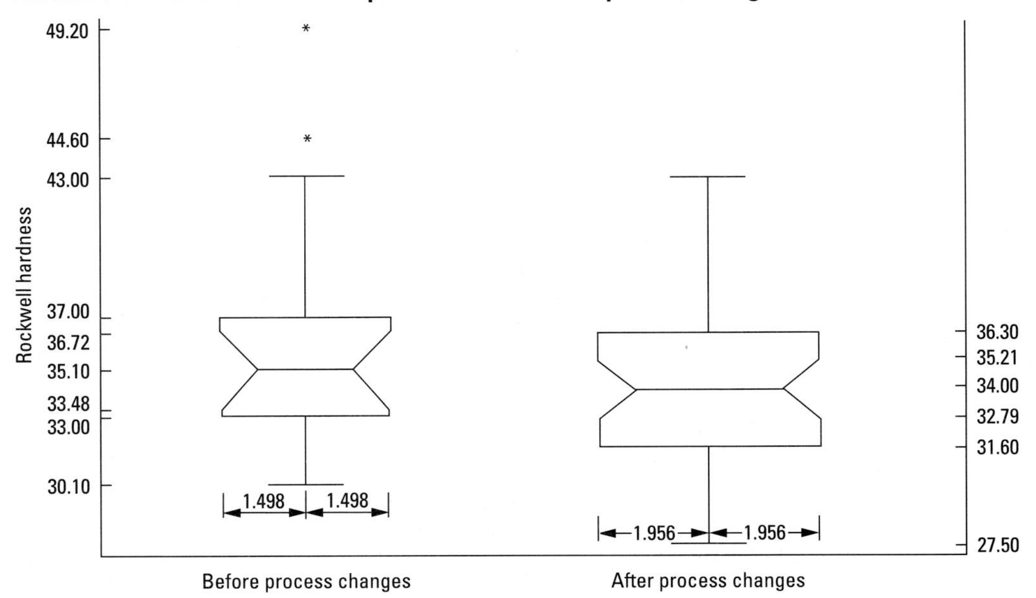

5-8 NORMAL PROBABILITY PLOTS

In many statistical techniques, the population from which the sample data is drawn is assumed to be normally distributed so that certain inferences can be made about the quality characteristic. Chapter 9 will examine three situations where the goal is to determine the proportion of a product occurring below or above certain values known as specification limits. If the assumption of normality holds, then this proportion can readily be found.

Plotting is often done on normal probability paper. The ordinate represents the theoretical percentile (assuming the data is from a normal distribution), while the abscissa represents the ordered data values. Basically, the vertical scale of normal probability paper is designed so that if the data is from a normal distribution, the observations will plot as a straight line. Deviations of the plotted points from a straight line indicate departures from normality. Analysts often use judgment to define a "significant" deviation from a straight line. Statistical methods including goodness-of-fit tests (such as chi-squared tests; see Duncan 1986) or Kolmogorov–Smirnov tests (see Massey 1951) can be used, but we will not discuss them here. Further discussion of probability plotting may be found in Michael (1983). Several software packages are available for normal probability plots (SAS 1990; Minitab 1996).

The following steps are used to construct a **normal probability plot:**

Step 1: Rank the observations in ascending order from smallest to largest. Assign a rank to each data value. The rank of the observations is denoted by i such that $i = 1, 2, \ldots, n$, where n is the number of data values in the sample.

Step 2: The probability plotting position of the observation with rank i is calculated from

$$F_i = \frac{i - 0.5}{n} \tag{5.3}$$

Step 3: On normal probability paper, plot the value of each observation (along the horizontal axis) versus its probability percentage ($100F_i$) along the ordinate.

Step 4: Estimate a straight line through the plotted points. A best-fit straight line through the points gives an estimate of its cumulative probability distribution function.

Step 5: Based on step 4, make a decision as to whether the data is from a normal distribution. Consider the closeness of fit to a straight line as well as any systematic deviations from the line.

While these steps demonstrate the construction of a normal probability plot, we now show an example using a software package.

Example 5-13: A sample of 50 coils to be used in an electrical circuit is randomly selected, and their resistances measured. The data values are shown in Table 5-6. A normal probability plot is constructed using Minitab. The data values on coil resistance are first entered. Choose **Stat > Basic Statistics > Normality Test.** Type the name of the variable, say Resistance, in **Variable** and click **OK.** The resulting normal probability plot is shown in Figure 5-12, where a straight line is fitted through the plotted points. Observe that a majority of the plotted points are in close proximity to the straight line. A few points on the extremes deviate from the line. Overall, we can conclude that the observations appear to come from a normal distribution.

If we accept the assumption of normality, then we can make several inferences from Figure 5-12. First, we can estimate the population mean by reading the value of the 50th percentile off the fitted straight line; it appears to be approximately 30.0 ohms. Second, we can estimate the population standard deviation as the difference between the 84th and 50th percentile data values. This is because the 84th percentile of a normal distribution is approximately 1 standard deviation away from the mean (the 50th percentile). From Figure 5-12, the population standard deviation is estimated to be 34.5 – 30.0 ohms = 4.5 ohms. Third, we can estimate the proportion of a process output that does not meet certain specification limits. For example, if the

TABLE 5-6 Coil Resistance

Observation	*Coil Resistance (ohms)*	*Observation*	*Coil Resistance (ohms)*	*Observation*	*Coil Resistance (ohms)*
1	35.1	18	25.8	35	31.4
2	35.4	19	26.4	36	28.5
3	36.3	20	25.6	37	28.4
4	38.8	21	33.1	38	27.6
5	39.0	22	33.6	39	27.6
6	22.5	23	32.3	40	28.2
7	23.7	24	32.6	41	30.8
8	25.0	25	32.2	42	30.6
9	25.3	26	27.5	43	30.4
10	25.0	27	26.5	44	30.5
11	34.7	28	26.9	45	30.5
12	34.2	29	26.7	46	28.5
13	34.4	30	27.2	47	30.2
14	34.7	31	31.8	48	30.1
15	34.3	32	32.1	49	30.0
16	26.4	33	31.5	50	28.9
17	25.5	34	31.2		

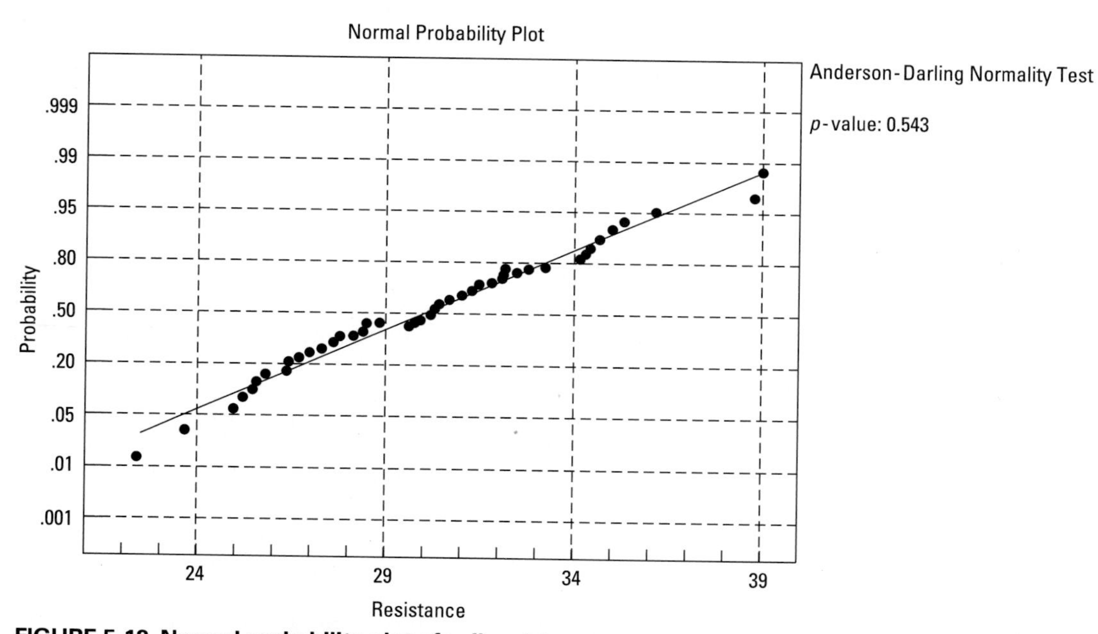

FIGURE 5-12 Normal probability plot of coil resistance.

lower specification limit of the resistance of cables is 25 ohms, from the figure we estimate that approximately 10% of the output is less than the specification limit.

The pattern of a normal probability plot gives us some idea of the process. Nelson (1979) describes the different types of patterns encountered in normal probability plots. These patterns and their possible interpretations are shown in Figure 5-13. In all instances, we must be aware of the subjectivity involved in interpretation. Experience and good judgment are essential to accurately determining whether a set of points deviates significantly from linearity. Using probability plots with statistical goodness-of-fit tests is a reliable way to make inferences regarding normality.

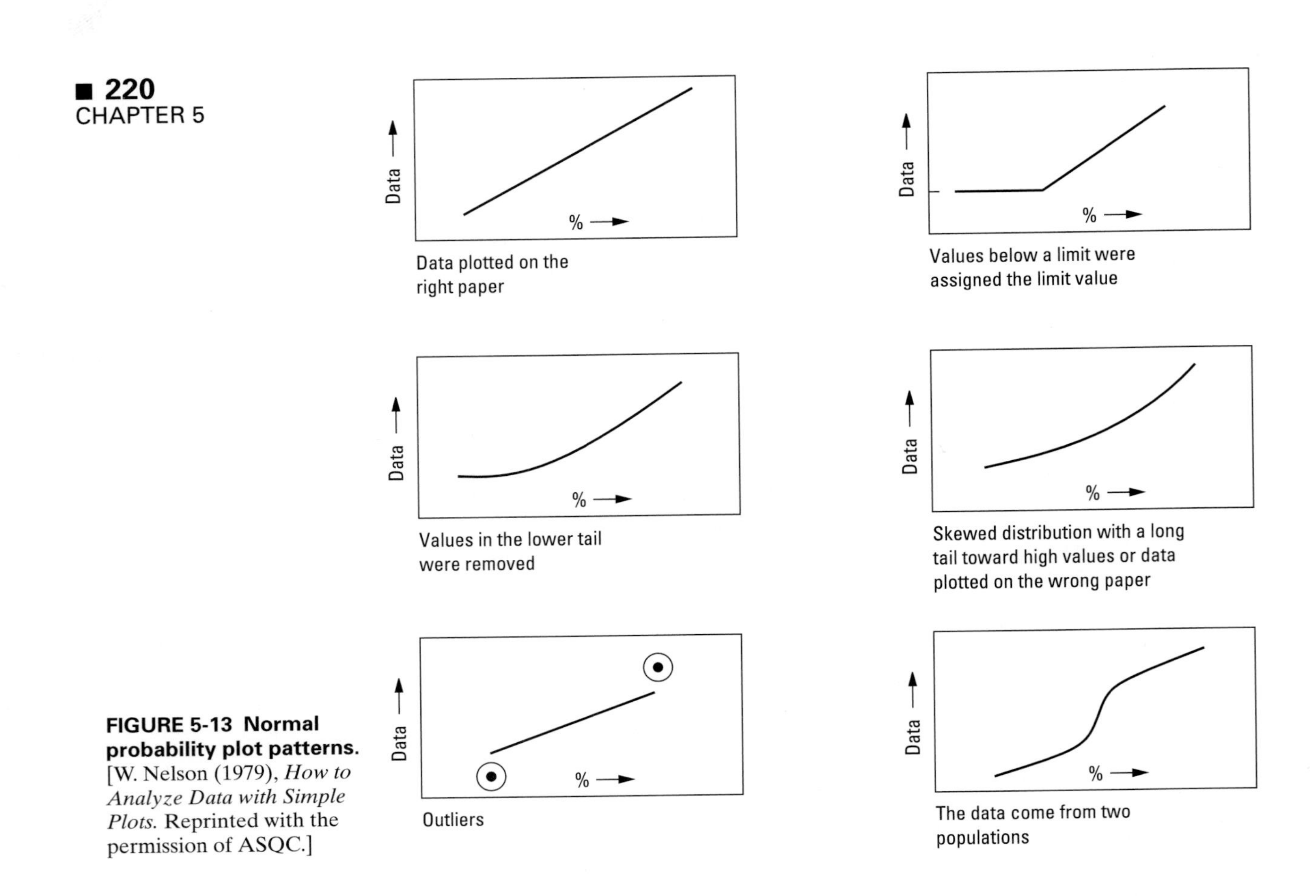

FIGURE 5-13 Normal probability plot patterns. [W. Nelson (1979), *How to Analyze Data with Simple Plots.* Reprinted with the permission of ASQC.]

5-9 EMPIRICAL QUANTILE-QUANTILE PLOTS

A comparison of the distributions of two data sets can be made through the use of empirical **quantile-quantile plots.** The notion of a quantile is similar to that of a percentile. The quantiles of a data set are simply the ordered data values themselves. The pth quantile, $Q(p)$, is the data value such that the proportion of data values that are less than or equal to it is p. Thus, the median is $Q(.5)$. For example, let's consider a company that uses two mixers to make a compound. The amount of polypropylene is measured from random samples taken from each mixer. We would like to know whether the distribution of polypropylene content is the same for the two mixers. Constructing a quantile-quantile plot will enable us to determine whether changes need to be made to the process. Later we can use plots constructed from data obtained before and after the changes to find out how the changes have influenced the process characteristic. Thus, the effects of actions can be investigated.

A quantile-quantile plot is simple to construct. Consider first the case in which the sample sizes of two data sets are equal to n. The values in the first data set are ranked in ascending order. The values in this ordered data set are denoted by X_i for $i = 1, 2, \ldots, n$. The values in the second data set are next ranked in ascending order. These ordered values are denoted by Y_i for $i = 1, 2, \ldots, n$. So, X_i is the $(i - .5)/n$ quantile of the first data set, and Y_i is the same quantile of the second data set. Bivariate points are created by grouping the pairs (X_i, Y_i). These points are plotted, and the resulting graph is known as an empirical quantile-quantile plot. If the plotted points lie on, or close to, a straight line, the samples are assumed to be from the same distribution.

In case the samples from the two data sets are unequal in size, a minor modification is necessary in which empirical quantiles are first found for all of the ordered observations from the smaller data set. Next, a corresponding set of quantiles (based on the smaller data set) is found through interpolation for the larger data set. The chosen points are then plotted to give the quantile-quantile plot. Thus, not all of the observations of the larger data set will be plotted.

Suppose the first data set has 10 observations, the ordered values of which are denoted by X_i, $i = 1, 2, \ldots, 10$. The second data set has 20 observations; these ordered values are denoted by Y_j, $j = 1, 2, \ldots, 20$. The smallest observation in the first data set is X_1, which is the $(1 - .5)/10$, or 0.05 quantile. Observe that for the second data set Y_1 is the $(1 - .5)/20$, or 0.025 quantile and that Y_2 is the $(2 - .5)/20$, or 0.075 quantile. So, to determine the corresponding value from the second data set for value X_1, we would linearly interpolate between Y_1 and Y_2 to obtain the 0.05 quantile. Since Y_1 is the 0.025 quantile and Y_2 is the 0.075 quantile, the 0.05 quantile from the second data set would be interpolated as $(Y_1 + Y_2)/2$. Thus a plotted point will be $[X_1, (Y_1 + Y_2)/2]$. This procedure is used to determine the other bivariate points for obtaining the plot.

Example 5-14: Samples of a size 30 are collected from each of two mixers A and B in a production facility. The polypropylene content is measured for each sample. Table 5-7 shows the data from the two mixers. We want to determine whether the distribution of polypropylene content is the same for the two mixers.

Solution. The two samples are ranked in ascending order and then merged to create a bivariate observation. Table 5-8 shows the ordered samples and the 30 bivariate points. These points are plotted as shown in Figure 5-14 to yield an empirical quantile-quantile plot. The plot shows that a straight line approximates the plotted observations. Most of the points lie close to the straight line. We may infer that distributions of polypropylene content from the two mixers are the same. There is some subjectivity in making this decision since determining the "closeness" of plotted points to a straight line involves judgment. However, your accuracy in these decisions improves with added experience in using these plots. Their ease of construction makes them an attractive tool for studying process characteristics.

TABLE 5-7 Polypropylene Content Data from Two Mixers

Data for Mixer A—Polypropylene Content, X					*Data for Mixer B—Polypropylene Content, Y*				
48.6	49.6	48.3	48.9	48.9	49.8	47.9	47.6	48.5	48.9
49.5	49.3	49.4	49.4	49.2	47.7	49.6	49.5	48.3	48.2
54.5	54.6	54.9	52.9	53.2	52.4	52.0	50.2	50.5	53.2
55.5	53.5	53.3	54.1	53.7	52.7	50.9	51.9	51.6	51.2
49.8	51.5	52.8	52.1	49.9	47.4	47.0	47.6	45.8	45.4
50.9	51.4	50.2	50.4	49.8	46.2	45.9	46.4	46.8	47.2

TABLE 5-8 Ordered Bivariate Data on Polypropylene Content from Two Mixers

Observation i	*Mixer A,* X_i	*Mixer B,* Y_i	*Observation* i	*Mixer A,* X_i	*Mixer B,* Y_i
1	48.3	45.4	16	50.9	48.5
2	48.6	45.8	17	51.4	48.9
3	48.9	45.9	18	51.5	49.5
4	48.9	46.2	19	52.1	49.6
5	49.2	46.4	20	52.8	49.8
6	49.3	46.8	21	52.9	50.2
7	49.4	47.0	22	53.2	50.5
8	49.4	47.2	23	53.3	50.9
9	49.5	47.4	24	53.5	51.2
10	49.6	47.6	25	53.7	51.6
11	49.8	47.6	26	54.1	51.9
12	49.8	47.7	27	54.5	52.0
13	49.9	47.9	28	54.6	52.4
14	50.2	48.2	29	54.9	52.7
15	50.4	48.3	30	55.5	53.2

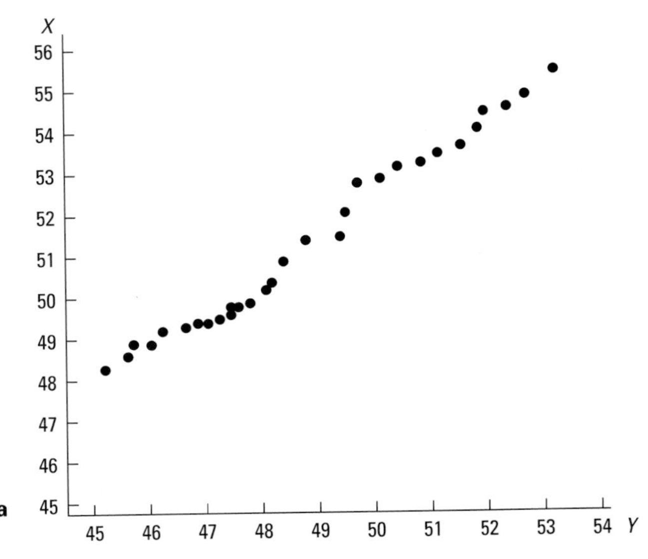

FIGURE 5-14 Quantile-quantile (empirical) plot of data from mixers.

5-10 SCATTER DIAGRAMS

The simplest form of a **scatter diagram** consists of plotting bivariate data to depict the relationship between two variables. When we analyze processes, the relationship between a controllable variable and a desired quality characteristic is frequently of importance. Knowing this relationship may help us decide how to set a controllable variable to achieve a desired level for the output characteristic.

Example 5-15: Suppose we are interested in determining the relationship between the depth of cut in a milling operation and the amount of tool wear. We take 40 observations from the process such that the depth of cut (in millimeters) is varied over a range of values and the corresponding amount of tool wear (also in millimeters) over 40 operation cycles is noted.

The data values are shown in Table 5-9. We can create scatter plots with Microsoft Excel software. Click the **Chart Wizard** button on the standard toolbar. In step 1 of the dialog box,

TABLE 5-9 Data on Depth of Cut and Tool Wear

Observation	*Depth of Cut (in mm)*	*Tool Wear (in mm)*	*Observation*	*Depth of Cut (in mm)*	*Tool Wear (in mm)*
1	2.1	0.035	21	5.6	0.073
2	4.2	0.041	22	4.7	0.064
3	1.5	0.031	23	1.9	0.030
4	1.8	0.027	24	2.4	0.029
5	2.3	0.033	25	3.2	0.039
6	3.8	0.045	26	3.4	0.038
7	2.6	0.038	27	2.8	0.040
8	4.3	0.047	28	2.2	0.031
9	3.4	0.040	29	2.0	0.033
10	4.5	0.058	30	2.9	0.035
11	2.6	0.039	31	3.0	0.032
12	5.2	0.056	32	3.6	0.038
13	4.1	0.048	33	1.9	0.032
14	3.0	0.037	34	5.1	0.052
15	2.2	0.028	35	4.7	0.050
16	4.6	0.057	36	5.2	0.058
17	4.8	0.060	37	4.1	0.048
18	5.3	0.068	38	4.3	0.049
19	3.9	0.048	39	3.8	0.042
20	3.5	0.036	40	3.6	0.045

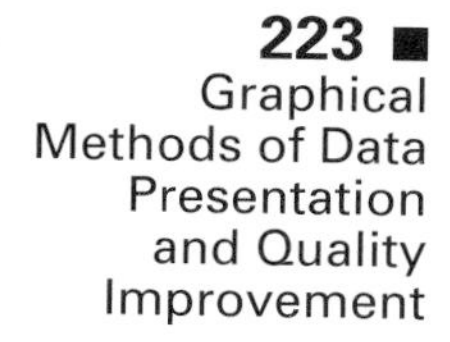

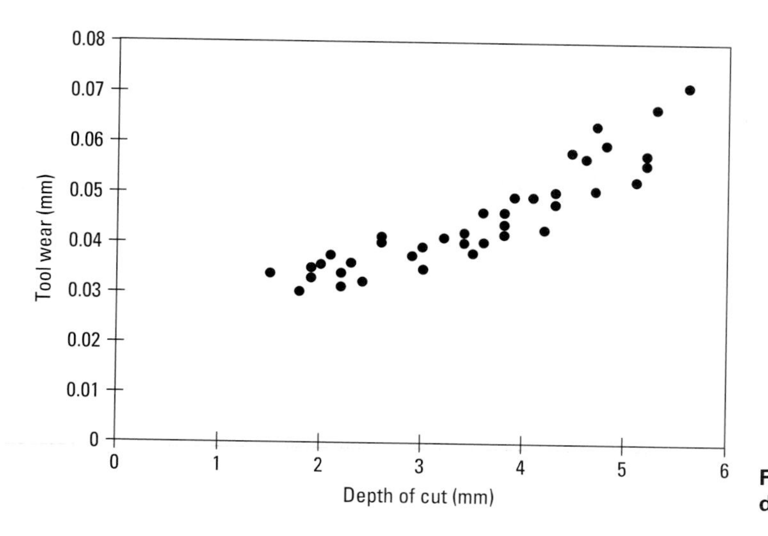

FIGURE 5-15 Scatter plot of tool wear versus depth of cut.

type the two columns whose range names you want to plot. Click **Next** to go to step 2. Double-click **XY(Scatter)** to select a scatter plot. In step 3, click **1** to show data values with no connecting lines, then click **Next.** In step 4, make sure that the **Columns Option** is selected, indicating that the variables are in columns, then click **Next.** In step 5, click the **No option** under "Add a Legend," because there is only one variable along the vertical axis. Click the **Chart Title** box, type a title, and click **Finish.**

The resulting scatter plot is shown in Figure 5-15. It gives us an idea of what relationship exists between depth of cut and amount of tool wear. In this case, the relationship is generally nonlinear. For depth-of-cut values less than 3.0 mm, tool wear seems to be constant, whereas with increases in depth of cut, tool wear starts increasing at an increasing rate. For depth-of-cut values above 4.5 mm, tool wear appears to increase drastically. This information will help us determine the depth of cut to use to minimize downtime due to tool changes.

5-11 MULTIVARIABLE CHARTS

In many manufacturing or service operations, there are usually several variables or attributes that affect product or service quality. The types of plots discussed up to this point have only dealt with two variables. Since realistic problems usually have more than two variables, **multivariable charts** are useful means of displaying collective information.

Several types of multivariable charts are available (Blazek et al. 1987; Kleiner and Hartigan 1981). One of these is known as a **radial plot,** or star, for which the different variables of interest correspond to different rays emanating from a star. The length of each ray represents the magnitude of the variable.

Example 5-16: Suppose the controllable variables in a process are temperature, pressure, manganese content, and silicon content. Figure 5-16 shows radial plots, or stars, for two samples of size 10, taken an hour apart. The sample means for the respective variables are calculated. These are represented by the length of the rays. A relative measure of quality performance is used to locate the center of a star vertically (in this case, percentage of nonconforming product), while the horizontal axis represents the two sampling times.

Several process characteristics can be observed from Figure 5-16. First, from time 1 to time 2, an improvement in the process performance is seen, as indicated by a decline in the percentage nonconforming. Next, we can examine what changes in the controllable variables led to this improvement. We see that a decrease in temperature, an increase in both pressure and manganese content, and a basically constant level of silicon caused this reduction in the percentage nonconforming.

Other forms of multivariable plots (such as standardized stars, glyphs, trees, faces, and weathervanes), are conceptually similar to radial plots. For details on these forms, refer to Friedman et al. (1972), Gnanadesikan (1977), Chernoff (1973), and Bruntz et al. (1974).

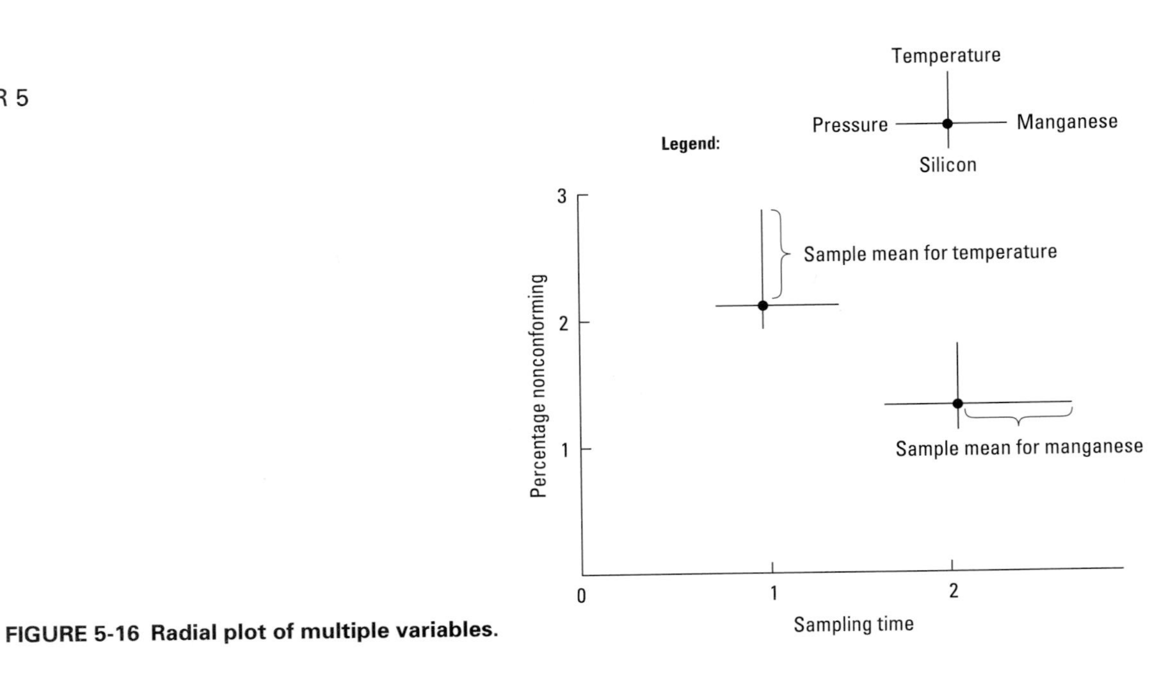

FIGURE 5-16 **Radial plot of multiple variables.**

5-12 MATRIX PLOTS AND THREE-DIMENSIONAL SCATTER PLOTS

Investigating quality improvement in products and processes often involves data that deal with more than two variables. The graphical methods discussed so far, with the exception of multivariable charts, only deal with one or two variables. The **matrix plot** is a graphical option for situations with more than two variables. This plot depicts two-variable relationships between a number of variables all in one plot. As a two-dimensional matrix of separate plots, it enables us to conceptualize relationships between the different variables. The Minitab software can produce matrix plots.

Example 5-17: Consider the data shown in Table 5-10 on temperature, pressure, and seal strength of plastic packages. Since temperature and pressure are process variables, we want to investigate their impact on seal strength, a product characteristic.

TABLE 5-10 Data on Temperature, Pressure, and Seal Strength for Plastic Packages

Observation	*Temperature (in °C)*	*Pressure (in kg/cm²)*	*Seal Strength (in kg)*	*Observation*	*Temperature (in °C)*	*Pressure (in kg/cm²)*	*Seal Strength (in kg)*
1	180	80	8.5	16	220	40	11.5
2	190	60	9.5	17	250	30	10.8
3	160	80	8.0	18	180	70	9.3
4	200	40	10.5	19	190	75	9.6
5	210	45	10.3	20	200	65	9.9
6	190	50	9.0	21	210	55	10.1
7	220	50	11.4	22	230	50	11.3
8	240	35	10.2	23	200	40	10.8
9	220	50	11.0	24	240	40	10.9
10	210	40	10.6	25	250	35	10.8
11	190	60	8.8	26	230	45	11.5
12	200	70	9.8	27	220	40	11.3
13	230	50	10.4	28	180	70	9.6
14	240	45	10.0	29	210	60	10.1
15	240	30	11.2	30	220	55	11.1

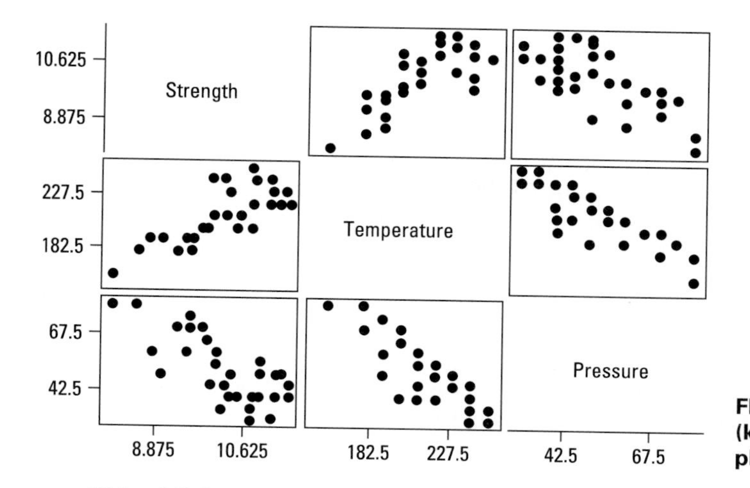

FIGURE 5-17 Matrix plot of the variables strength (kg), temperature (°C), and pressure (kg/cm²) for plastic packages.

Using Minitab, the data is entered for the three variables. Next, choose **Graph > Matrix Plot.** Type in the variable names, say Temperature, Pressure, Strength in **Graph variables,** then click **OK.** The resulting matrix plot is shown in Figure 5-17. Observe that seal strength tends to increase linearly with temperature up to a certain point, which is about 210°C. Beyond 210°C, seal strength tends to decrease. The relationship between seal strength and pressure decreases with pressure. Also, the existing process conditions exhibit a relationship between temperature and pressure that decreases with pressure. Such graphical aids provide us with some insight on the relationship between the variables, taken two at a time.

Another plot of interest is the **three-dimensional scatter plot,** which depicts the joint relationship of a dependent variable with two independent variables. While matrix plots demonstrate two-variable relationships, they do not show the joint effect of more than one variable on a third variable. Since interactions do occur between variables, the three-dimensional scatter plot is useful in identifying optimal process parameters based on a desired level of an output characteristic.

Example 5-18: Let's reconsider the plastic package data shown in Table 5-10. Suppose we want to identify the joint relationship of the process variables, temperature and pressure, on seal strength of packages, an output characteristic. Using Minitab, we choose **Graph > 3D Plot.** Type in the variable names, say Strength in **Z,** Temperature in **Y,** and Pressure in **X,** then click **OK.** Figure 5-18 shows the resulting three-dimensional scatter plot of strength versus temperature and pressure. This scatter plot helps us identify optimal process parameter values that will maximize a variable. For example, a temperature around 230°C and a pressure around 30 kg/cm^2 appear to be desirable process parameter values for maximizing seal strength.

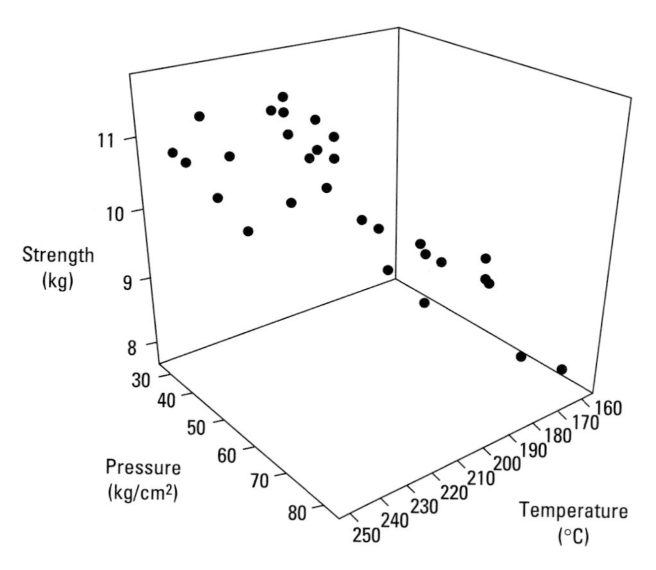

FIGURE 5-18 Three-dimensional scatter plot of strength versus temperature and pressure for plastic packages.

5-13 SUMMARY

This chapter presents a number of graphical methods that are useful for analyzing the existing status of product and process characteristics. The variability of product or process characteristics is always a concern, and the techniques presented here provide an idea of the degree of variability. Methods such as frequency histograms, stem-and-leaf plots, box plots, and normal probability plots are useful for interpreting measures of both central tendency and dispersion. They can be used to compare product and process performance before and after certain changes are made in the process. In addition to methods of data presentation, the chapter discusses several graphical methods that are useful in the analysis of quality improvement.

Neverending improvements in products and their corresponding processes are a necessity. Thus, after achieving control in a process, management will seek means for further improvement in order to remain competitive. The methods discussed in this chapter provide guidelines for determining possible areas where such changes should be investigated. Pareto diagrams identify major problem areas. Cause-and-effect diagrams go a step further and analyze the various subcauses associated with each major cause. A graphical display of the effect of a certain variable on an output characteristic is obtained through a scatter plot, while the output of two different process conditions can be compared through an empirical quantile-quantile plot. Real-world problems usually involve more than one variable, and multivariable charts, matrix plots, and three-dimensional scatter plots are useful for displaying the relative impacts of such variables on an output characteristic. All of these methods are easy to understand and apply.

CASE STUDY

Accelerating Improvement*

The need to improve quality continues to capture the attention of American industry. The Seven-Step Method and the Project Team Review Process are related techniques that, in the proper management setting, can accelerate process improvement.

The Seven-Step Method is a structured approach to problem solving and process improvement. It leads a team through a logical sequence of steps that force a thorough analysis of the problem, its potential causes, and possible solutions. The structure imposed by the Seven-Step Method helps a team focus on the correct issues rather than diffuse its energy on tangential or even counterproductive undertakings.

The Seven-Step Method is most successful when accompanied by regular project reviews performed by managers with a vested interest in the project's outcome. In many organizations, project teams are not reviewed until a solution or recommendation is to be presented—the notion of a status review is foreign. However, there is a formal review process in which peers and superiors guide, support, and monitor project teams while they are working on problems. This Project Team Review Process structures a session so that it becomes a productive meeting with positive consequences, thereby providing teams with support and focus.

THE SEVEN-STEP METHOD

The value of the Seven-Step Method lies in the discipline and logic that it imposes. The seven steps are now briefly described.

Step 1: Define the Problem

1. Define the problem in terms of a gap between what is and what should be. (For example, "Customers report an excessive number of errors. The team's objective is to reduce the number of errors.")

*Adapted from M. Gaudard, R. Coates, and L. Freeman (1991), "Accelerating Improvement," *Quality Progress,* Vol. 24 (No. 10): 81–88.

2. Document why it is important to be working on this particular problem:
   - Explain how you know it is a problem, providing any data you might have that supports this.
   - List the customer's key quality characteristics. State how closing the gap will benefit the customer in terms of these characteristics.
3. Determine what data you will use to measure progress:
   - Decide what data you will use to provide a baseline against which improvement can be measured.
   - Develop an operational definition you will need to collect the data.

Step 2: Study the Current Situation

1. Collect the baseline data and plot them. (Sometimes historical data can be used for this purpose.) A run chart or control chart is usually used to exhibit baseline data. Decide how you will exhibit these data on the run chart. Decide how you will label your axes.
2. Develop flow charts of the processes.
3. Provide any helpful sketches or visual aids.
4. Identify any variables that might have a bearing on the problem. Consider the variables of what, where, to what extent, and who. Data will be gathered on these variables to localize the problem.
5. Design data collection instruments.
6. Collect the data and summarize what you have learned about the variables' effects on the problem.
7. Determine what additional information would be helpful at this time. Repeat substeps 2 through 7 until there is no additional information that would be helpful at this time.

Step 3: Analyze the Potential Causes

1. Determine potential causes of the current conditions:
   - Use the data collected in step 2 and the experience of the people who work in the process to identify conditions that might lead to the problem.
   - Construct cause-and-effect diagrams for these conditions of interest.
   - Decide on the most likely causes by checking against the data from step 2 and the experience of the people working in the process.
2. Determine whether more data are needed. If so, repeat substeps 2 through 7 of step 2.
3. If possible, verify the causes through observation or by directly controlling variables.

Step 4: Implement a Solution

1. Develop a list of solutions to be considered. Be creative.
2. Decide which solutions should be tried:
   - Carefully assess the feasibility of each solution, the likelihood of success, and potential adverse consequences.
   - Clearly indicate why you are choosing a particular solution.
3. Determine how the preferred solution will be implemented. Will there be a pilot project? Who will be responsible for the implementation? Who will train those involved?
4. Implement the preferred solution.

Step 5: Check the Results

1. Determine whether the actions in step 4 were effective:
   - Collect more data on the baseline measure from step 2.
   - Collect any other data related to the conditions at the start that might be relevant.
   - Analyze the results. Determine whether the solution tested was effective. Repeat prior steps as necessary.
2. Describe any deviations from the plan and what was learned.

Step 6: Standardize the Improvement

1. Institutionalize the improvement:
   - Develop a strategy for institutionalizing the improvement, and assign responsibilities.
   - Determine whether the improvement should be applied elsewhere, and plan for its implementation.

Step 7: Establish Future Plans

1. Determine your plans for the future:
   - Decide whether the gap should be narrowed further and, if so, how another project should be approached and who should be involved.
   - Identify related problems that should be addressed.
2. Summarize what you learned about the project team experience, and make recommendations for future project teams.

APPLICATION

A restaurant caters to business travelers and has a self-service breakfast buffet. Interested in customer satisfaction, the manager constructs a survey, distributes it to customers over a three-month period, and summarizes the results in a Pareto chart (see Figure 5-19). The Pareto chart indicates that the restaurant's major problem is that customers have to wait too long to be seated. A team of employees is formed to work on this problem.

Step 1: Define the project. With the survey as the background, the team undertakes the first step. The problem is that customers wait too long to be seated. They should not have to wait at all. The problem is important because customers have complained, and this is supported by the Pareto chart constructed from the survey data. Most of the customers are business travelers who want either a speedy breakfast or a chance to conduct business during breakfast. Decreasing the wait to be seated will increase the restaurant's ability to respond to these key quality characteristics. Progress can be measured by the percent of customers each day who have to wait in excess of, say, one minute to be seated. The team develops an operational definition of "waiting to be seated" to answer such questions as: When does the wait start? When does it end? How is it measured?

Step 2: Study the current situation. The team collects baseline data and plots it (see Figure 5-20). At the same time, the team develops a flowchart of seating a party. The team members feel that a floor diagram might be helpful, so they produce one (see Figure 5-21). The variables they identify as potentially affecting the problem are day of the week, size of the party, reason for waiting, and time of the morning. Data relating to these variables are collected.

From the baseline data, the team learns that the percent of people served who have to wait is higher early in the week and decreases during the week, with only a small percent waiting on weekends. This is reasonable, since the restaurant's clientele primarily consists of business travelers. The size of the party does not appear to be a factor, because parties of all sizes wait in approximately the same proportions. A histogram of the number of people waiting by the time of the morning reveals nothing surprising: more people wait during the busy hours than during the slow hours (see Figure 5-22). The reason for waiting, however, is interesting. Most people wait because a table is not available or because they have a seating preference (as opposed to the host or hostess not

FIGURE 5-19 Pareto chart of complaints.

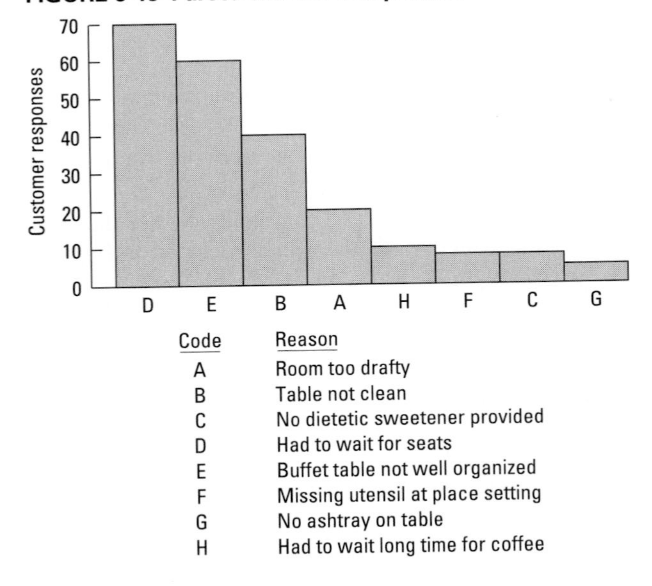

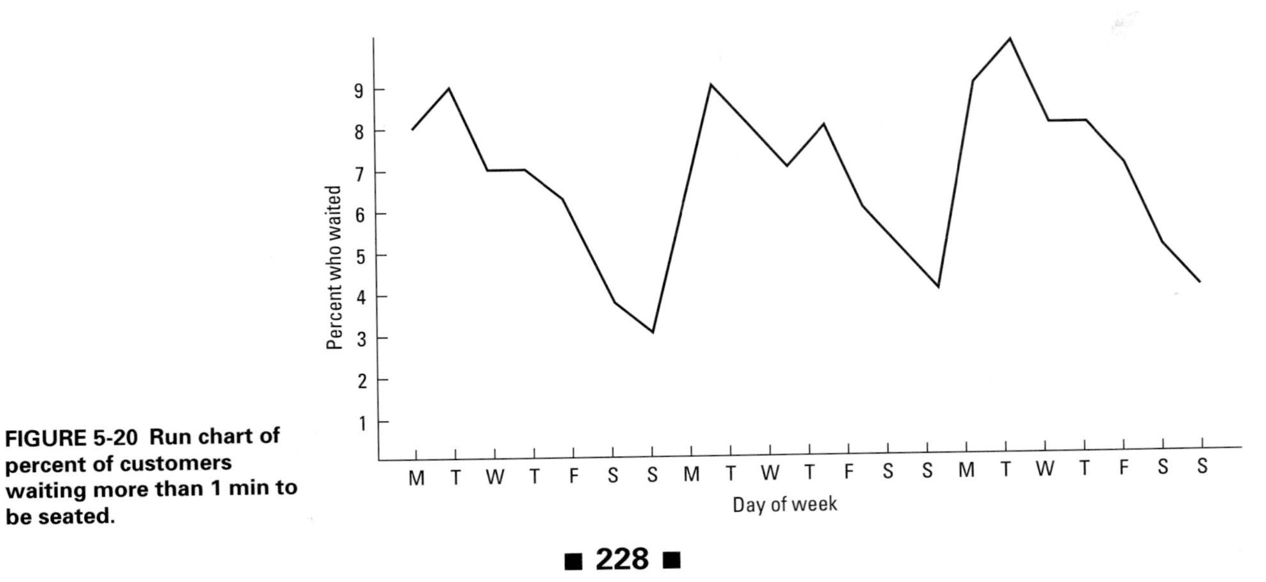

FIGURE 5-20 Run chart of percent of customers waiting more than 1 min to be seated.

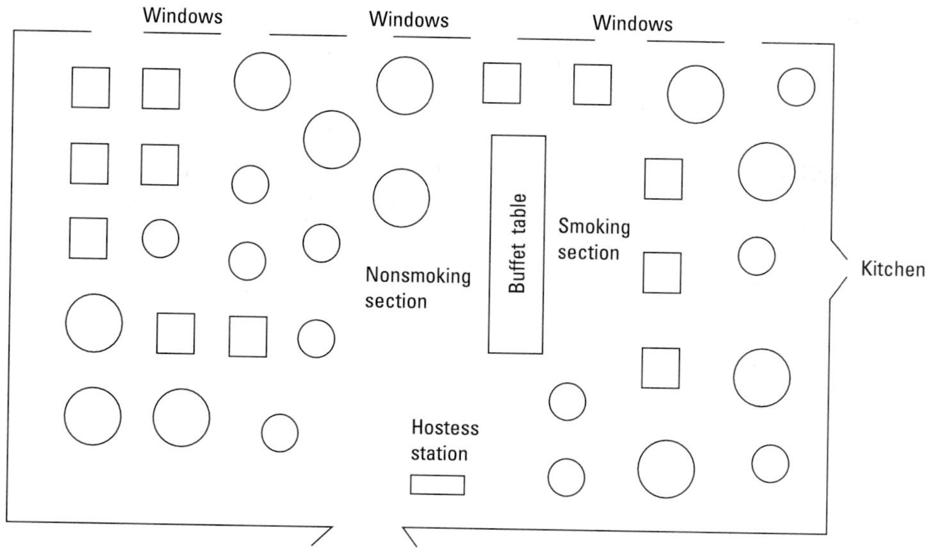

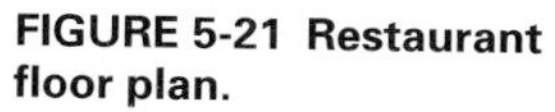

FIGURE 5-21 Restaurant floor plan.

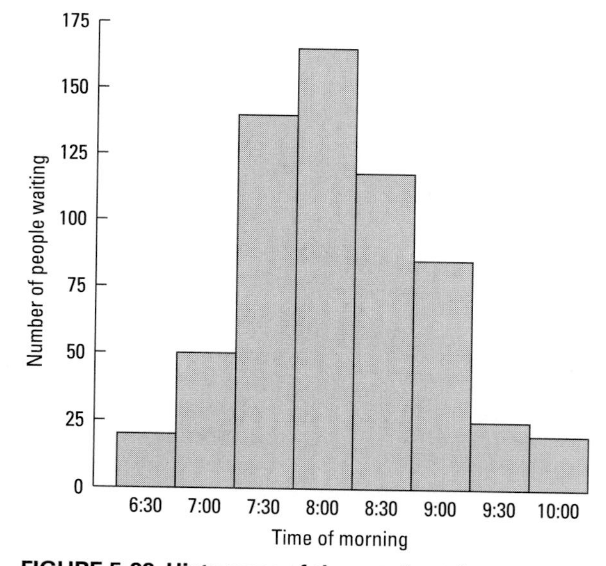

FIGURE 5-22 Histogram of the number of customers waiting more than 1 min for particular times of morning.

being around to seat customers or customers waiting for friends to join them).

At this point, it would be easy for the team to jump to the solution of putting more staff on early in the week and during busy hours in the morning—but analyzing causes is not done until the next step.

The team decides additional information is needed on why tables are not available and how seating preferences affect waiting. After data are collected, it learns that tables are generally unavailable because they are not cleared (as opposed to being occupied) and that most of the people who have a seating preference wait for a table in the nonsmoking area.

Step 3: Analyze the potential causes. A cause-and-effect diagram is constructed showing why tables are not cleared quickly, with particular emphasis on identifying root causes (see Figure 5-23). This diagram, together with the rest of the data the team has gathered, leads the team to conclude that the most likely cause is the distance from the tables to the kitchen, particularly in the nonsmoking area.

Step 4: Implement a solution. The team develops a list of possible solutions. Since the team has not been able to verify the cause by controlling the variables, it chooses a solution that can be easily tested: Set up temporary workstations in the nonsmoking area. No other changes are made. The team continues to collect data on the percent of people waiting longer than one minute to be seated.

Step 5: Check the results. After a month, the team analyzes the data collected in step 4. As Figure 5-24 shows, the improvement is dramatic.

Step 6: Standardize the improvement. The temporary workstations are replaced with permanent ones.

Step 7: Establish future plans. The team decides that the next highest bar in the Pareto chart of customer complaints—the buffet table being not well organized—should be addressed.

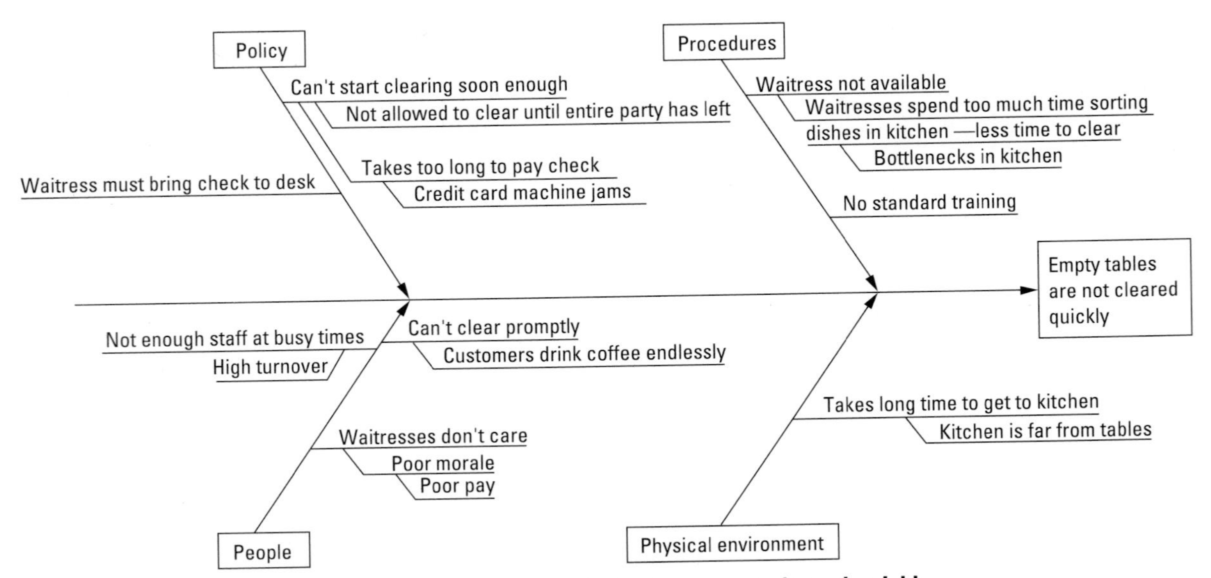

FIGURE 5-23 Cause-and-effect diagram describes why tables are not cleared quickly.

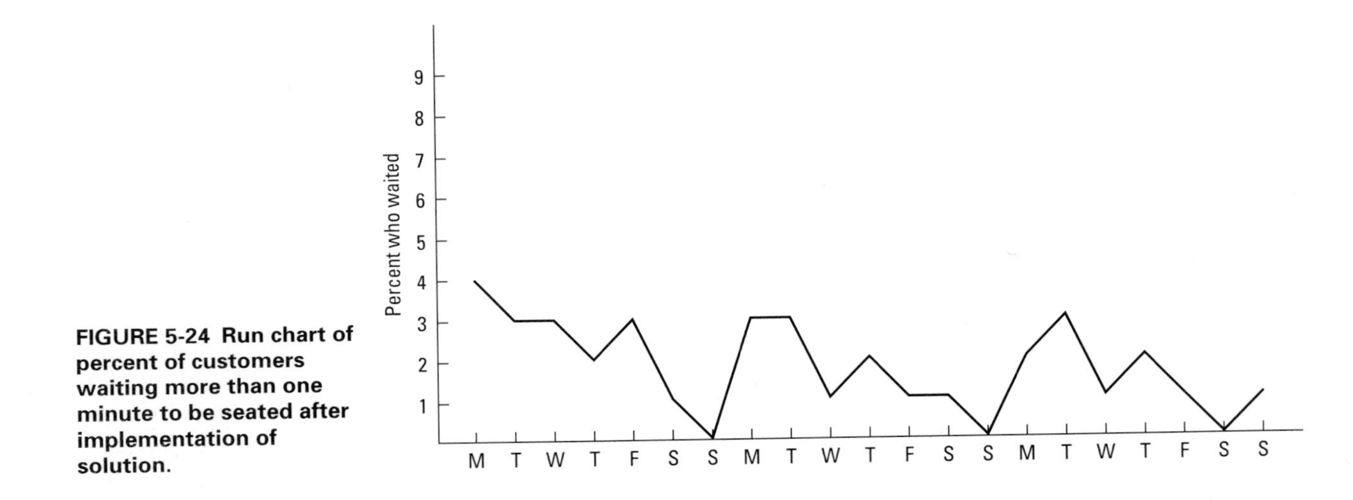

FIGURE 5-24 Run chart of percent of customers waiting more than one minute to be seated after implementation of solution.

THE PURPOSE OF THE SEVEN STEPS

The overall purpose of the Seven-Step Method is to facilitate process improvement.

Since the plan-do-check-act (PDCA) process, also known as the Deming cycle, also has process improvement as its goal, it seems natural to ask how it and the Seven-Step Method are related. The *plan* step consists of planning a change or test aimed at improvement; the *do* step consists of carrying out the change or test, preferably on a small scale; the *check* step involves studying the results to understand what has been learned; and the *act* step consists of adopting the change, abandoning it, or repeating the cycle.

It seems clear that much of the PDCA thinking is present in the Seven-Step Method. The PDCA cycle, however, is a broad paradigm for process improvement that applies to situations where the Seven-Step Method does not. The Seven-Step Method is appropriate when a deep understanding of the problem is needed to determine and plan an effective solution. A team acquires this understanding through the data-based localization and cause analysis in steps 2 and 3. In these steps, the team is continually restrained from jumping to solutions. Only in step 4 does the team formulate and implement a solution. This step is similar to the plan and do steps in the PDCA cycle. In step 5, which is comparable to the PDCA cycle's check step, the team checks its results. After checking, the team either standardizes its findings (step 6) or returns to a prior step to obtain an even deeper understanding of the problem or possible solutions—a process much like the act step of the PDCA cycle. In addition to reflecting on its experience in terms of tasks, group processes, and organizational issues, in step 7 the team identifies future needs. Identification of these needs continues the PDCA cycle, and if appropriate, those needs are addressed using the Seven-Step Method.

The Seven-Step Method is directed at analytic rather than enumerative studies. In an enumerative study, an existing population is studied; an analytic study focuses on prediction. In analytic situations, the key is to learn about the cause systems that underlie the processes of interest to understand the effects of various conditions. Since organizations are usually interested in learning about the future, almost all problems in industry—and certainly the most important ones—are of an analytic nature. The Seven-Step Method helps solve analytic problems efficiently because it focuses on understanding causal relationships, as evidenced in steps 2 through 5.

THE METHOD'S VALUE

After management had been using the Seven-Step Method in their project teams for about three months, we asked them at a training session to brainstorm regarding what they had learned as a result of the method. The managers overwhelmingly found the method's focus and restraint to be difficult but valuable. They also valued the way the method provides organization, logic, and thoroughness. They were impressed by the use of data instead of opinions. A number of people commented on how they were listening to each other more carefully and respecting each other's ideas more and on how, perhaps because of the focus on data, there had been a lowering of territorial fences and a promotion of cooperation and trust. The manager's perceptions concerning the method were also shaped by the group process skills they had been practicing and the project reviews they had undertaken.

DIFFICULT ISSUES

The three project teams found several concepts in the first two steps extremely difficult. The first was arriving at a problem statement. The initial tendency was to frame a solution as a problem. The analog of this in the restaurant case study would be to state the problem as "There aren't enough tables" or "The waitresses and waiters need to work harder" instead of "The customers wait too long." Once this hurdle was crossed, there were others. One team needed to agree on several operational definitions before it could even begin to formulate a problem statement. Another team kept revising its problem statement during the first four months before settling on one that was consistent with what the team was doing. The third team arrived at a problem statement relatively easily by comparison, but even they struggled.

Localization—the process of focusing on smaller and smaller vital pieces of the problem—is another task that the teams found difficult. Localization is usually achieved by stratifying data using Pareto charts with categorical data and run charts with continuous data. Localization is what makes a problem tractable. Although team members could see the value of localization in solving sample problems, they found it hard to internalize this in solving their own problems. Realizing that their own problems were not overwhelming but instead tractable through localization was an important achievement for the project teams.

There were other issues that proved difficult. It was much easier for the teams to justify the importance of the problem in terms of internal considerations rather than in terms of the customer's key quality characteristics. We sensed that this occurred because the team members had not yet internalized the idea that improvement should be driven by customer requirements, not internal indicators. Some team members could not see the benefits of collecting data accurately and consistently, so they resisted devising ways to accomplish this. The teams had trouble understanding how baseline data would be used to validate a solution. Causes of the problem often crept into discussions where they did not belong. The teams had difficulty keeping an open mind about potential causes. For example, they resisted investigating the effects of variables that they felt were not causes. The teams needed a significant amount of coaching in how to obtain information in a nonthreatening way (through interviews or surveys) from the people who work in the system. The teams also faced several organizational challenges, such as finding the time to work on their projects, arranging meetings, and getting support from workers who were to collect the data.

QUESTIONS FOR DISCUSSION

1. Discuss the Seven-Step Method used for problem solving and process improvement.
2. Discuss the similarities between the Seven-Step Method and Deming's plan-do-check-act (PDCA) cycle.
3. In addition to the waiting time of customers, if you had one or more characteristics that you decided to examine, what method would you use?

4. Develop a cause-and-effect diagram to conduct dispersion analysis for one of the main causes identified in the restaurant case study.
5. What are some of the difficulties faced in applying the Seven-Step Method along with a Project Team Review?

Key Terms

- box plot
  - notched
  - variable-width
- cause-and-effect diagram
- cause enumeration
- dispersion analysis
- fishbone diagram
- frequency
  - distribution
  - histogram
- interquartile range
- Ishikawa diagram
- kurtosis
- matrix plot
  - two-dimensional
- multivariable chart
- normal probability plot
- outlier
- Pareto diagram
- process analysis
- quantile-quantile plot
- quartile
- radial plot
- run chart
- scatter plot
  - three-dimensional
- skewness
- stem-and-leaf plot
- tail length
- whisker

Exercises

Discussion Questions

1. What is the advantage of stem-and-leaf plots over frequency histograms?
2. Explain the importance of cause-and-effect diagrams in process improvement. Select a problem of your choice, and construct a cause-and-effect diagram.
3. Describe the importance of Pareto diagrams in process improvement.
4. What is a box plot used for? How does it help in the assimilation of product or process information?
5. Explain the importance of normal probability plots in quality control and improvement.
6. Compare and contrast quantile-quantile plots and scatter plots. Explain their roles in product or process improvement.
7. What are the advantages of multivariable charts over scatter diagrams in quality improvement?

Problems

8. A random sample of 50 observations of the mileage per gallon (mpg) of a particular brand of gasoline is shown:

33.2	29.4	36.5	38.1	30.0
29.1	32.2	29.5	36.0	31.5
34.5	33.6	27.4	30.4	28.4
32.6	30.4	31.8	29.8	34.6
30.7	31.9	32.3	28.2	27.5
34.9	32.8	27.7	28.4	28.8
30.2	26.8	27.8	30.5	28.5
31.8	29.2	28.6	27.5	28.5
30.8	31.8	29.1	26.9	34.2
33.5	27.4	28.5	34.8	30.5

   a. Construct a frequency histogram.
   b. Construct a relative frequency histogram and a cumulative frequency histogram.
   c. What conclusions can you draw regarding the product?
   d. If the company has a goal of a gas mileage of 31 mpg, is it achieving its objective?
9. Construct a stem-and-leaf plot for the data shown in Exercise 8. What inferences can you make from the plot?
10. Construct a normal probability plot for the data in Exercise 8. What inferences can you make from the plot?
11. Construct a box plot for the data in Exercise 8. What insights do you get from this plot?

12. Construct a notched box plot using a 95% level of confidence for the data in Exercise 8.
13. The following waiting times (in minutes) before being served in a local post office are observed for 50 randomly chosen customers:

```
2.1 0.5 3.6 1.4 2.0
0.8 0.4 4.2 3.5 2.5
4.8 2.8 1.9 1.2 3.2
1.6 2.5 2.4 1.9 2.0
3.5 5.2 3.1 1.6 1.5
1.9 2.4 2.7 2.1 1.8
4.6 3.8 1.5 4.5 3.9
5.5 2.5 3.8 5.0 4.6
2.1 2.8 1.6 3.8 4.2
3.5 5.2 4.8 3.9 2.6
```

a. Construct a frequency histogram and a relative frequency histogram.
b. What conclusions can you draw?
c. If the post office's goal is to have a waiting time of less than 4.0 mins, has this goal been achieved? Comment on your conclusion.

14. Construct a stem-and-leaf plot for the data in Exercise 13. What inferences can you make?
15. Construct a normal probability plot for the data in Exercise 13, and comment on the conclusions that you draw from this plot.
16. Construct a box plot for the data in Exercise 13. What do you conclude from the plot?
17. Construct a notched box plot using a 95% level of confidence for the data in Exercise 13.
18. Consider the situation of Exercise 8. The company, on further experimentation with product development, has come up with a new brand of gasoline. A random sample of 30 observations yields the following values for mileage per gallon:

```
32.9 31.5 34.3 36.8 35.0
29.4 33.2 37.8 35.0 32.7
28.5 30.4 32.6 31.5 30.6
35.8 36.4 34.2 35.0 33.5
31.8 32.5 28.4 33.8 35.1
30.2 33.0 34.6 32.4 32.0
```

a. Construct a relative frequency histogram.
b. Has an improvement taken place in the product?
c. Is the company meeting its goal of a gas mileage of 31 mpg?

19. Construct a quantile-quantile plot using the data in Exercises 8 and 18. What can you conclude from this plot?
20. Construct notched box plots using a 95% level of confidence of appropriate widths for the data in Exercises 8 and 18. Can you draw any conclusions on product improvement?
21. An analysis of defects of the output from a job shop produced the following results:

Type of Defect	*Frequency*	*Dollar Value (in $1000)*
Nonconforming diameter	40	3
Rough surface	80	7
Warped flange	50	2
Nonconforming length	20	1
Nonconforming ream	60	2.5

a. Construct a Pareto diagram and discuss the results.
b. If management has an allocation of $10,000, which problem areas should they tackle?

22. The pH values of a dye for 30 samples taken consecutively over time are listed row-wise as follows:

```
20.3 15.5 18.2 18.0 20.5 22.8
21.6 21.0 22.5 23.8 23.9 24.2
23.6 24.9 27.4 25.5 20.9 25.8
24.6 25.5 27.3 26.4 26.8 27.5
26.4 26.8 27.2 27.1 27.4 27.8
```

Construct a run chart. What conclusions can you draw?

23. An insurance company is interested in determining whether life insurance coverage is influenced by disposable income. A randomly chosen sample of size 20 produced the following data. Construct a scatter plot. What conclusions can you draw?

Disposable Income (in $1000)	*Life Insurance Coverage (in $1000)*	*Disposable Income (in $1000)*	*Life Insurance Coverage (in $1000)*
45	60	65	80
40	58	60	90
65	100	45	50
50	50	40	50
70	120	55	70
80	100	55	60
70	80	60	80
40	50	75	100
50	70	45	50
45	60	65	70

24. In a chemical process, the parameters of temperature, pressure, proportion of catalyst, and pH

value of the mixture influence the acceptability of the batch. The following data gives the process parameter values and the proportion of nonconforming output based on a sample of 10 values. Construct a multivariable chart. What inferences can you make regarding the desirable values of the process parameters?

Observation	*Temperature (°C)*	*Pressure (kg/cm²)*	*Proportion of Catalyst*	*Acidity (pH value)*	*Proportion Nonconforming*
1	300	100	.03	10	.080
2	350	90	.04	20	.070
3	400	80	.05	15	.040
4	500	70	.06	25	.060
5	550	60	.04	10	.070
6	500	50	.06	15	.050
7	450	40	.05	15	.055
8	450	30	.04	20	.060
9	350	40	.04	15	.054
10	400	40	.04	15	.052

References

Blazek, L. W., B. Novic, and D. M. Scott (1987). "Displaying Multivariate Data Using Polyplots," *Journal of Quality Technology* 19(2): 69–74.

Bruntz, S. M., W. S. Cleveland, B. Kleiner, and S. L. Warner (1974). "The Dependence of Ambient Ozone on Solar Radiation, Wind, Temperature, and Mixing a Height," *Proceedings of the Symposium of Atmospheric Diffusion of Air Pollution.* American Meteorological Society, pp. 125–128.

Chambers, J. M. (1977). *Computational Methods for Data Analysis.* New York: John Wiley.

Chambers, J. M., W. S. Cleveland, B. Kleiner, and P. A. Tukey (1983). *Graphical Methods for Data Analysis.* Belmont, Calif.: Wadsworth.

Chernoff, H. (1973). "The Use of Faces to Represent Points in *K*-Dimensional Space Graphically," *Journal of the American Statistical Association* (68): 361–368.

Duncan, A. J. (1986). *Quality Control and Industrial Statistics,* 5th ed. Homewood, Ill.: Richard D. Irwin.

Friedman, H. P., E. S. Farrell, R. M. Goldwyn, M. Miller, and S. H. Siegel (1972). "A Graphic Way of Describing Changing Multivariate Patterns," *Proceedings of the Sixth Interface Symposium on Computer Science and Statistics.* Berkeley, Calif.: University of California, pp. 56–59.

Gaudard, M., R. Coates, and L. Freeman (1991). "Accelerating Improvement," *Quality Progress* 24(10): 81–88.

Gnanadesikan, R. (1977). *Methods of Statistical Data Analysis of Multivariate Observations.* New York: John Wiley.

Ishikawa, K. (1976). *Guide to Quality Control.* Asian Productivity Organization, Nordica International Limited, Hong Kong. (Available in the United States from UNIPUB, New York.)

Kendall, M. G., and A. Stuart (1967). *The Advanced Theory of Statistics,* 2nd ed., Vol. 1. New York: Hafner.

Kleiner, B., and J. A. Hartigan (1981). "Representing Points in Many Dimensions by Trees and Castles," *Journal of the American Statistical Association* (76): 260–276.

McGill, R., J. W. Tukey, and W. A. Larsen (1978). "Variations of Box Plots," *The American Statistician* 32(1): 12–16.

Massey, F. J., Jr. (1951). "The Kolmogorov–Smirnov Test of Goodness of Fit," *Journal of the American Statistical Association* (46): 68–78.

Michael, J. R. (1983). "The Stabilized Probability Plot," *Biometrika* 70(1): 11–17.

Microsoft Corporation (1994). *Microsoft Excel User's Guide,* Version 5.0. Redmond, Wash.: Microsoft.

Minitab, Inc. (1996). *Minitab Reference Manual,* Release 11, State College, Penn.: Minitab.

Nelson, W. (1979). *How to Analyze Data with Simple Plots.* Milwaukee, Wis.: American Society for Quality Control.

SAS Institute, Inc. (1990). *SAS/STAT User's Guide: Volumes 1 and 2,* Version 6, 4th ed. Cary, N.C.: SAS Institute.

CHAPTER

6

Statistical Process Control Using Control Charts

Chapter Outline

Symbols	
θ	Parameter
$\hat{\theta}$	Estimator
α	Probability of a Type I error
β	Probability of a Type II error
$\sigma_{\overline{X}}$	Standard deviation of sample mean
σ	Process standard deviation
n	Subgroup or sample size

6-1 INTRODUCTION

We have discussed at length the importance of satisfying the customer by improving the product or service. A process capable of meeting or exceeding customer requirements is a key part of this endeavor. Part III of this book deals with the topic of process control and improvement. It provides the necessary background for understanding statistical process control through control charts. In this chapter we build the foundation for using control charts. Chapter 7 examines control charts for variables, and Chapter 8 discusses control charts for attributes. Process capability analysis is covered in Chapter 9.

A **control chart** is a graphical tool for monitoring the activity of an ongoing process. Control charts are sometimes referred to as **Shewhart control charts,** because Walter A. Shewhart first proposed their general theory. The values of the quality characteristic are plotted along the vertical axis, and the horizontal axis represents the samples, or subgroups (in order of time), from which the quality characteristic is found. Samples of a certain size (say, 4 or 5 observations) are selected, and the quality characteristic (say, average length) is calculated based on the number of observations in the sample. These characteristics are then plotted in the order in which the samples were taken. Figure 6-1 shows a typical control chart.

Examples of quality characteristics include average length, average diameter, average tensile strength, average resistance, and average service time. These characteristics are *variables,* and numerical values can be obtained for each. The term *attribute* applies to such quality characteristics as the proportion of nonconforming items, the number of nonconformities in a unit, and the number of demerits per unit.

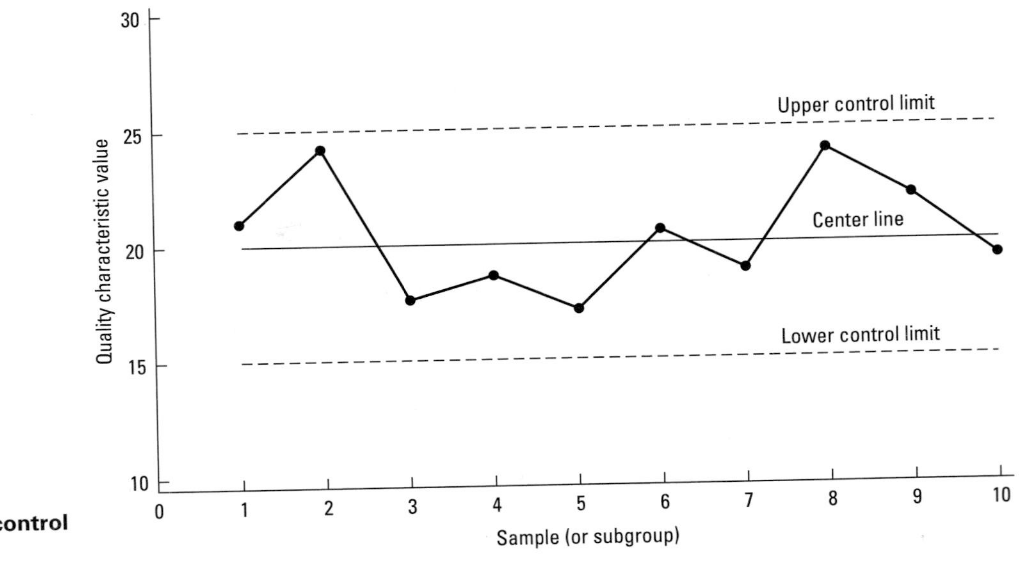

FIGURE 6-1 Typical control chart.

Three lines are indicated on the control chart. The **center line,** which typically represents the average value of the characteristic being plotted, is an indication of where the process is centered. Two limits, the **upper control limit** and the **lower control limit,** are used to make decisions regarding the process. If the points plot within the control limits and do not exhibit any identifiable pattern, the process is said to be in *statistical control.* If a point plots outside the control limits or if an identifiable nonrandom pattern exists (such as 12 out of 14 successive points plotting above the center line), the process is said to be out of statistical control. Details are given in Section 6-5 on the rules for identifying out-of-control conditions.

Several benefits can be realized by using control charts. Such charts indicate the following:

1. *When to take corrective action.* A control chart indicates when something may be wrong so that corrective action can be taken.
2. *Type of remedial action necessary.* The pattern of the plot on a control chart diagnoses possible causes and hence indicates possible remedial actions.
3. *When to leave a process alone.* Variation is part of any process. A control chart shows when an exhibited variability is normal and inherent such that no corrective action is necessary. As explained in Chapter 2, inappropriate overcontrol through frequent adjustments only increases process variability.
4. *Process capability.* If the control chart shows a process to be in statistical control, we can estimate the capability of the process and hence its ability to meet customer requirements. This helps product and process design.
5. *Possible means of quality improvement.* The control chart provides a baseline for instituting and measuring quality improvement. Control charts also provide useful information regarding actions to take for quality improvement.

6-2 CAUSES OF VARIATION

Variability is a part of any process, no matter how sophisticated, so management and operators must understand it. Several factors over which we have some control, such as methods, equipment, people, materials, and policies, influence variability. Environmental factors also contribute to variability. The causes of variation can be subdivided into two groups—common causes and special causes. Control of a process is achieved through the elimination of special causes. Improvement of a process is accomplished through the reduction of common causes.

Special Causes

Variability caused by **special** or assignable **causes** is something that is not inherent in the process. That is, it is not part of the process as designed and does not affect all items. Special causes can be the use of a wrong tool, an improper raw material, or an operator error. If an observation falls outside the control limits or a nonrandom pattern is exhibited, special causes are assumed to exist, and the process is said to be out of control. One objective of a control chart is to detect the presence of special causes as soon as possible to allow appropriate corrective action. Once the special causes are eliminated through remedial actions, the process is again brought to a state of statistical control.

Deming believed that 15% of all problems are due to special causes. Actions on the part of both management and workers will reduce the occurrence of such causes.

Common Causes

Variability due to **common** or chance **causes** is something inherent to a process. It exists as long as the process is not changed and is referred to as the natural variation in a process. It is an inherent part of the process *design* and affects all items. This variation is the effect of many small causes and cannot be totally eliminated. When this variation is constant, we have what is known as a stable system of common causes. A process operating under a stable system of common causes is said to be in **statistical control.** Examples include inherent variation in incoming raw material from a qualified vendor, a lack of adequate supervision skills, the vibration of machines, and fluctuations in working conditions.

Management alone is responsible for common causes. Deming believed that about 85% of all problems are due to common causes and hence can be solved only by action on the part of management. In a control chart, if quality characteristic values are within the control limits and no nonrandom pattern is visible, it is assumed that a system of common causes exists and that the process is in a state of statistical control.

6-3 STATISTICAL BASIS FOR CONTROL CHARTS

Basic Principles

A control chart has a center line and lower and upper control limits. The center line is usually found in accordance with the data in the samples. It is an indication of the mean of a process and is usually found by taking the average of the values in the sample. However, the center line can also be a desirable target or standard value.

Normal distributions play an important role in the use of control charts (Duncan, 1986). The values of the statistic plotted on a control chart (for example, average diameter) are assumed to have an approximately normal distribution. For large sample sizes or for small sample sizes with a population distribution that is unimodal and close to symmetric, the Central Limit Theorem states that if the plotted statistic is a sample average, it will tend to have a normal distribution. Thus, even if the parent population is not normally distributed, control charts for averages and other related statistics are based on normal distributions.

The control limits are two lines, one above and one below the center line, that aid in the decision-making process. These limits are chosen so that the probability of the sample points falling between them is almost 1 (usually about 99.7% for 3σ limits) if the process is in statistical control. As discussed previously, if a system is operating under a stable system of common causes, it is assumed to be in statistical control. Typical control limits are placed at 3 standard deviations away from the mean of the statistic being plotted. Normal distribution theory states that a sample statistic will fall within the limits 99.74% of the time if the process is in control. If a point falls outside the control limits, there is a reason to believe that a special cause exists in the system. We must then try to identify the special cause and take corrective action to bring the process back to control.

The most common basis for deciding whether a process is out of control is the presence of a sample statistic outside the control limits. Other rules exist for determining out-of-control process conditions, and are discussed in Section 6-5. These rules focus on nonrandom or systematic behavior of a process as evidenced by a nonrandom plot pattern. For example, if seven successive points plot above the center line but within the upper control limit, there is a reason to believe that something might be wrong with the process. If the process were in control, the chances of this happening would be extremely small. Such a pattern might suggest that the process mean has shifted upward. Hence, appropriate actions would need to be identified in order to lower the process mean.

A control chart is a means of **on-line process control.** Data values are collected for a process, and the appropriate sample statistics (such as sample mean, sample range, or sample standard deviation) based on the quality characteristic of interest (such as diameter, length, or strength) are obtained. These sample statistics are then plotted on a control chart. If they fall within the control limits and do not exhibit any systematic or nonrandom pattern, the process is judged to be in statistical control. If the control limits are calculated from current data, the chart tells us whether the process is presently in control. If the control limits were calculated from previous data based on a process that was in control, the chart can be used to determine whether the current process has drifted out of control.

Control charts are important management control tools. If management has some target value in mind for the process mean (say, average part strength), a control chart can be constructed with that target value as the center line. Sample statistics, when plotted on the control chart, will show how close the actual process output comes to the desired standard. If the deviation is unsatisfactory, management will have to come up with remedial actions.

Control charts help management set realistic goals. For example, suppose the output of a process shows that the average part strength is 3000 kg, with a standard deviation of 100 kg. If management has a target average strength of at least 3500 kg, the control chart will indicate that such a goal is unrealistic and may not be feasible for the existing process. Major changes in the system and process, possibly only through action on the part of management, will be needed to create a process that will meet the desired goal.

If a process is under statistical control, then control chart information can estimate such process parameters as the mean, standard deviation, and the proportion of nonconforming items (also known as fallout). These estimates can then be used to determine the capability of the process. **Process capability** refers to the ability of the process to produce within desirable specifications (ASQC 1987; Montgomery 1996). Conclusions drawn from studies on process capability have a tremendous influence on major management decisions such as whether to make or buy, how to direct capital expenditures for machinery, how to select and control vendors, and how to implement process improvements to reduce variability. Process capability is discussed in Chapter 9.

For variables, the value of a quality characteristic is measurable numerically. Control charts for variables are constructed to show measures of central tendency as well as dispersion. Variable control charts display such information as sample mean, sample range, sample standard deviation, cumulative sum, individual values, and moving average. Control charts for variables are described in Chapter 7. Attributes, on the other hand, indicate the presence or absence of a condition. Typical attribute charts deal with the fraction of nonconforming items, the number of nonconforming items, the total number of nonconformities, the number of nonconformities per unit, or the number of demerits per unit. Control charts for attributes are described in Chapter 8.

There are several issues pertinent to the construction of a control chart: the number of items in a sample, the frequency with which data is sampled, how to minimize errors in making inferences, the analysis and interpretation of the plot patterns, and rules for determining out-of-control conditions. We will discuss these issues in the following sections.

Selection of Control Limits

Let θ represent a quality characteristic of interest and $\hat{\theta}$ represent an estimate of θ. For example, if θ is the mean diameter of parts produced by a process, $\hat{\theta}$ would be the sample mean diameter of a set of parts chosen from the process. Let $E(\hat{\theta})$ represent the mean, or expected value, and let $\sigma(\hat{\theta})$ be the standard deviation of the estimator $\hat{\theta}$.

The center line and **control limits** for this arrangement are given by

$$\begin{aligned} \text{CL} &= E(\hat{\theta}) \\ \text{UCL} &= E(\hat{\theta}) + k\,\sigma(\hat{\theta}) \\ \text{LCL} &= E(\hat{\theta}) - k\,\sigma(\hat{\theta}) \end{aligned} \quad \textbf{(6.1)}$$

where k represents the number of standard deviations of the sample statistic that the control limits are placed from the center line. Typically, the value of k is chosen to be 3 (hence the name 3σ limits). If the sample statistic is assumed to have an approximately normal distribution, a value of $k = 3$ implies that there is a probability of only .0026 of a sample statistic falling outside the control limits if the process is in control.

Sometimes, the selection of k in eq. (6.1) is based on a desired probability of the sample statistic falling outside the control limits when the process is in control. Such limits are known as **probability limits.** For example, if we want the probability that the sample statistic will fall outside the control limits to be .002, then Appendix A-3 gives $k = 3.09$ (assuming the sample statistic is normally distributed). The probabilities of the sample statistic falling above the upper control limit and below the lower control limit are each equal to .001. Using this principle, the value of k and hence the control limits can be found for any desired probability.

The choice of k is influenced by error considerations also. As discussed in the next section, two types of errors (Type I and Type II) can be made in making inferences from control charts. The choice of a value k is influenced by how significant we consider the impact of such errors to be.

Example 6-1: A semiautomatic turret lathe machines the thickness of a part that is subsequently used in an assembly. The process mean is known to be 30 mm with a standard deviation of 1.5 mm. Construct a control chart for the average thickness using 3σ limits if samples of size 5 are randomly selected from the process. Table 6-1 shows the average thickness of 15 samples selected from the process. Plot these on a control chart, and make inferences.

Solution. The center line is

$$\text{CL} = 30 \text{ mm}$$

The standard deviation of the sample mean $\overline{X}$ is given by

$$\sigma_{\overline{X}} = \frac{\sigma}{\sqrt{n}} = \frac{1.5}{\sqrt{5}} = 0.671 \text{ mm}$$

Assuming a normal distribution of the sample mean thickness, the value of k in eq. (6.1) is selected as 3. The control limits are calculated as follows:

$$\text{UCL} = 30 + 3(0.671) = 32.013$$
$$\text{LCL} = 30 - 3(0.671) = 27.987$$

The center line and control limits are shown in Figure 6-2. The sample means for the 15 samples shown in Table 6-1 are plotted on this control chart. Figure 6-2 shows that all of the sample means are within the control limits. Also, the pattern of the plot does not exhibit any nonrandom behavior. Thus, we conclude that the process is in control.

Errors in Making Inferences from Control Charts

Making inferences from a control chart is analogous to testing a hypothesis. Suppose we are interested in testing the null hypothesis that the average diameter θ of a part from a particular process is 25 mm. This situation is represented by the null hypothesis H_0: $\theta = 25$; the alternative hypothesis is H_a: $\theta \neq 25$. The rejection region of the null hypothesis is thus two-tailed. The control limits are the critical points that separate the rejection and acceptance regions. If a sample value (sample average diameter, in this case) falls above the upper control limit or below the lower control limit, we reject the null hypothesis. In such a case, we conclude that the process mean differs from 25 mm and the process is therefore out of control. There are two types of errors—Type I and Type II—that can occur when making inferences from control charts.

TABLE 6-1 Average Part Thickness Values

Sample	*Average Part Thickness* $\overline{X}$ *(in mm)*	*Sample*	*Average Part Thickness* $\overline{X}$ *(in mm)*	*Sample*	*Average Part Thickness* $\overline{X}$ *(in mm)*
1	31.56	6	31.45	11	30.20
2	29.50	7	29.70	12	29.10
3	30.50	8	31.48	13	30.85
4	30.72	9	29.52	14	31.55
5	28.92	10	28.30	15	29.43

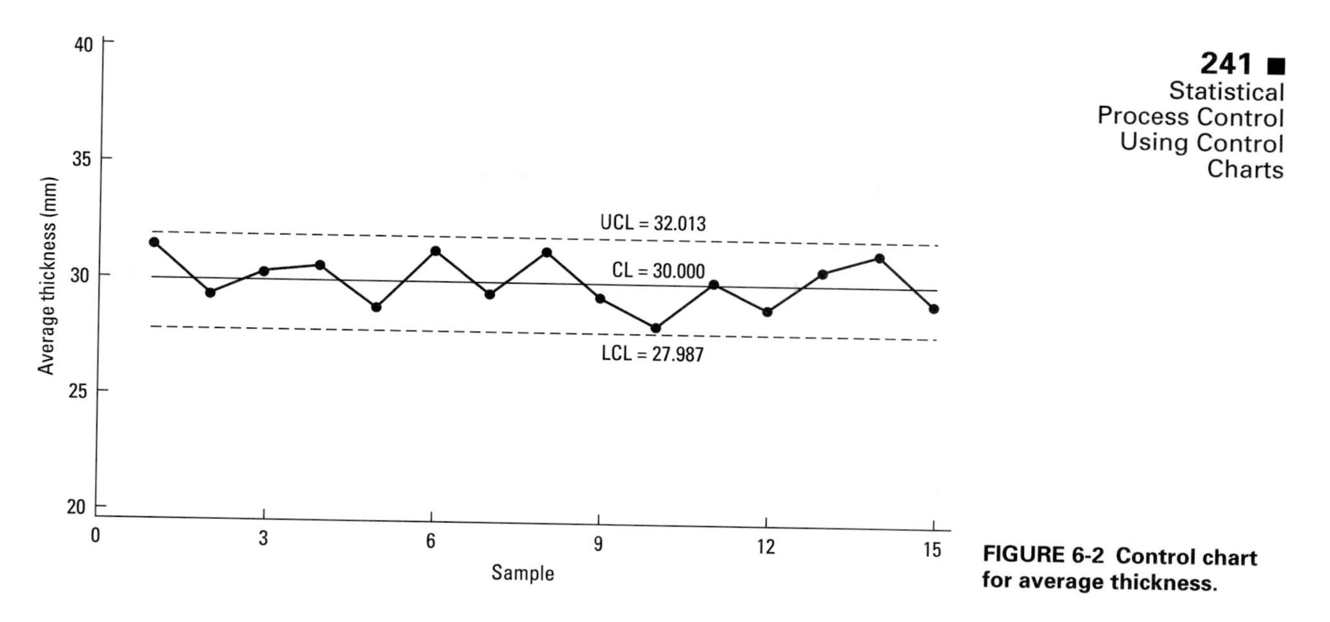

FIGURE 6-2 Control chart for average thickness.

Type I Errors

Type I errors result from inferring that a process is out of control when it is actually in control. The probability of a Type I error is denoted by α. Suppose a process is in control. If a point on the control chart falls outside the control limits, we assume that the process is out of control. However, since the control limits are a finite distance (usually 3 standard deviations) from the mean, there is a small chance (about .0026) of a sample statistic falling outside the control limits. In such instances, inferring that the process is out of control is a wrong conclusion. Figure 6-3 shows the probability of making a Type I error in control charts. It is the sum of the two tail areas outside the control limits.

Type II Errors

Type II errors result from inferring that a process is in control when it is really out of control. If no observations fall outside the control limits, we conclude that the process is in control. Suppose, however, that a process is actually out of control. Perhaps the process mean has changed (say, an operator has inadvertently changed a depth of

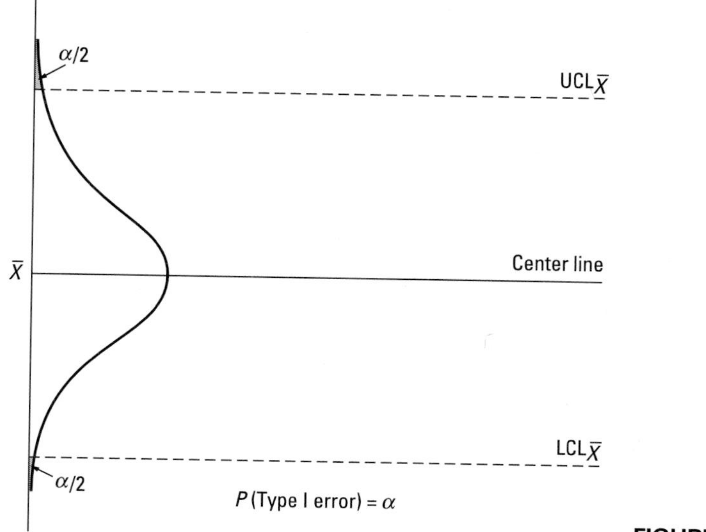

FIGURE 6-3 Type I error in control charts.

cut or the quality of raw materials has decreased). Or, the process could go out of control because the process variability has changed (due to the presence of a new operator). Under such circumstances, a sample statistic could fall within the control limits, yet the process would be out of control—this is a Type II error.

Let's consider Figure 6-4, which depicts a process going out of control due to a change in the process mean from A to B. For this situation, the correct conclusion is that the process is out of control. However, there is a strong possibility of the sample statistic falling within the control limits (as indicated by the shaded area), in which case we would conclude that the process is in control and thus make a Type II error.

Example 6-2: A control chart is to be constructed for the average breaking strength of nylon fibers. Samples of size 5 are randomly chosen from the process. The process mean and standard deviation are estimated to be 120 kg and 8 kg, respectively.

a. If the control limits are placed 3 standard deviations from the process mean, what is the probability of a Type I error?

Solution. From the problem statement, $\hat{\mu} = 120$ and $\hat{\sigma} = 8$. The center line for the control chart is at 120 kg. The control limits are

These limits are shown in Figure 6-5a.

Since the control limits are 3 standard deviations from the mean, the standardized normal value at the upper control limit is

$$Z = \frac{\overline{X} - \mu}{\sigma_{\overline{X}}}$$

$$= \frac{130.733 - 120}{8/\sqrt{5}} = 3.00$$

Similarly, the Z-value at the lower control limit is –3.00. For these Z-values in the standard normal table in Appendix A-3, each tail area is found to be .0013. The probability of a Type I error, as shown by the shaded tail areas in Figure 6-5a, is therefore .0026.

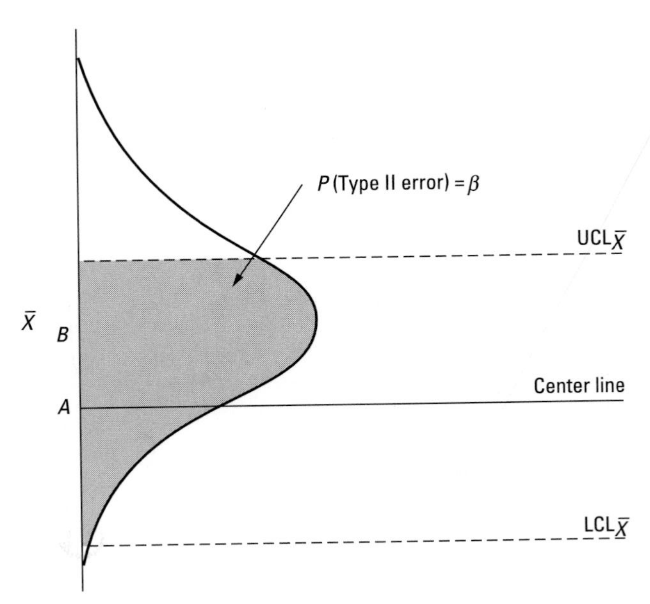

FIGURE 6-4 Type II error in control charts.

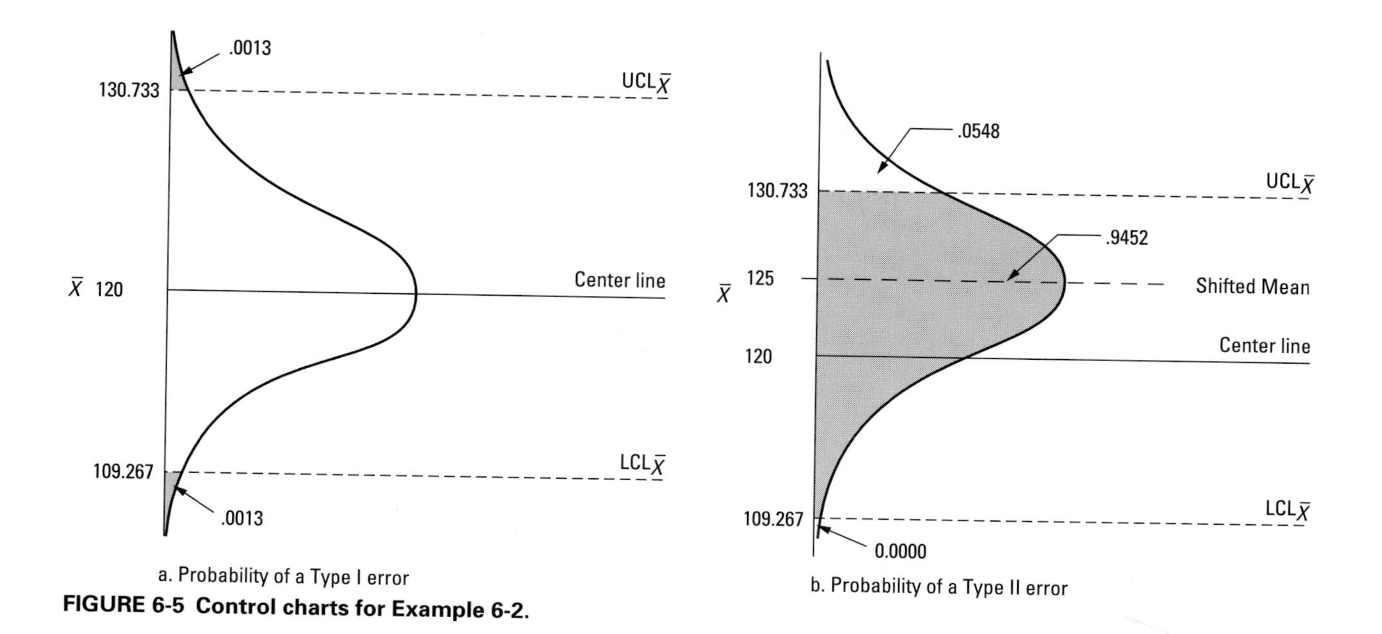

FIGURE 6-5 Control charts for Example 6-2.

b. If the process mean shifts to 125 kg, what is the probability of concluding that the process is in control and hence making a Type II error on the first sample plotted after the shift?

Solution. The process mean shifts to 125 kg. Assuming that the process standard deviation is the same as before, the distribution of the sample means is shown in Figure 6-5b. The probability of concluding that the process is in control is equivalent to finding the area between the control limits under the distribution shown in Figure 6-5b. We find the standardized normal value at the upper control limit as

$$Z_1 = \frac{130.733 - 125}{8/\sqrt{5}} = 1.60$$

From the standard normal table in Appendix A-3, the tail area above the upper control limit is .0548. The standardized normal value at the lower control limit is

$$Z_2 = \frac{109.267 - 125}{8/\sqrt{5}} = -4.40$$

From Appendix A-3, the tail area below the lower control limit is approximately .0000. The area between the control limits is $1 - (.0548 + .0000) = .9452$. Hence, the probability of concluding that the process is in control and making a Type II error is .9452 or 94.52%. This implies that for a shift of this magnitude, there is a pretty good chance of not detecting it in the first sample drawn after the shift.

c. What is the probability of detecting the shift by the second sample plotted after the shift if the samples are chosen independently?

Solution. The probability of detecting the shift by the second sample is *P*(Detecting shift on sample 1) + *P*(Not detecting shift in sample 1 and detecting shift in sample 2). This first probability was found in part b to be .0548. The second probability, using eqs. (4.2) and (4.5), is found to be $(1 - .0548)(.0548) = .0518$, assuming independence of the two samples. The total probability is $.0548 + .0518 = .1066$. Thus, there is a 10.66% chance of detecting a shift in the process by the second sample.

Operating Characteristic Curve

An **operating characteristic (OC) curve** is a measure of goodness of a control chart's ability to detect changes in process parameters. Specifically, it is a plot of the probability of the Type II error versus the shifting of a process parameter value from its in-control value. OC curves enable us to determine the chances of not detecting a shift of a certain magnitude in a process parameter on a control chart.

A typical OC curve is shown in Figure 6-6. The shape of an OC curve is similar to an inverted S. For small shifts in the process mean, the probability of nondetection is high. As the change in the process mean increases, the probability of nondetection decreases; that is, it becomes more likely that we will detect the shift. For large changes, the probability of nondetection is very close to zero. The ability of a control chart to detect changes quickly is indicated by the steepness of the OC curve and the quickness with which the probability of nondetection approaches zero. Calculations for constructing an operating characteristic curve are identical to those for finding the probability of a Type II error.

Example 6-3: Refer to the data in Example 6-2 involving the control chart for the average breaking strength of nylon fibers. Samples of size 5 are randomly chosen from a process whose mean and standard deviation are estimated to be 120 kg and 8 kg, respectively. Construct the operating characteristic curve for increases in the process mean from 120 kg.

Solution. A sample calculation for the probability of not detecting the shift when the process mean increases to 125 kg is given in Example 6-2. This same procedure is used to calculate the probabilities of nondetection for several values of the process mean. Table 6-2 displays some sample calculations. The vertical axis of the operating characteristic curve in Figure 6-6 represents the probabilities of nondetection given in Table 6-2 (these values are also the probabilities of a Type II error). The graph shows that for changes in the process mean exceeding 15 kg, the probability of nondetection is fairly small (less than 10%), while shifts of 5 kg have a high probability (over 85%) of nondetection.

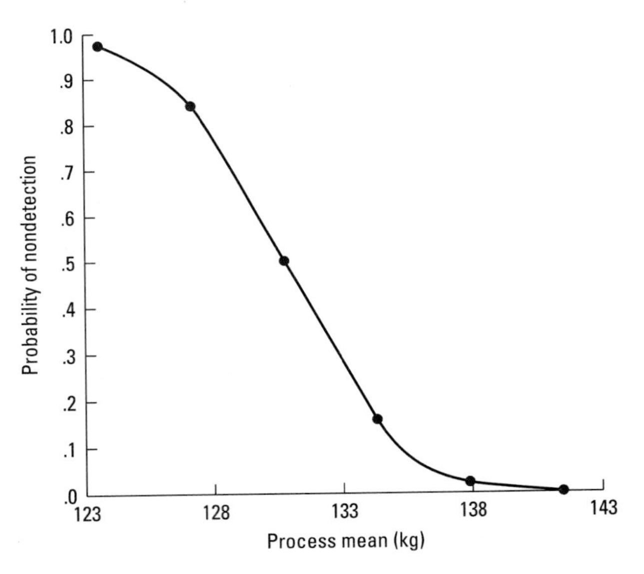

FIGURE 6-6 Operating characteristic curve for a control chart.

TABLE 6-2 Probabilities for OC Curve

Process Mean	*Z-value at UCL,* Z_1	*Area Above UCL*	*Z-value at LCL,* Z_2	*Area Below LCL*	*Probability of Nondetection,* β
123.578	2.00	.0228	−4.00	.0000	.9772
127.156	1.00	.1587	−5.00	.0000	.8413
130.733	0.00	.5000	−6.00	.0000	.5000
134.311	−1.00	.8413	−7.00	.0000	.1587
137.888	−2.00	.9772	−8.00	.0000	.0228
141.466	−3.00	.9987	−9.00	.0000	.0013

Effect of Control Limits on Errors in Inference Making

The choice of the control limits influences the likelihood of the occurrence of Type I and Type II errors. As the control limits are placed farther apart, the probability of a Type I error decreases (refer to Figure 6-3). For control limits placed 3 standard deviations from the center line, the probability of a Type I error is about .0026. For control limits placed 2.5 standard deviations from the center line, Appendix A-3 gives the probability of a Type I error as .0124. On the other hand, for control limits placed 4 standard deviations from the mean, the probability of a Type I error is negligible. If a process is in control, the chance of a sample statistic falling outside the control limits decreases as the control limits expand. Hence, the probability of making a Type I error decreases, too. The control limits could be placed sufficiently far apart, say 4 or 5 standard deviations on each side of the center line, to reduce the probability of a Type I error, but doing so affects the probability of making a Type II error.

Moving the control limits has the opposite effect on the probability of a Type II error. As the control limits are placed farther apart, the probability of a Type II error increases (refer to Figure 6-4). Ideally, to reduce the probability of a Type II error, we would tend to have the control limits placed closer to each other. But this, of course, has the detrimental effect of increasing the probability of a Type I error. Thus, the two types of errors are inversely related to each other as the control limits change. As the probability of a Type I error decreases, the probability of a Type II error increases.

If all other process parameters are held fixed, the probability of a Type II error will decrease with an increase in sample size. As n increases, the standard deviation of the sampling distribution of the sample mean decreases. Thus, the control limits will be drawn closer, and the probability of a Type II error will be reduced. Figure 6-7 demonstrates this effect. The new sample size is larger than the old sample size. The sampling distribution of the new sample mean has a reduced variance, so the new control limits are closer to each other. As can be seen from the figure, the probability of a Type II error is smaller for the larger sample.

Because of the inverse relationship between Type I and Type II errors, a judicious choice of control limits is desirable. In the majority of uses, the control limits are placed at 3 standard deviations from the center line, thereby restricting the probability of a Type I error to .0026. The reasoning behind this choice of limits is that the chart user does not want to look unnecessarily for special causes in a process when there are none. By placing the control limits at 3 standard deviations, the probability of a false alarm is small, and minimal resources will be spent on locating nonexistent problems with the process. However, the probability of a Type II error may be large for small shifts in the process mean.

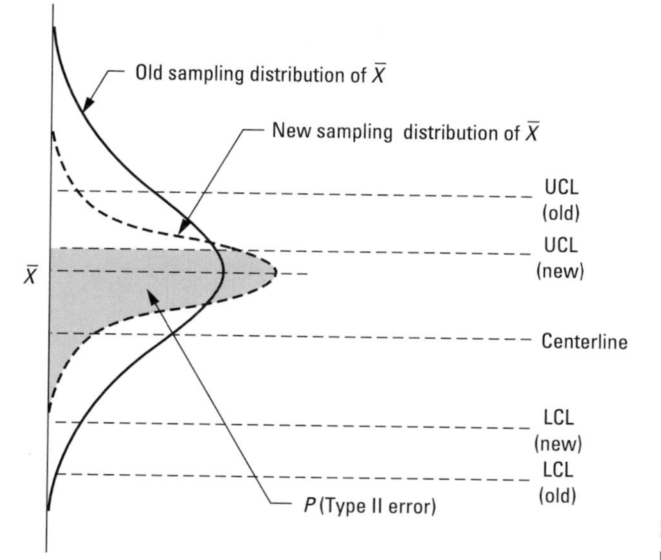

FIGURE 6-7 Effect of an increased sample size on the probability of a Type II error.

If it is more important to detect small changes in the process than to avoid spending time looking for nonexistent problems, it may be desirable to place the control limits closer (at, say, 2 or 2.5 standard deviations). For sophisticated processes, it is often crucial to detect small changes as soon as possible, because the impact on downstream activities is enormous if they are not detected right away. In this case, tighter control limits are preferable, even if this means incurring some costs for unnecessary investigation of problems when the process is in control.

Warning Limits

Warning limits are usually placed at 2 standard deviations from the center line. When a sample statistic falls outside the warning limits but within the control limits, the process is not considered to be out of control, but the users are now alerted that the process may be going out of control. For a normally distributed sample statistic, Appendix A-3 gives the probability of it falling in the band between the warning limit and the control limit to be .043 (that is, there is about a 4.3% chance of this happening). Thus, a sample statistic outside the warning limits is reason to be wary. If two out of three successive sample statistics fall within the warning/control limit on a given side, the process may indeed be out of control, because the probability of this happening in an in-control process is very small (.0027, obtained from $2 \times 3 \times .0215 \times .0215 \times .9785$).

Effect of Sample Size on Control Limits

The **sample size** usually has an influence on the standard deviation of the sample statistic being plotted on the control chart. For example, consider a control chart for the sample mean $\overline{X}$. The standard deviation of $\overline{X}$ is given by

$$\sigma_{\overline{X}} = \frac{\sigma}{\sqrt{n}}$$

where σ represents the process standard deviation and n is the sample size. We see that the standard deviation of $\overline{X}$ is inversely related to the square root of the sample size. Since the control limits are placed a certain number of standard deviations (say, 3) from the center line, an increase in the sample size causes the control limits to be drawn closer. Similarly, decreasing the sample size causes the limits to expand. Increasing the sample size provides more information, intuitively speaking, and causes the sample statistics to have less variability. A lower variability reduces the frequency with which errors occur in making inferences.

Average Run Length

An alternative measure of the performance of a control chart, in addition to the OC curve, is the **average run length (ARL).** This denotes the number of samples, on average, required to detect an out-of-control signal. Suppose the rule used to detect an out-of-control condition is a point plotting outside the control limits. Let P_d denote the probability of an observation plotting outside the control limits. Then, the run length is 1 with a probability of P_d, 2 with a probability of $(1 - P_d)P_d$, 3 with a probability of $(1 - P_d)^2 P_d$, and so on. The average run length is given by

$$\text{ARL} = \sum_{j=1}^{\infty} j(1 - P_d)^{j-1} P_d$$

$$= P_d \sum_{j=1}^{\infty} j(1 - P_d)^{j-1} \qquad \textbf{(6.2)}$$

The infinite series inside the summation is obtained from $1/[1-(1-P_d)]^2$. Hence, we have

$$\text{ARL} = \frac{P_d}{[1-(1-P_d)]^2} = \frac{1}{P_d} \tag{6.3}$$

For a process in control, P_d is equal to α, the probability of a Type I error. Thus, for three-sigma control charts with the selected rule for the detection of an out-of-control condition, ARL is $1/.0026 \simeq 385$. This indicates that an observation will plot outside the control limits every 385 samples, on average. For a process in control, we prefer the ARL to be large because an observation plotting outside the control limits represents a false alarm.

For an out-of-control process, it is desirable for the ARL to be small because we want to detect the out-of-control condition as soon as possible. Let's consider a control chart for the process mean and suppose that a change takes place in this parameter. In this situation, $P_d = 1 - \beta$, where β is the probability of a Type II error. So, $\text{ARL} = 1/(1-\beta)$.

We have computed β, the probability of a Type II error, for a control chart on changes in the process mean (see Figure 6-4). Because it is straightforward to develop a general expression for β in terms of the shift in the process mean (expressed in units of the process standard deviation, σ), we can also construct ARL curves for the control chart. Figure 6-8 shows ARL curves for sample sizes of 1, 2, 3, 4, 5, 7, 9, and 16 for a control chart for the mean where the shifts in the mean are shown in units of σ. Note that if we wish to detect a shift of 1.0σ in the process mean, using a sample of size 5, the

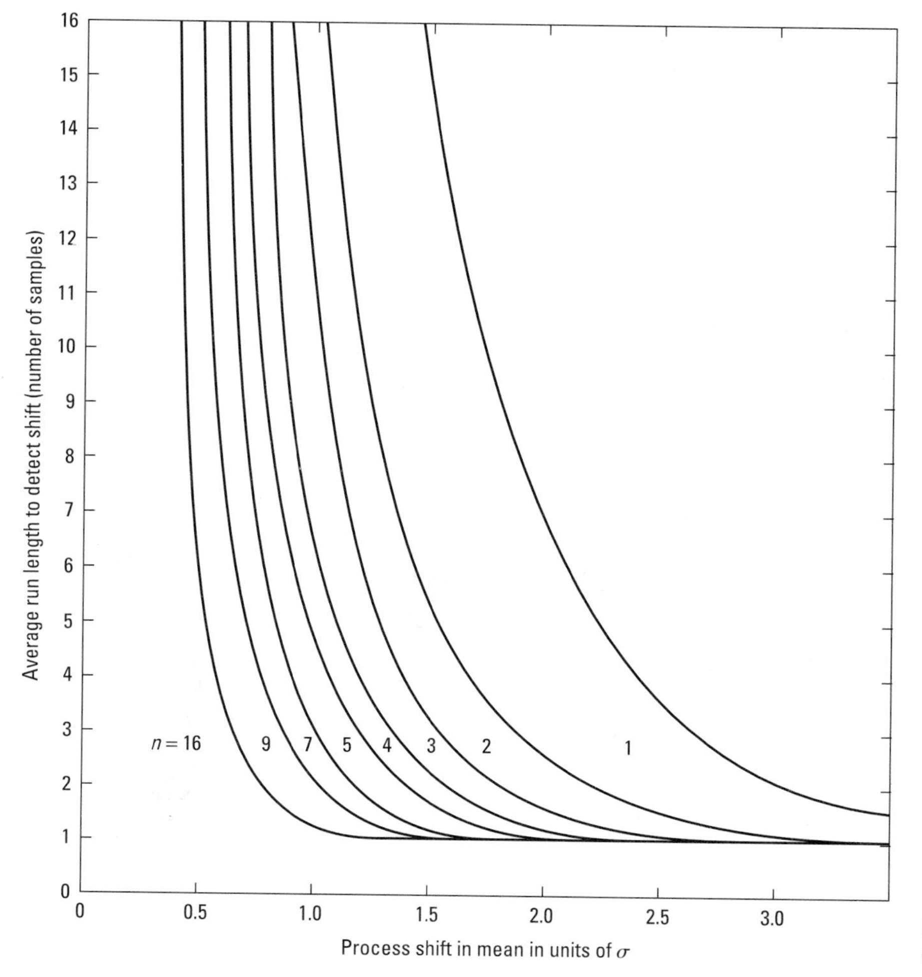

FIGURE 6-8 ARL curves for control charts for the mean.

TABLE 6-3 Computation of ARL for Changes in Process Mean

Process Mean	*Shift in Process Mean in Units of $\sigma_{\overline{X}}$*	*Shift in Process Mean in Units of σ*	P_d	*ARL*
123.578	1	0.7156	.0228	43.86
127.156	2	1.4312	.1587	6.30
130.733	3	2.1468	.5000	2.00
134.311	4	2.8624	.8413	1.19
137.888	5	3.5780	.9772	1.02
141.466	6	4.2936	.9987	1.00

average number of samples required will be about 4. In case this ARL is not suitable and we wish to reduce it, the sample size could be increased to 7, whereupon the ARL is reduced to approximately 3. When we need to express ARL in terms of the expected number of individual units sampled, *I,* the expression is

$$I = n(\text{ARL}) \tag{6.4}$$

where n denotes the sample size.

Example 6-4: Let's reconsider Example 6-3 on the average breaking strength of nylon fibers. Our sample size is 5. Table 6-2 shows calculations for β, the probability of nondetection, for different values of the process mean. Table 6-3 displays the values of ARL for each change in the process mean. The change in the process mean, from the in-control value of 120, is shown in multiples of the process standard deviation, σ.

From Table 6-3, we find that for shifts in the process mean of 3 or more standard deviations, the control chart is quite effective because the ARL is slightly above 1. This indicates that, on average, the out-of-control condition will be detected on the first sample drawn after the shift takes place. For a shift in the process mean of 2.15σ, the ARL is 2, while for a smaller shift in the process mean of 1.43σ, the ARL is above 6. These values of ARL represent a measure of the strength of the control chart in its ability to detect process changes quickly. For a small shift in the process mean of about 0.72σ, about 44 samples, on average, will be required to detect the shift.

6-4 SELECTION OF RATIONAL SAMPLES

Shewhart described the fundamental criteria for the selection of rational subgroups, or rational samples, the term we will use in this text. The premise is that a **rational sample** is chosen in such a manner that the variation within it is considered to be due only to common causes. So, samples are selected such that if special causes are present they will occur between the samples. Therefore, the differences *between* samples will be maximized, and differences *within* samples will be minimized.

In most cases, the sampling is done by time order. Let's consider a job shop with several machines. Samples are collected at random times from each machine. Control charts for the average value of the characteristic for each machine are plotted separately. If two operators are producing output, samples are formed from the output of each operator, and a separate control chart is plotted for each operator. If output between two shifts differs, then the two outputs should not be mixed in the sampling process. Rather, samples should first be selected from shift 1 and a control chart constructed to determine the stability of that shift's output. Next, rational samples are selected from shift 2 and a control chart constructed for this output.

Selection of the sample observations is done by the **instant-of-time method** (Besterfield 1990). Observations are selected at approximately the same time for the population under consideration. This method provides a time frame for each sample, which makes the identification of problems simpler. The instant-of-time method minimizes variability within a sample and maximizes variability between samples if special causes are present.

Sample Size

Selecting sample size—the number of items in each sample—is a necessity in using control charts. The degree of shift in the process parameter expected to take place will influence the choice of sample size. As noted in the discussion of operating characteristic curves, large shifts in a process parameter (say, the process mean) can be detected by smaller sample sizes than those needed to detect smaller shifts. Having an idea of the degree of shift we wish to detect enables us to select an appropriate sample size. If we can tolerate smaller changes in the process parameters, then a small sample size might suffice. Alternatively, if it is important to detect slight changes in process parameters, a larger sample size will be needed.

Frequency of Sampling

The **sampling frequency** must be decided prior to the construction of control charts. Choosing large samples very frequently is the sampling scheme that provides the most information. However, this is not always feasible because of resource constraints. Other options include choosing small sample sizes at frequent intervals or choosing large sample sizes at infrequent intervals. In practice, the former is usually adopted.

Other factors also influence the frequency of sampling and the sample size. The type of inspection needed to obtain the measurement—that is, destructive or nondestructive—can be a factor. The current state of the process (in control or out of control) is another factor. If the process is stable, we might get by with sampling at infrequent intervals. However, for processes that indicate greater variability, we would need to sample more frequently.

The cost of sampling and inspection per unit is another area of concern. The choice of sample size is influenced by the loss incurred due to a nonconforming item being passed on to the consumer. These intangible costs are sometimes hard to identify and quantify. Because larger sample sizes detect shifts in process parameters sooner than smaller sample sizes, they can be the most cost-effective choice.

6-5 ANALYSIS OF PATTERNS IN CONTROL CHARTS

One of the main objectives of using control charts is to determine when a process is out of control so that necessary actions may be taken. Criteria other than a plotted point falling outside the control limits are also used to determine whether a process is out of control. We discuss some **rules for out-of-control processes** next.

Later, we examine some typical control chart patterns and the reasons for their occurrence. As mentioned previously, plot patterns often indicate whether the process is in control or not; a systematic or nonrandom pattern suggests an out-of-control process. Analyzing these patterns is more difficult than plotting the chart. Identifying the causes of nonrandom patterns requires a knowledge of the process, equipment, and operating conditions as well as their impact on the characteristic of interest.

Some Rules for Identifying an Out-of-Control Process

RULE 1

A process is assumed to be out of control if a single point plots outside the control limits.

This is the most commonly used rule. If the control limits are placed at 3 standard deviations from the mean of the quality characteristic being plotted (assuming a normal distribution), the probability of a point falling outside these limits if the process is in control is very small (about .0026). Figure 6-9 depicts this situation.

RULE 2

A process is assumed to be out of control if two out of three consecutive points fall outside the 2σ warning limits on the same side of the center line.

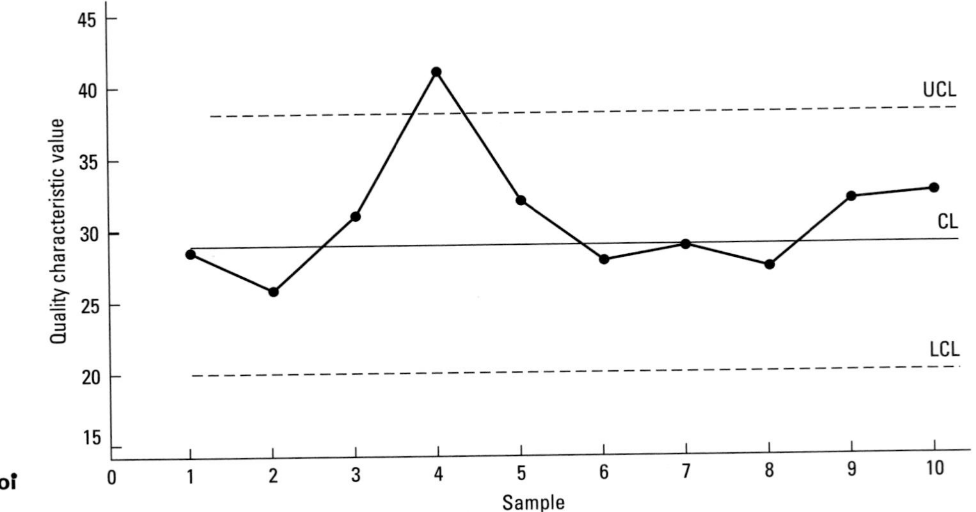

FIGURE 6-9 Out-of-control patterns: Rule 1.

As noted in Section 6-3, warning limits at 2 standard deviations of the quality characteristic from the center line can be constructed. These are known as 2σ limits. If the process is in control, the chance of two out of three points falling outside the warning limits is small. In Figure 6-10, observe that samples 7 and 9 fall above the upper 2σ limit. We can infer that this process has gone out of control, so special causes should be investigated.

RULE 3

A process is assumed to be out of control if four out of five consecutive points fall beyond the 1σ limit on the same side of the center line.

If the control limits are first determined, then the standard deviation can be calculated. Note that the distance between the center line and the upper control limit is 3 standard deviations (assuming 3σ limits). Dividing this distance by 3 gives the standard deviation of the characteristic being plotted. Adding and subtracting this standard deviation from the center line value gives the 1σ limits. Consider Figure 6-11, for which samples 4, 5, 6, and 8 plot below the lower 1σ limit. Based on Rule 3, this process would be considered out of control.

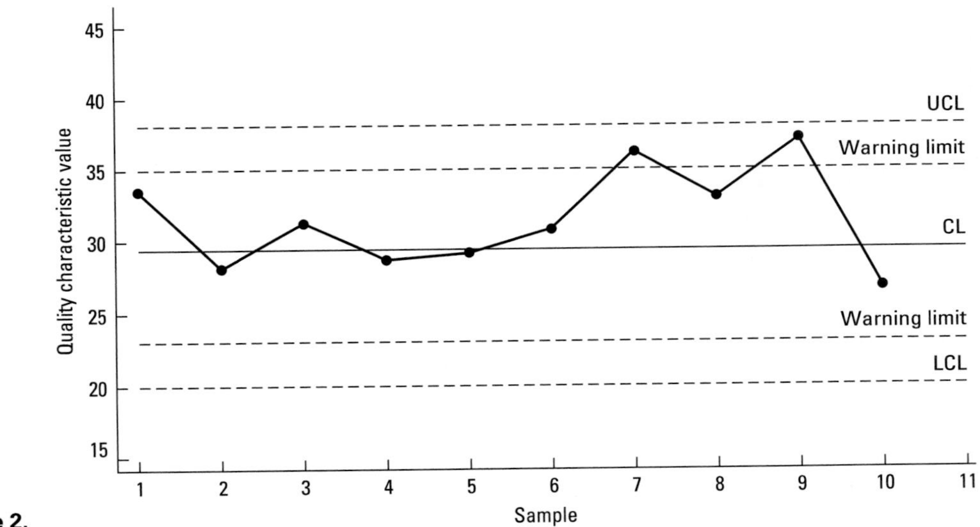

FIGURE 6-10 Out-of-control patterns: Rule 2.

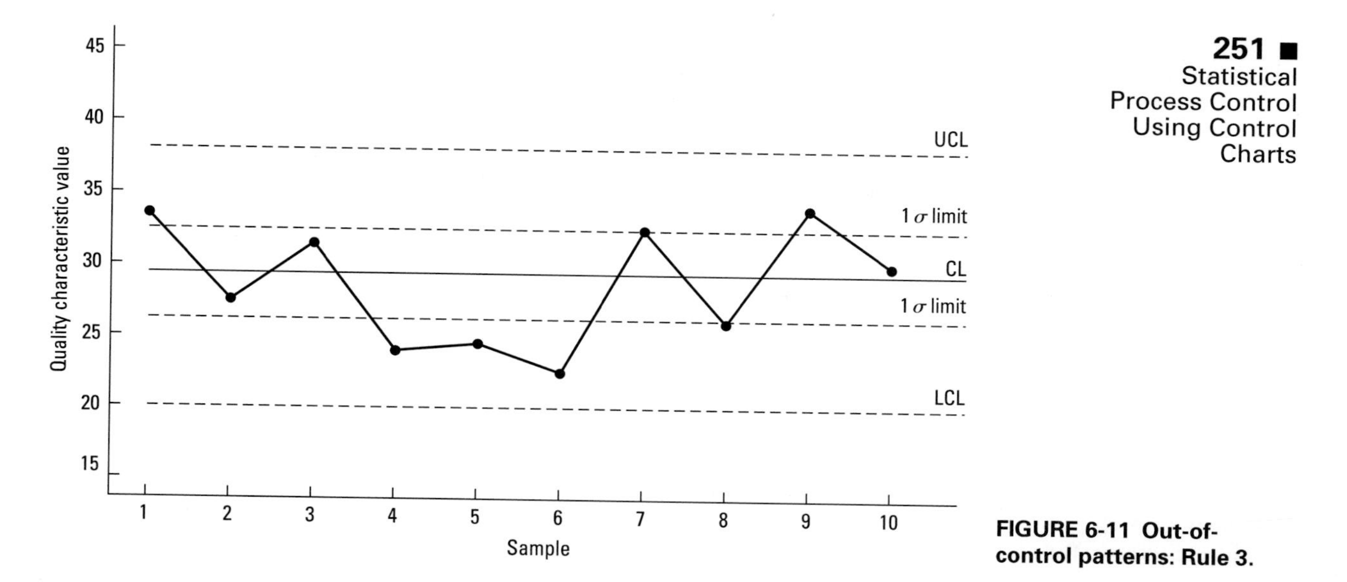

FIGURE 6-11 Out-of-control patterns: Rule 3.

RULE 4

A process is assumed to be out of control if nine or more consecutive points fall to one side of the center line.

For a process in control, a roughly equal number of points should be above or below the center line, with no systematic pattern visible. The condition stated in Rule 4 is highly unlikely if a process is in control. For instance, if nine or more consecutive points plot above the center line on an $\overline{X}$-chart, an upward shift in the process mean may have occurred. In Figure 6-12, samples 2, 3, 4, 5, 6, 7, 8, 9, and 10 plot above the center line. The process is assumed to be out of control.

RULE 5

A process is assumed to be out of control if there is a run of six or more consecutive points steadily increasing or decreasing.

A run is a sequence of like observations. Thus, if three successive points increase in magnitude, we would have a run of three points. In Figure 6-13, samples 2–8 show a continual increase; so this process would be deemed out of control.

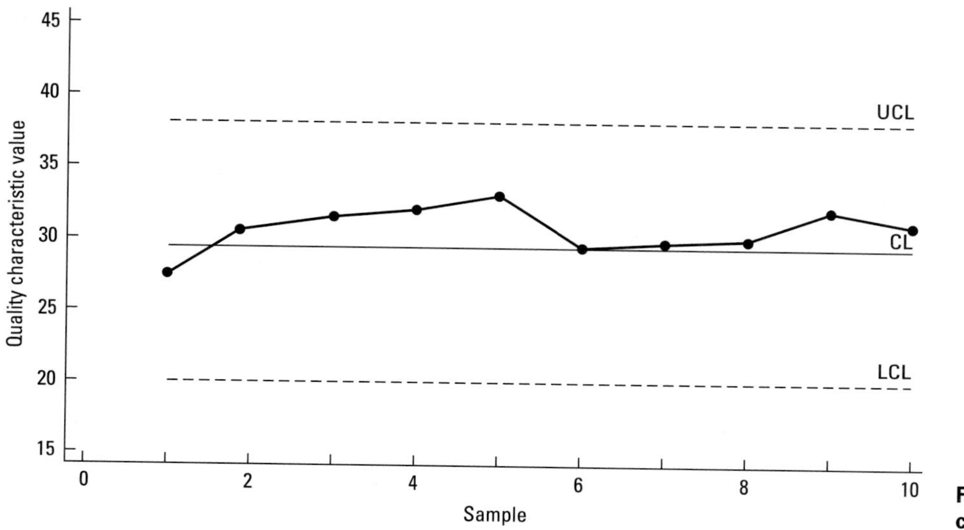

FIGURE 6-12 Out-of-control patterns: Rule 4.

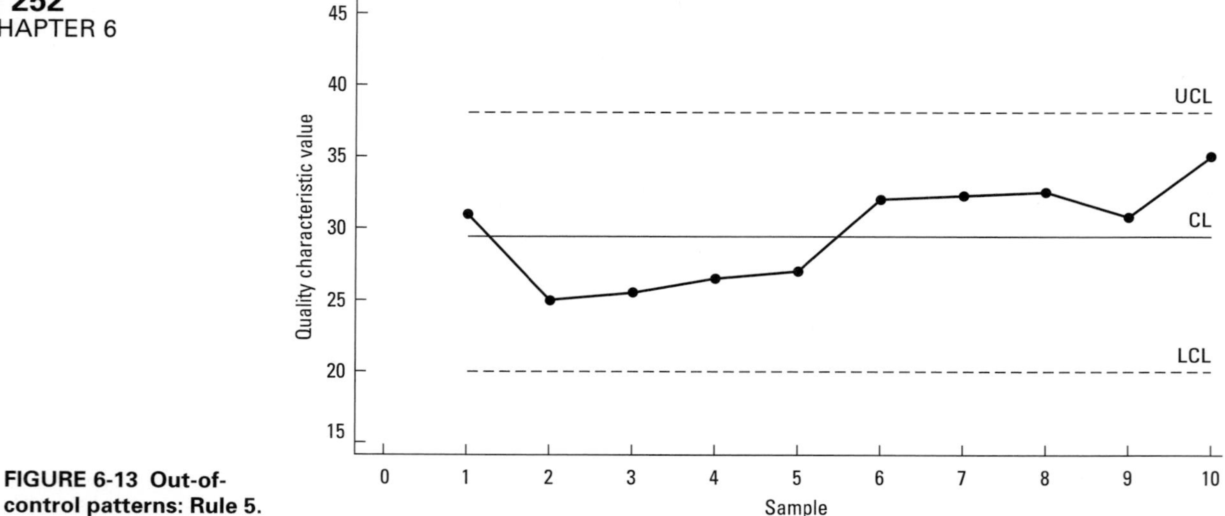

FIGURE 6-13 Out-of-control patterns: Rule 5.

Interpretation of Plots

The five rules for determining out-of-control conditions are not all used simultaneously. Rule 1 is routinely used along with a couple of the other rules (say, Rules 2 and 3). The reason for not using all of them simultaneously is that doing so increases the chance of a Type I error. In other words, the probability of a false alarm increases as more rules are used to determine an out-of-control state. Even though the probability of the stated condition occurring is rather small for any one rule with an in-control process, the **overall Type I error rate,** based on the number of rules that are used, may not be small.

Suppose the number of independent rules used for out-of-control criteria is *k*. Let α_i be the probability of a Type I error of rule *i*. Then, the overall probability of a Type I error is

$$\alpha = 1 - \prod_{i=1}^{k} (1 - \alpha_i) \tag{6.5}$$

Suppose four independent rules are being used to determine whether a process is out of control. Let the probability of a Type I error for each rule be given by $\alpha_1 = .005$, $\alpha_2 = .02$, $\alpha_3 = .03$, and $\alpha_4 = .05$. The overall false alarm rate, or the probability of a Type I error, would be

$$\alpha = 1 - (.995)(.98)(.97)(.95) = .101$$

If several more rules were used simultaneously, the probability of Type I error would become too large to be acceptable. Note that the relationship in eq. (6.5) is derived under the assumption that the rules are independent. The rules, however, are not independent, so eq. (6.5) is only an approximation to the probability of a Type I error. For more information, see Walker, Philpot, and Clement (1991).

Using many rules for determining out-of-control conditions complicates the decision process and sabotages the purpose of using control limits. One of the major advantages of control charts is that they are easy to construct, interpret, and use.

In addition to the rules we've been discussing, there are many other nonrandom patterns that a control chart user has to judiciously interpret. It is possible for the process to be out of control yet none of the five rules to be applicable. This is where experience, judgment, and interpretive skills come into play. Consider, for example, Figure 6-14. None of the five rules for out-of-control conditions apply even though the pattern is clearly nonrandom. The systematic nature of the plot and the somewhat

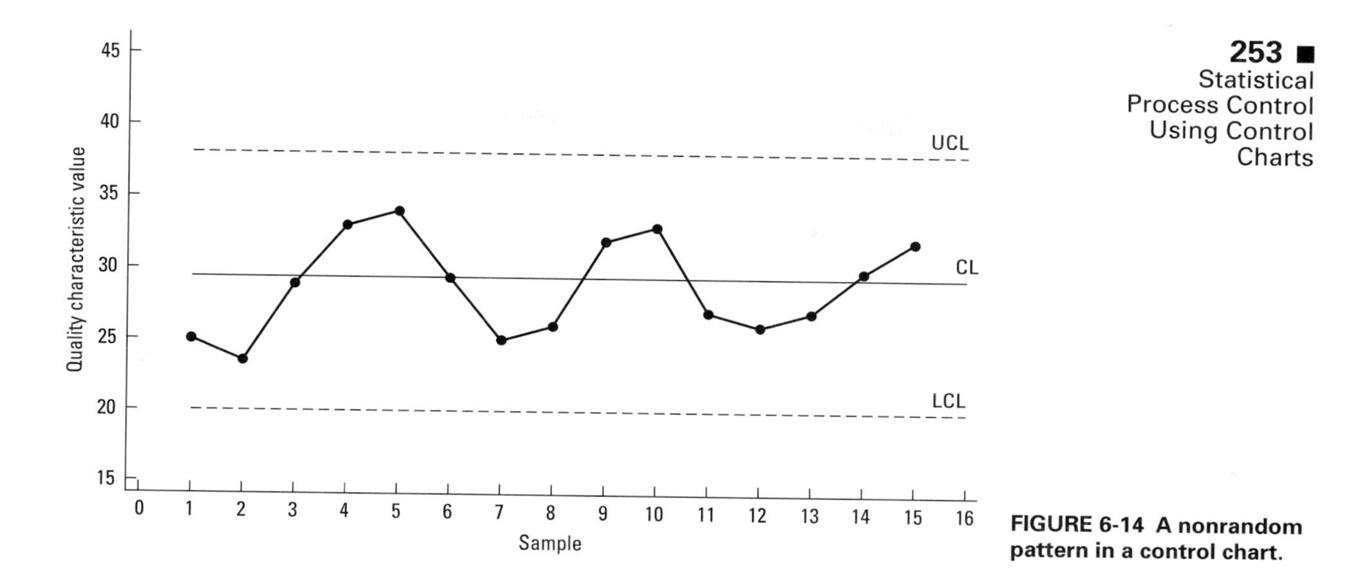

FIGURE 6-14 A nonrandom pattern in a control chart.

cyclic behavior are important clues. This pattern probably means that special causes are present and the process is out of control. You should always keep an eye out for nonrandom patterns when you examine control charts.

Determination of Causes of Out-of-Control Points

The task of the control chart user does not end with the identification of out-of-control points. In fact, the difficult part begins when out-of-control points have been determined. Now we must pinpoint the causes associated with these points—not always a trivial task. This requires a thorough knowledge of the process and the sensitivity of the output quality characteristic to the process parameters.

Determination of cause is usually a collective effort, with people from product design, process design, tooling, production, and purchasing and vendor control involved. A cause-and-effect chart is often an appropriate tool here. Once special causes have been identified, appropriate remedial actions need to be proposed. Typical control chart patterns for out-of-control processes, along with their possible causes, are discussed in Chapter 7. These include a sudden shift in the pattern level, a gradual shift in the pattern level, a cyclic pattern, or a mixture pattern, among others.

6-6 MAINTENANCE OF CONTROL CHARTS

Although the construction of control charts is an important step in statistical process control, it should be emphasized that quality control and improvement are an ongoing process. Therefore, implementation and **control chart maintenance** are a vital link in the quality system. When observations plotted on control charts are found to be out of control, the center line and control limits need to be revised; this, of course, will eliminate these out-of-control points. There are some exceptions to the elimination of out-of-control points, however, especially for points below the lower control limit. These will be discussed in Chapters 7 and 8. Once the computation of the revised center line and control limits is completed, these lines are drawn on charts where future observations are to be plotted. The process of revising the center line and control limits is ongoing.

Proper placement of the control charts on the shop floor is important. Each person who is associated with a particular quality characteristic should have easy access to the chart. When a statistical process control system is first implemented, the chart is usually placed in a conspicuous place where operators can look at it. The involvement of everyone, from the operator to the manager to the chief executive officer, is essential to the success of the program. The control charts should be attended to by everyone involved. Proper maintenance of these charts on a regular ongoing basis helps ensure the success of the quality systems approach. If a particular quality characteristic becomes insignificant, its control chart can be replaced by others that are relevant. Products, processes, vendors, and equipment change with time. Similarly, the different control charts that are kept should be chosen to reflect important characteristics of the current environment.

6-7 SUMMARY

This chapter has introduced the basic concepts of control charts for statistical process control. The benefits that can be derived from using control charts have been discussed. This chapter explains the statistical background for the use of control charts, the selection of the control limits, and the manner in which inferences can be drawn from the charts. It provides a discussion of the two types of errors that can be encountered in making inferences from control charts. Guidelines for the proper selection of sample size and rules for determining out-of-control conditions have been explored. Several control chart patterns have been studied with a focus on identifying possible special causes. Since this chapter is intended solely to explain the fundamentals of control charts, such technical details as formulas for various types of control charts have intentionally been omitted here. We will discuss these in Chapters 7 and 8, and you may also find them in Banks (1989), Montgomery (1996), and Wadsworth, Stephens, and Godfrey (1986).

CASE STUDY

SPC: What Data Should I Collect? What Charts Should I Use?*

Statistical process control (SPC) can help companies cut costs, improve quality, and pursue continuous improvement. However, those people unfamiliar with statistics who try to implement SPC usually have two questions: "What data should I collect?" and "What control chart can I use under some specific circumstances?" Employees in Boeing's Fabrication Division were confronted with these two questions and were able to answer them using two straightforward tools: the data definition process and the control chart decision tree.

WHAT DATA SHOULD I COLLECT?

In one of Boeing's foundries, a team was formed to ensure that the shop would meet its goals of continuous improvement and producing the highest quality castings using the most cost-effective methods. The team initially developed a process flow diagram (see Figure 6-15) to analyze and measure plan development.

Realizing that decisions are only as good as the data collected for analysis, the measurement plan started with the data definition process. The work sheet shown in Table 6-4 helped the team with this task.

The problem statement was "The defect rate in the foundry casting process results in unwanted scrap and rework." The foundry team used this statement to identify the quality characteristics that needed to be measured to improve the process: casting defect rate, scrap, and rework.

Then the team used the data definition model to define the data requirements. The model provides a plan to identify and collect the data required to ensure proper analysis and achieve optimal process improvement. The model focuses on four important aspects of data collection:

*Adapted from J. Munoz and C. Nielson (1991), "SPC: What Data Should I Collect? What Charts Should I Use?" *Quality Progress* 24(1): 50–52.

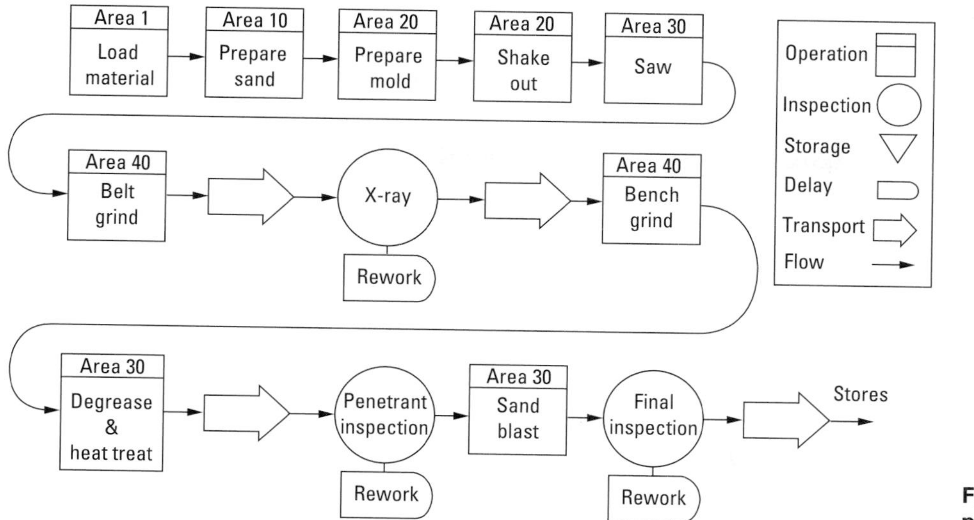

FIGURE 6-15 Simplified process flow diagram.

TABLE 6-4 Data Definition Work Sheet

Problem statement: The defect rate in foundry casting process results in unwanted scrap and rework.		
Quality characteristic(s): Defect rate, scrap, and rework.		
Scope of Study	*Data Definition*	*Parameters to Measure*
Boundaries? Starts: Shop load Ends: Final shop Population? All part orders processed on all shifts Time period? Three months to determine current status Continuously monitor performance	Attribute(s): In-process: Defects/part (identify defect type) Q.C.: Same as above Variable(s): n/a	Where: Saw, belt grind, bench grind Where: n/a
Charts/graphs? Run chart, u control chart, and Pareto chart Charting frequency? Monthly (quality characteristics) Daily (attribute) Type of comparisons? Part types, shift to shift Means of summarizing? Quality char.: Monthly average and totals Attribute: Weekly defect totals and types Statistics to calculate? Process average, totals, control limits, percentages Data analysis requirements	Who? In-process: Operators and shop leaders (recorded during shift and collected at end) How? Two automatic data collectors In-house developed software Sampling? Only at QC inspections Cost? Data collectors plus 0.5 labor hours/day DATA COLLECTION RESOURCES	

1. *Scope of study.* The scope defines the boundary and population of the process to be measured.
2. *Parameters to measure.* The parameters define the types of attributes and variables to be measured and identify where in the process they will be measured.
3. *Data analysis requirements.* The requirements define the methods and tools needed to analyze the data.
4. *Data collection resources.* The resources required to collect the data are defined.

Using the process flow diagram, the team defined each section of the model.

SCOPE OF STUDY

Boundaries: The casting process beginning with shop load and ending with the final inspection of the part (see Figure 6-15).

Population: Includes all part orders processed on all shifts.

Time period: Data will be collected for three months to determine which attribute has the greatest effect on the defect rate, scrap, and rework. Data collection will continue in order to monitor improvements and determine the next attribute to be eliminated.

PARAMETERS TO MEASURE

Attributes: Casting defects will be tracked at each foundry operation previously identified. Defects will be measured by defect types and defects per part. The same defect data will also be collected at three inspection points (X-ray, penetrant, and final).

Variables: No variable measures were needed.

DATA ANALYSIS REQUIREMENTS

Charts/graphs: Charts will include the run (trend) chart to display the monthly defect rate, scrap, and rework; the Pareto chart to display occurrence of defects; and the u-chart to monitor the number of defects per unit.

Charting frequency: The run chart will be compiled and reviewed monthly. Pareto charts will be produced weekly and monthly. The u-chart will be compiled daily.

Types of comparisons: Defect rates and types will be compared between part types. Shift-to-shift variations will also be compared.

Means of summarization: Data for the quality characteristics will be summarized as monthly averages and totals. Attribute data will be summarized weekly by total defects and types.

Statistics to calculate: Process average (defects per unit), defect totals (occurrence and type), control limits, and Pareto analysis.

DATA COLLECTION RESOURCES

Who: In-process data will be recorded by operators at defined data collection points. Shop leaders will collect data at shift end. QC data will be provided by QC personnel.

How: Automated data collectors will gather in-process and QC data; software developed in-house will chart and analyze the data.

Sampling: Sampling will be used only at QC inspection points. *Cost:* Data collection costs include the expense of two automated units and 0.5 labor hours per day.

Having completed the data definition process, the foundry team had a documented plan to collect data for measuring and improving the foundry casting process. All that was left was to get the resources and begin collecting data.

What Control Chart Should I Use?

During the data definition process, the foundry team struggled in deciding what type of statistical control chart it should use to monitor the variation in the process. The team was then introduced to the decision tree process (see Figure 6-16). In using the decision tree, the first item that must be identified is the type of data being collected: variable (measurable quality characteristics) or attribute (nonmeasurable but countable quality characteristics).

If variable data are used, two control charts are required: one to monitor the process average and one to monitor process variation. The control chart used to monitor the process average is the $\overline{X}$-chart. The chart used to monitor process variation is determined by the sample size n. When n is small, the R-chart (range) is generally preferred over the s-chart (sample standard deviation) because there is little difference in efficiency between the two charts and the R-chart is much simpler to use. As n increases, the efficiency of the R-chart as

FIGURE 6-16 Control chart decision tree.

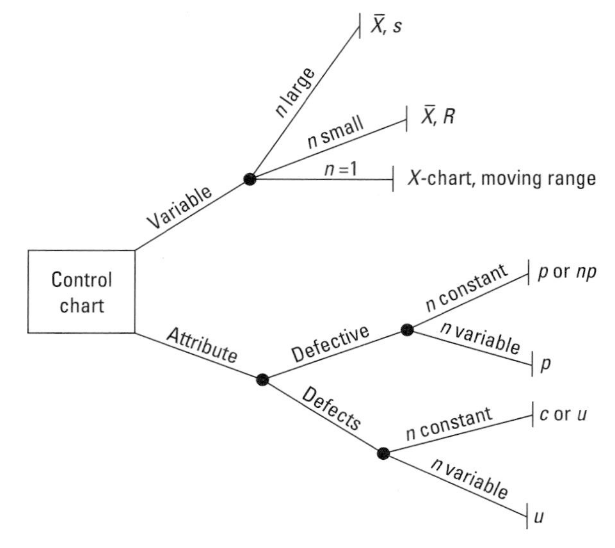

compared to that of the s-chart decreases. So, if n is large, the s-chart should be used instead of the R-chart.

Thus, n determines which variable control charts to use. If n is large (10 or more observations per sample), $\overline{X}$ and s-charts are desirable. If n is small (fewer than 10 observations per sample), $\overline{X}$ and R-charts are preferred. If only one observation per sample is possible, then the X-chart (control chart for individuals) and the moving range chart are appropriate.

If attribute data are collected, just one control chart is required in most cases. It must be decided whether defective units or number of defects are being counted. If the process is being monitored through the collection of data on nonconforming (defective) parts, the question is whether the sample size is constant. If it is constant, then a p-chart (fraction defective) or an np-chart (number of defective units) would work. But if the sample size varies, only the p-chart can be used, and the control limits will vary with sample size.

If the attribute is monitored through the collection of data on the number of defects (nonconformances or discrepancies), the condition of the sample size will again determine the type of chart. If n is constant, c-charts (number of defects per sample) or u-charts (number of defects per unit) can be used, but if n is variable, only the u-chart is acceptable.

The foundry team used this simple decision tree, making its search to find the most appropriate control chart (the u-chart) a lot easier.

QUESTIONS FOR DISCUSSION

1. Describe how information from the Pareto chart could influence the control chart to monitor the number of defects per unit.
2. What is the purpose of constructing a run chart and a u-chart?
3. Describe some characteristics, in this setting, for which a variables chart for the mean and range would be used.
4. Describe some attributes, in this setting, for which a control chart for the proportion nonconforming would be appropriate.
5. If casting nonconformities are to be observed by defect type, which may be prioritized, discuss the type of data to be collected and the control chart that should be constructed.
6. Discuss how shift-to-shift variability in defect rates may be monitored. Describe a statistical test that could be performed to determine whether there is a difference in the process average between two shifts.
7. What method would you suggest for identifying causes behind the production of scrap and rework?
8. Given the inspection points shown in the process flow diagram, discuss how easy or difficult it would be to identify the specific operation in the process that may cause a particular defect type.

Key Terms

- average run length (ARL)
- center line
- common cause
- control chart
- control chart maintenance
- control limits
  - upper control limit
  - lower control limit
- instant-of-time method
- on-line process control
- operating characteristic (OC) curve
- probability limits
- process capability
- rational samples
- remedial actions
- rules for out-of-control processes
- sample size
- sampling frequency
- Shewhart control charts
- special cause
- statistical control
- Type I error
  - overall rate
- Type II error
- warning limits

Exercises

Discussion Questions

1. What are the benefits of using control charts?
2. Explain the difference between common causes and special causes. Give examples of each.
3. Explain the rationale behind placing the control limits at 3 standard deviations from the mean.
4. Define and explain Type I and Type II errors in the context of control charts. Are they related?

How does the choice of control limits influence these two errors?

5. What are warning limits, and what purpose do they serve?
6. What is the utility of the operating characteristic curve? How can the discriminatory power of the curve be improved?
7. Describe the role of the average run length (ARL) in the selection of control chart parameters. Explain how ARL influences sample size.
8. Discuss the relationship between ARL and Type I and Type II errors.
9. How are rational samples selected? Explain the importance of this in the total quality systems approach.
10. State and explain each rule for determining out-of-control points.
11. What are some reasons for a process to go out of control due to a sudden shift in the level?
12. Explain some causes that would make the control chart pattern follow a gradually increasing trend.

Problems

13. What is meant by an overall Type I error rate? If Rules 1, 2, and 3 of this chapter are simultaneously used, assuming independence, what is the probability of an overall Type I error if 3σ control limits are used?
14. The diameter of cotter pins produced by an automatic machine is a characteristic of interest. Based on historical data, the process average diameter is 15 mm with a process standard deviation of 0.8 mm. If samples of size 4 are randomly selected from the process:
    a. Find the 1σ and 2σ control limits.
    b. Find the 3σ control limits for the average diameter.
    c. What is the probability of a false alarm?
    d. If the process mean shifts to 14.5 mm, what is the probability of not detecting this shift on the first sample plotted after the shift? What is the ARL?
    e. What is the probability of failing to detect the shift by the second sample plotted after the shift?
    f. Construct the OC curve for this control chart.
    g. Construct the ARL curve for this control chart.
15. The length of industrial filters is a quality characteristic of interest. Thirty samples, each of size 5, are chosen from the process. The data yields an average length of 110 mm, with the process standard deviation estimated to be 4 mm.
    a. Find the warning limits for a control chart for the average length.
    b. Find the 3σ control limits. What is the probability of a Type I error?
    c. If the process mean shifts to 112 mm, what are the chances of detecting this shift by the third sample drawn after the shift?
    d. What is the chance of detecting the shift for the first time on the second sample point drawn after the shift?
    e. What is the ARL for a shift in the process mean to 112 mm? How many samples, on average, would it take to detect a change in the process mean to 116 mm?
16. The tensile strength of nonferrous pipes is of importance. Samples of size 5 are selected from the process output, and their tensile strength values are found. After 30 such samples, the process mean strength is estimated to be 3000 kg with a standard deviation of 50 kg.
    a. Find the 1σ and 2σ control limits. For the 1σ limits, what is the probability of concluding that the process is out of control when it is really in control?
    b. Find the 3σ limits.
    c. If Rule 1 and Rule 2 are simultaneously used to detect out-of-control conditions, assuming independence, what is the overall probability of a Type I error if 3σ control limits are used?
17. Suppose 3σ control limits are constructed for the average temperature in a furnace. Samples of size 4 were selected with the average temperature being 5000°C and a standard deviation of 50°C.
    a. Find the 3σ control charts.
    b. Suppose Rule 2 and Rule 3 are used simultaneously to determine out-of-control conditions. What is the overall probability of a Type I error assuming independence of the rules?
    c. Approximately how many samples, on average, will be analyzed before detecting a change when Rules 2 and 3 are used simultaneously?
    d. If the process average temperature drops to 4960°C, what is the probability of failing to detect this change by the third sample point drawn after the change?
    e. What is the probability of the shift being detected within the first two samples?
18. A manager is contemplating using Rules 1 and 4 for determining out-of-control conditions. Suppose the manager constructs 3σ limits.
    a. What is the overall Type I error probability assuming independence of the rules?
    b. On average, how many samples will be analyzed before detecting a change in the process

mean? Assume that the process mean is now at 110 mm (having moved from 105 mm) and that the process standard deviation is 6 mm. Samples of size 4 are selected from the process.

References

American Society for Quality Control. (1987). ANSI/ASQC A1-1987. *Definitions, Symbols, Formulas, and Tables for Control Charts.* Milwaukee, Wis.: ASQC.

Banks, J. (1989). *Principles of Quality Control.* New York: John Wiley.

Besterfield, D. H. (1990). *Quality Control.* 3rd ed. Upper Saddle River, N.J.: Prentice Hall.

Duncan, A. J. (1986). *Quality Control and Industrial Statistics,* 5th ed. Homewood, Ill.: Irwin.

Montgomery, D. C. (1996). *Introduction to Statistical Quality Control.* 3rd ed. New York: John Wiley.

Munoz, J., and J. Nielsen (1991). "SPC: What Data Should I Collect? What Charts Should I Use?" *Quality Progress,* 24(1): 50–52.

Nelson, L. S. (1984). "The Shewhart Control Chart—Tests for Special Causes," *Journal of Quality Technology,* 16(4): 237–239.

Wadsworth, H. M., K. S. Stephens, and A. B. Godfrey (1986). *Modern Methods for Quality Control and Improvement.* New York: John Wiley.

Walker, E., J. W. Philpot, and J. Clement (1991). "False Signal Rates for the Shewhart Control Chart with Supplementary Runs Tests," *Journal of Quality Technology,* 23(3): 247–252.

CHAPTER

7

Control Charts for Variables

Chapter Outline

Symbols

μ	Process (or population) mean	k_i	Standardized value for range of sample number i
σ	Process (or population) standard deviation	σ_0	Target or standard value of process standard deviation
$\hat{\sigma}$	Estimate of process standard deviation	$\sigma_{\overline{X}}$	Standard deviation of the sample mean
$\overline{X}$	Sample average	S_m	Cumulative sum at sample number m
R	Sample range	w	Span, or width, in calculation of moving average
s	Sample standard deviation	$\overline{\overline{X}}$	Mean of sample means
n	Sample or subgroup size	$\overline{R}$	Mean of sample ranges
X_i	ith observation	G_t	Geometric moving average at time t
W	Relative range	M_t	Arithmetic moving average at time t
g	Number of samples or subgroups	T^2	Hotelling's T^2 multivariate statistic
$\overline{X}_0$	Target or standard value of process mean	MR	Moving range
Z_i	Standardized value for average of sample number i		

7-1 INTRODUCTION

Chapter 6 introduced the fundamentals of control charts. In this chapter we look at the details of **control charts for variables**—quality characteristics that are measurable on a numerical scale. Examples of **variables** include length, thickness, diameter, breaking strength, temperature, acidity, and viscosity. We must be able to control the mean value of a quality characteristic as well as its variability. The **mean** gives an indication of the central tendency of a process, and the variability provides an idea of the process dispersion. Therefore, we need information about both these statistics to keep a process in control.

Let's consider Figure 7-1. A change in the **process mean** of a quality characteristic (say, length of a part) is shown in Figure 7-1a, where the **mean shifts** from μ_0 to μ_1. It is, of course, important that this change be detected because if the **specification limits** are as shown in Figure 7-1a, a change in the process mean would change the proportion of parts that do not meet specifications. Figure 7-1b shows a change in the dispersion of the process; the **process standard deviation** has changed from σ_0 to σ_1, with the process mean remaining stationary at μ_0. Note that the proportion of the output that does not meet specifications has increased. Control charts aid in detecting such changes in process parameters.

Variables provide more information than attributes. **Attributes** deal with qualitative information such as whether an item is nonconforming or what the number of nonconformities in an item is. Thus, attributes do not show the degree to which a quality characteristic is nonconforming. For instance, if the specifications on the length of a part are 40 ± 0.5 mm and a part has length 40.6 mm, attribute information would indicate as nonconforming both this part *and* a part of length 42 mm. The degree to which these two lengths deviate from the specifications is lost in attribute information. This is not so with variables, however, because the numerical value of the quality characteristic (length, in this case) is used in the control chart.

The cost of obtaining variable data is usually higher than for attributes because attribute data is collected by means such as go/no-go gages, which are easier to use and therefore less costly. The total cost of data collection is the sum of two components: the fixed cost and the variable unit cost. Fixed costs include the cost of the inspection equipment; variable unit costs include the cost of inspecting units. The more units inspected, the higher the variable cost, whereas the fixed cost is unaffected. As the use of automated devices for measuring quality characteristic values spreads, the difference in the variable unit cost between variables and attributes may not be much. However, the fixed costs, such as investment costs, may increase.

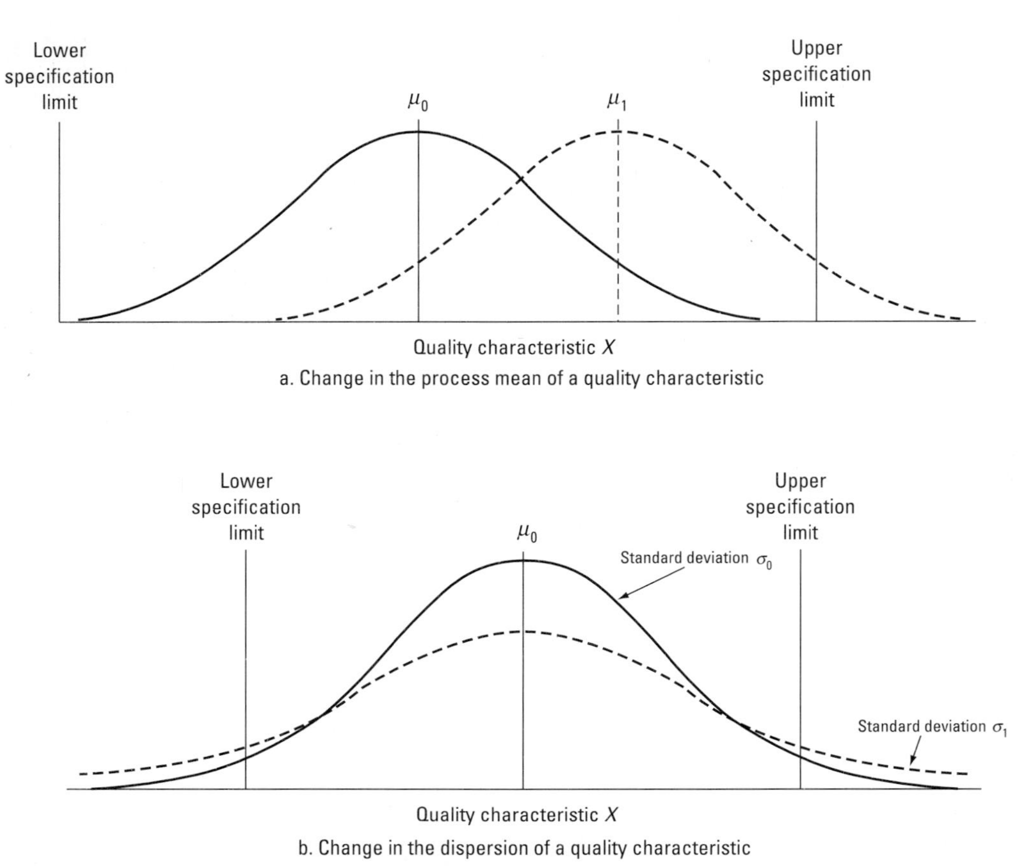

FIGURE 7-1 Changes in the mean and dispersion of a process.

At the plant level, when information is aggregated, we lose the specific identity of the constituent parts. That is, the output of several machines taken collectively does not allow us to distinguish between their performance. While an attribute chart for the proportion of nonconforming product may represent the general operational level of the plant, a variables chart is not appropriate at this level. To pinpoint problems, data for variable charts and attribute charts must be collected at a more specific level.

7-2 SELECTION OF CHARACTERISTICS FOR INVESTIGATION

In small organizations as well as in large ones, many possible product and process quality characteristics exist. A single component usually has several quality characteristics such as length, width, height, surface finish, and elasticity. In fact, the number of quality characteristics that affect a product is usually quite large. Now multiply such a number by even a small number of products and the total number of characteristics quickly increases to an unmanageable value. It is normally not feasible to maintain a control chart for each possible variable.

With each additional control chart, the decision-making process becomes more complicated; thus, the primary advantage of using the control charts—ease of use—diminishes. Operators find it difficult to look at 10 or 15 control charts simultaneously, so only a manageable number of control charts should be maintained at a given time. Balancing feasibility and completeness of information is an ongoing task. Accomplishing it involves selecting a few vital quality characteristics from the many candidates.

Selecting which quality characteristics to maintain control charts on requires giving higher priority to those that cause more nonconforming items and that increase costs. The goal is to select the "vital few" from among the "trivial many." In most industries, there are numerous variables to choose from. This is where **Pareto analysis** comes in because it clarifies which are the "important" quality characteristics.

When nonconformities occur because of different defects, the frequency of each defect can be tallied. Table 7-1 shows the Pareto analyses for various defects in an assembly. Alternatively, the cost of producing the nonconformity could be collected. Table 7-1 shows that the three most important defects are the inside hub diameter, the hub length, and the slot depth.

Using the percentages given in Table 7-1, we can construct a Pareto diagram like the one shown in Figure 7-2. The defects are thus shown in a nonincreasing order of occurrence. From the figure we can see that if we have only enough resources to construct three variable charts, we will choose inside hub diameter (code 4), hub length (code 3), and slot depth (code 7).

Once quality characteristics for which control charts are to be maintained have been identified, a scheme for obtaining the data should be set up. Quite often, it is desirable to measure process characteristics that have a causal relationship to product quality characteristics. Process characteristics are typically controlled directly through

TABLE 7-1 Pareto Analysis of Defects for Assembly Data

Defect Code	*Defect*	*Frequency*	*Percentage*
1	Outside diameter of hub	30	8.82
2	Depth of keyway	20	5.88
3	Hub length	60	17.65
4	Inside diameter of hub	90	26.47
5	Width of keyway	30	8.82
6	Thickness of flange	40	11.77
7	Depth of slot	50	14.71
8	Hardness (measured by Brinell hardness number)	20	5.88
		340	100.00

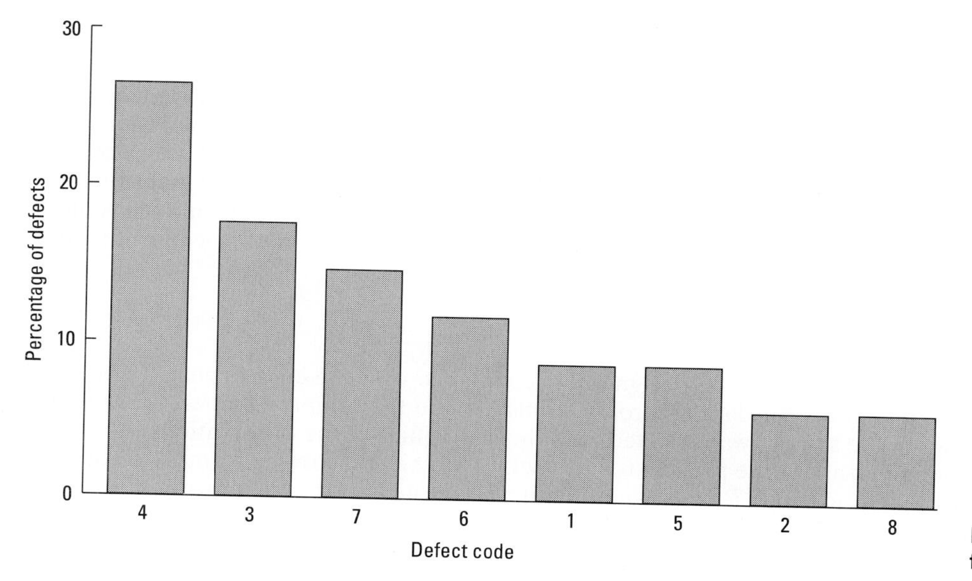

FIGURE 7-2 Pareto diagram for assembly data.

control charts. In the assembly example of Table 7-1, we might decide to monitor process variables (cutting speed, depth of cut, and coolant temperature) that have an impact on hub diameter, hub length, and slot depth. Monitoring process variables through control charts implicitly controls product characteristics.

7-3 PRELIMINARY DECISIONS

Certain decisions must be made before we can construct control charts. Several of these were discussed in detail in Chapter 6.

Selection of Rational Samples

The manner in which we sample the process deserves our careful attention. The sampling method should maximize differences between samples and minimize differences within samples. This means that separate control charts may have to be kept for different operators, machines, or vendors.

Lots from which samples are chosen should be homogeneous. As mentioned in Chapter 6, if our objective is to determine shifts in process parameters, then samples should be made up of items produced at nearly the same time. This gives us a time reference and will be helpful if we need to determine special causes. Alternatively, if we are interested in the nonconformance of items produced since the previous sample was selected, then samples should be chosen from items produced since that time.

Sample Size

Sample sizes are normally between 4 and 10, and it is quite common in industry to have sample sizes of 4 or 5. The larger the sample size, the better the chance of detecting small shifts. Other factors such as cost of inspection or cost of shipping a nonconforming item to the customer also influence the choice of sample size.

Frequency of Sampling

The sampling frequency depends on the cost of obtaining information compared to the cost of not detecting a nonconforming item. As processes are brought into control, the frequency of sampling will likely diminish.

Choice of Measuring Instruments

The accuracy of the measuring instrument directly influences the quality of the data collected. Measuring instruments should be calibrated and tested for dependability under controlled conditions. Low-quality data leads to erroneous conclusions. The characteristic being controlled and the desired degree of measurement precision both have an impact on the choice of measuring instrument. In measuring dimensions such as length, height, or thickness, something as simple as a set of calipers or a micrometer may be acceptable. On the other hand, measuring the thickness of silicon wafers may require complex optical sensory equipment.

Design of Data Recording Forms

Recording forms should be designed in accordance with the control chart to be used. Common features for data recording forms include the sample number, the date and time when the sample was selected, and the raw values of the observations. A column for comments about the process is also useful. A typical recording form for a chart for the mean $\overline{X}$ and range R is shown in Figure 7-3. Note that the top portion of the form contains information about part name, lot number, operation, operator, machine, gage used, unit of measurement, and specifications. The next segment of the chart contains

MEAN AND RANGE CHART

LOT NO.	P/N

PART NAME		OPERATION	SPEC.
OPERATOR	MACHINE	GAGE	UNIT OF MEASURE

DATE																
TIME																
SAMPLE VALUES 1																
2																
3																
4																
5																
SUM																
MEAN, $\bar{X}$																
RANGE, R																
INSP.																
	1	2	3	4	5	6	7	8	9	10	11	12	13	14	15	16
MEANS																
RANGES																

FIGURE 7-3 Data recording form for $\overline{X}$- and R-charts.

information about the date and time the sample is taken, the raw observations, and summary information such as the sum, mean, and range, as well as space for comments. Below this is space for the control chart for the mean and range.

7-4 CONTROL CHARTS FOR THE MEAN AND RANGE

After preliminary decisions, the following steps are used to develop the control charts.

Development of the Charts

Step 1: Using a preselected sampling scheme and sample size, record measurements of the selected quality characteristic on the appropriate forms.

Step 2: For each sample, calculate the sample mean and range using the following formulas:

$$\overline{X} = \frac{\sum_{i=1}^{n} X_i}{n} \tag{7.1}$$

$$R = X_{\max} - X_{\min} \tag{7.2}$$

where X_i represents the ith observation, n is the sample size, $X_{\max}$ is the largest observation, and $X_{\min}$ is the smallest observation.

Step 3: Obtain and draw the **center line** and the **trial control limits** for each chart. For the $\overline{X}$-chart, the center line $\overline{\overline{X}}$ is given by

$$\overline{\overline{X}} = \frac{\sum_{i=1}^{g} \overline{X}_i}{g} \tag{7.3}$$

where g represents the number of samples. For the R-chart, the center line $\overline{R}$ is found from

$$\overline{R} = \frac{\sum_{i=1}^{g} R_i}{g} \tag{7.4}$$

Conceptually, the **3σ control limits** for the $\overline{X}$-chart are

$$\overline{\overline{X}} \pm 3\sigma_{\overline{X}} \tag{7.5}$$

Rather than compute $\sigma_{\overline{X}}$ from the raw data, we can use the relation between the process standard deviation σ (or the standard deviation of the individual items) and the mean of the ranges ($\overline{R}$). Multiplying factors used to calculate the center line and control limits are given in Appendix A-7. When sampling from a population that is normally distributed, the distribution of the statistic $W = R/\sigma$ (known as the relative range) is dependent on the sample size n. The mean of W is represented by d_2 and is tabulated in Appendix A-7. Thus, an estimate of the process standard deviation is

$$\hat{\sigma} = \frac{\overline{R}}{d_2} \tag{7.6}$$

The control limits for an **$\overline{X}$-chart** are therefore estimated as

$$(\text{UCL}_{\overline{X}}, \text{LCL}_{\overline{X}}) = \overline{\overline{X}} \pm \frac{3\hat{\sigma}}{\sqrt{n}}$$

$$= \overline{\overline{X}} \pm \frac{3\overline{R}}{\sqrt{n}\,d_2}$$

$$(\text{UCL}_{\overline{X}}, \text{LCL}_{\overline{X}}) = \overline{\overline{X}} \pm A_2\overline{R} \tag{7.7}$$

where $A_2 = 3/\sqrt{n}d_2$ and is tabulated in Appendix A-7. Equation (7.7) is the working equation for determining the $\overline{X}$-chart control limits, given $\overline{R}$.

The control limits for the R-chart are conceptually given by

$$(\text{UCL}_R, \text{LCL}_R) = \overline{R} \pm 3\sigma_R \tag{7.8}$$

Since $R = \sigma W$, we have $\sigma_R = \sigma\sigma_W$. In Appendix A-7, σ_W is tabulated as d_3. Using eq. (7.6), we get

$$\hat{\sigma}_R = \left(\frac{\overline{R}}{d_2}\right)d_3$$

The control limits for the ***R*-chart** are estimated as

$$\begin{aligned} \text{UCL}_R &= \overline{R} + 3d_3\left(\frac{\overline{R}}{d_2}\right) = D_4\overline{R} \\ \text{LCL}_R &= \overline{R} - 3d_3\left(\frac{\overline{R}}{d_2}\right) = D_3\overline{R} \end{aligned} \tag{7.9}$$

where

$$D_4 = 1 + \frac{3d_3}{d_2} \quad \text{and} \quad D_3 = \max\left(0, 1 - \frac{3d_3}{d_2}\right)$$

Equation (7.9) is the working equation for calculating the control limits. Values of D_4 and D_3 are tabulated in Appendix A-7.

Step 4: Plot the values of the range on the control chart for range, with the center line and the control limits drawn. Determine whether the points are in statistical control. If not, investigate the special causes associated with the **out-of-control** points (see the **rules** for this in Chapter 6) and take appropriate remedial action to eliminate special causes.

Typically, only some of the rules are used simultaneously. The most commonly used criterion for determining an out-of-control situation is the presence of a point outside the control limits.

An R-chart is usually analyzed before the $\overline{X}$-chart to determine out-of-control situations. An R-chart reflects process variability, which should be brought to control first. As shown by eq. (7.7), the control limits for an $\overline{X}$-chart involve the process variability and hence $\overline{R}$. Therefore, if an R-chart shows an out-of-control situation, the limits on the $\overline{X}$-chart may not be meaningful.

Let's consider Figure 7-4. On the R-chart, sample 12 plots above the upper control limit and so is out of control. The $\overline{X}$-chart, however, does not show the process to be out of control. Suppose the special cause is identified as a problem with a new vendor who supplies raw materials and components. The task is to eliminate the cause perhaps by choosing a new vendor or requiring evidence of statistical process control at the vendor's plant.

Step 5: Delete the out-of-control point(s) for which **remedial actions** have been taken to remove special causes (in this case, sample 12) and use the remaining samples (here they are samples 1–11 and 13–15) to determine the revised center line and control limits for the $\overline{X}$- and R-charts.

These limits are known as the **revised control limits.** The cycle of obtaining information, determining the trial limits, finding out-of-control points, identifying and correcting special causes, and determining revised control limits then continues. The revised control limits will serve as trial control limits for the immediate future until the limits are revised again. This ongoing process is a critical component of continuous improvement.

A point of interest regarding the revision of R-charts concerns observations that plot below the lower control limit when the lower control limit is greater than zero. Such points that fall below LCL_R are, statistically speaking, out of control; however,

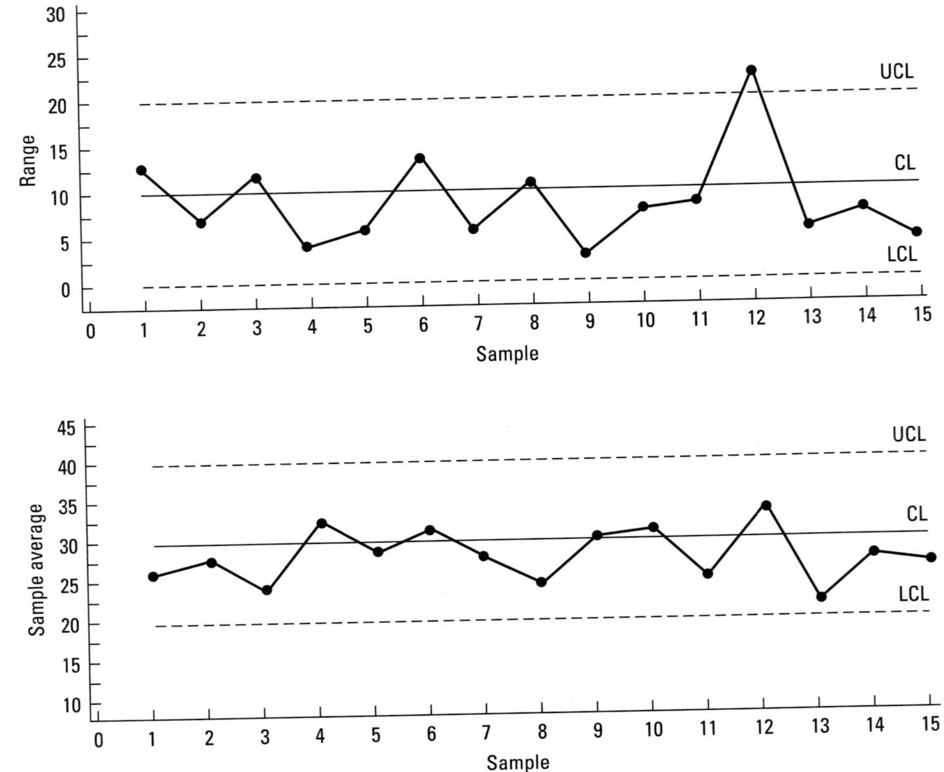

FIGURE 7-4 Plot of sample values on an $\overline{X}$- and R-chart.

they are also desirable because they indicate unusually small variability within the sample, which is, after all, one of our main objectives. Such small variability is most likely due to special causes.

If the user is convinced that the small variability does indeed represent the operating state of the process during that time, an effort should be made to identify the causes. If such conditions can be created consistently, process variability will be reduced. The process should be set to match those favorable conditions, and the observations should be retained for calculating the revised center line and the revised control limits for the R-chart.

Step 6: Implement the control charts.

The $\overline{X}$- and R-charts should be implemented for future observations, using the revised center line and control limits. The charts should be displayed in a conspicuous place where they will be visible to operators, supervisors, and managers. Statistical process control will be effective only if everyone is committed to it—from the operator to the chief executive officer.

Example 7-1: Consider a process by which coils are manufactured. Samples of size 5 are randomly selected from the process, and the resistance values (in ohms) of the coils are measured. The data values are given in Table 7-2, as are the sample mean $\overline{X}$ and the range R. First, the sum of the ranges is found and then the center line $\overline{R}$. We have

$$\overline{R} = \frac{\sum_{i=1}^{g} R_i}{g} = \frac{87}{25} = 3.48$$

For a sample of size 5, Appendix A-7 gives $D_4 = 2.114$ and $D_3 = 0$. The trial control limits for the R-chart are calculated as follows:

$$\text{UCL}_R = D_4\overline{R} = (2.114)(3.48) = 7.357$$
$$\text{LCL}_R = D_3\overline{R} = (0)(3.48) = 0$$

TABLE 7-2 Coil Resistance Data

Sample	*Observations (ohms)*	$\overline{X}$	*R*	*Comments*
1	20, 22, 21, 23, 22	21.60	3	
2	19, 18, 22, 20, 20	19.80	4	
3	25, 18, 20, 17, 22	20.40	8	New vendor
4	20, 21, 22, 21, 21	21.00	2	
5	19, 24, 23, 22, 20	21.60	5	
6	22, 20, 18, 18, 19	19.40	4	
7	18, 20, 19, 18, 20	19.00	2	
8	20, 18, 23, 20, 21	20.40	5	
9	21, 20, 24, 23, 22	22.00	4	
10	21, 19, 20, 20, 20	20.00	2	
11	20, 20, 23, 22, 20	21.00	3	
12	22, 21, 20, 22, 23	21.60	3	
13	19, 22, 19, 18, 19	19.40	4	
14	20, 21, 22, 21, 22	21.20	2	
15	20, 24, 24, 23, 23	22.80	4	
16	21, 20, 24, 20, 21	21.20	4	
17	20, 18, 18, 20, 20	19.20	2	
18	20, 24, 22, 23, 23	22.40	4	
19	20, 19, 23, 20, 19	20.20	4	
20	22, 21, 21, 24, 22	22.00	3	
21	23, 22, 22, 20, 22	21.80	3	
22	21, 18, 18, 17, 19	18.60	4	High temperature
23	21, 24, 24, 23, 23	23.00	3	Wrong die
24	20, 22, 21, 21, 20	20.80	2	
25	19, 20, 21, 21, 22	20.60	3	
		Sum = 521.00	Sum = 87	

The center line on the $\overline{X}$-chart is obtained as follows:

$$\overline{\overline{X}} = \frac{\sum_{i=1}^{g} \overline{X}_i}{g} = \frac{521.00}{25} = 20.840$$

Appendix A-7, for $n = 5$, gives $A_2 = 0.577$. Hence, the trial control limits on the $\overline{X}$-charts are

$$\text{UCL}_{\overline{X}} = \overline{\overline{X}} + A_2\overline{R} = 20.84 + (0.577)(3.48) = 22.848$$
$$\text{LCL}_{\overline{X}} = \overline{\overline{X}} - A_2\overline{R} = 20.84 - (0.577)(3.48) = 18.832$$

We can use Minitab to construct trial $\overline{X}$- and R-charts for the data in Table 7-2. Choose **Stat > Control Charts > Xbar-R.** Indicate whether the subgroups will be arranged in a single column or in rows, and click **OK.** Figure 7-5 shows the Minitab $\overline{X}$- and R-charts with three-sigma limits (represented by 3.0SL). Observe that sample 3 is above the upper control limits on the R-chart and samples 22 and 23 are below and above the $\overline{X}$-chart control limits, respectively. When the special causes for these three samples were investigated, operators found that the large value for the range in sample 3 was due to the quality of raw materials and components purchased from a new vendor. Management decided to require the new vendor to provide documentation showing that adequate control measures are being implemented at the vendor's plant and that the subsequent deliveries of raw materials and components will conform to standards.

When the special causes for samples 22 and 23 were examined, operators found that the oven temperature was too high for sample 22 and the wrong die was used for sample 23. Remedial actions were taken to rectify these situations.

With samples 3, 22, and 23 deleted, the revised center line on the R-chart is

$$\overline{R} = \frac{72}{22} = 3.273$$

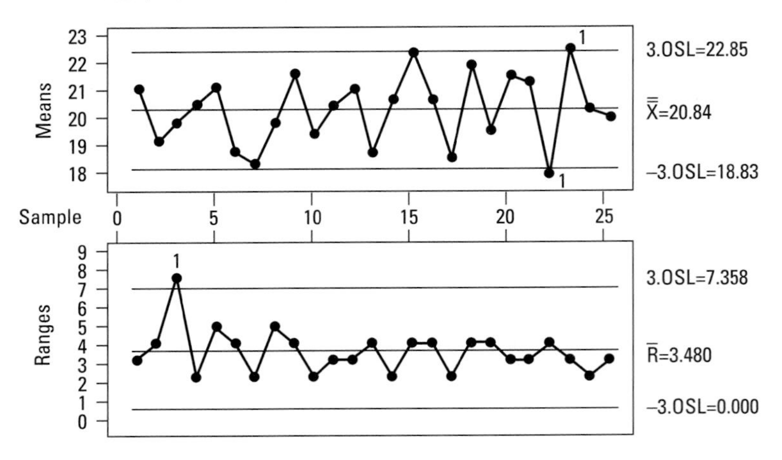

FIGURE 7-5 $\overline{X}$- and R-charts for data on coil resistance using Minitab.

The revised control limits on the R-chart are

$$\text{UCL}_R = D_4\overline{R} = (2.114)(3.273) = 6.919$$
$$\text{LCL}_R = D_3\overline{R} = (0)(3.273) = 0$$

The revised center line on the $\overline{X}$-chart is

$$\overline{\overline{X}} = \frac{459}{22} = 20.864$$

The revised control limits on the $\overline{X}$-chart are

$$\text{UCL}_{\overline{X}} = \overline{\overline{X}} + A_2\overline{R} = 20.864 + (0.577)(3.273) = 22.753$$
$$\text{LCL}_{\overline{X}} = \overline{\overline{X}} - A_2\overline{R} = 20.864 - (0.577)(3.273) = 18.975$$

Note that sample 15 falls slightly above the upper control limit on the $\overline{X}$-chart. On further investigation, no special causes could be identified for this sample. So, the revised limits will be used for future observations until a subsequent revision takes place.

Variable Sample Size

So far, our sample size has been assumed to be constant. A change in the sample size has an impact on the control limits for the $\overline{X}$- and R-charts. It can be seen from eqs. (7.7) and (7.9) that an increase in the sample size n reduces the width of the control limits. For an $\overline{X}$-chart, the width of the control limits from the center line is inversely proportional to the square root of the sample size. Appendix A-7 shows the pattern in which the values of the control chart factors A_2, D_4, and D_3 decrease with an increase in sample size.

Standardized Control Charts

When the sample size varies, the control limits on an $\overline{X}$- and an R-chart will change, as discussed previously. With fluctuating control limits, the rules for identifying out-of-control conditions we discussed in Chapter 6 become difficult to apply—that is, except for Rule 1 (which assumes a process to be out of control when an observation plots outside the control limits). One way to overcome this drawback is to use a standardized control chart. When we standardize a statistic, we subtract its mean from its value and divide this value by its standard deviation. The standardized values then represent the deviation from the mean in units of standard deviation. They are dimensionless and have a mean of zero. The control limits on a standardized chart are at ±3 and are therefore constant. It's easier to interpret shifts in the process from a standardized chart than from a chart with fluctuating control limits.

Let the sample size for sample i be denoted by n_i, and let $\overline{X}_i$ and s_i denote its average and standard deviation, respectively. The mean of the sample averages is found as

$$\overline{\overline{X}} = \frac{\sum_{i=1}^{g} n_i \overline{X}_i}{\sum_{i=1}^{g} n_i} \tag{7.10}$$

An estimate of the process standard deviation, $\hat{\sigma}$, is the square root of the weighted average of the sample variances, where the weights are 1 less the corresponding sample sizes. So,

$$\hat{\sigma} = \sqrt{\frac{\sum_{i=1}^{g} (n_i - 1)s_i^2}{\sum_{i=1}^{g} (n_i - 1)}} \tag{7.11}$$

Now, for sample i, the standardized value for the mean, Z_i, is obtained from

$$Z_i = \frac{\overline{X}_i - \overline{\overline{X}}}{\hat{\sigma}/\sqrt{n_i}} \tag{7.12}$$

where $\overline{\overline{X}}$ and $\hat{\sigma}$ are given by eqs. (7.10) and (7.11), respectively. A plot of the Z_i values on a control chart, with the center line at 0, the upper control limit at 3, and the lower control limit at –3, represents a standardized control chart for the mean.

To standardize the range chart, the range R_i for sample i is first divided by the estimate of the process standard deviation, $\hat{\sigma}$, given by eq. (7.11), to obtain

$$r_i = \frac{R_i}{\hat{\sigma}} \tag{7.13}$$

The values of r_i are then standardized by subtracting its mean d_2 and dividing by its standard deviation d_3 (Nelson 1989). The factors d_2 and d_3 are tabulated for various sample sizes in Appendix A-7. So, the standardized value for the range, k_i, is given by

$$k_i = \frac{r_i - d_2}{d_3} \tag{7.14}$$

These values of k_i are plotted on a control chart with a center line at 0, and upper and lower control limits at 3 and –3, respectively.

Control Limits for a Given Target or Standard

Management sometimes wants to specify values for the process mean and standard deviation. These values may represent goals or desirable standard or **target values.** Control charts based on these target values help determine whether the existing process is capable of meeting the desirable standards. Furthermore, they also help management set realistic goals for the existing process.

Let $\overline{X}_0$ and σ_0 represent the target values of the process mean and standard deviation, respectively. The center line and control limits based on these standard values for the $\overline{X}$-chart are given by

$$\begin{aligned} CL_{\overline{X}} &= \overline{X}_0 \\ UCL_{\overline{X}} &= \overline{X}_0 + 3\frac{\sigma_0}{\sqrt{n}} \\ LCL_{\overline{X}} &= \overline{X}_0 - 3\frac{\sigma_0}{\sqrt{n}} \end{aligned} \tag{7.15}$$

Let $A = 3/\sqrt{n}$. Values for A are tabulated in Appendix A-7. Equation (7.15) may be rewritten as

$$\begin{aligned} CL_{\overline{X}} &= \overline{X}_0 \\ UCL_{\overline{X}} &= \overline{X}_0 + A\sigma_0 \\ LCL_{\overline{X}} &= \overline{X}_0 - A\sigma_0 \end{aligned} \tag{7.16}$$

For the R-chart, the center line is found as follows. Since $\hat{\sigma} = \overline{R}/d_2$, we have

$$CL_R = d_2\sigma_0 \tag{7.17}$$

where d_2 is tabulated in Appendix A-7. The control limits are

$$\begin{aligned} UCL_R &= \overline{R} + 3\sigma_R = d_2\sigma_0 + 3d_3\sigma_0 \\ &= (d_2 + 3d_3)\sigma_0 = D_2\sigma_0 \end{aligned} \tag{7.18}$$

where $D_2 = d_2 + 3d_3$ (Appendix A-7) and $\sigma_R = d_3\sigma$.

Similarly,

$$\begin{aligned} LCL_R &= \overline{R} - 3\sigma_R = d_2\sigma_0 - 3d_3\sigma_0 \\ &= (d_2 - 3d_3)\sigma_0 = D_1\sigma_0 \end{aligned} \tag{7.19}$$

where $D_1 = d_2 - 3d_3$ (Appendix A-7).

We must be cautious when we interpret control charts based on target or standard values. Sample observations can fall outside the control limits even though no special causes are present in the process. This is because these desirable standards may not be consistent with the process conditions. Thus, we could waste time and resources looking for special causes that do not exist.

On an $\overline{X}$-chart, plotted points can fall outside the control limits because a target process mean is specified as too high or too low compared to the existing process mean. Usually, it is easier to meet a desirable target value for the process mean than it is for the process variability. For example, adjusting the mean diameter or length of a part can often be accomplished by simply changing controllable process parameters. However, correcting for R-chart points that plot above the upper control limit is generally much more difficult.

An R-chart based on target values can also indicate excessive process variability without special causes present in the system. Therefore, meeting the target value σ_0 may involve drastic changes in the process. Such an R-chart may be implying that the existing process is not capable of meeting the desired standard. This information enables management to set realistic goals.

Example 7-2: Refer to the coil resistance data in Example 7-1. Let's suppose the target values for the average resistance and standard deviation are 21.0 and 1.0 ohms, respectively. The sample size is 5. The center line and the control limits for the $\overline{X}$-chart are as follows:

$$\begin{aligned} CL_{\overline{X}} &= \overline{X}_0 = 21.0 \\ UCL_{\overline{X}} &= \overline{X}_0 + A\sigma_0 = 21.0 + (1.342)(1.0) = 22.342 \\ LCL_{\overline{X}} &= \overline{X}_0 - A\sigma_0 = 21.0 - (1.342)(1.0) = 19.658 \end{aligned}$$

The center line and control limits for the R-chart are

$$\begin{aligned} CL_R &= d_2\sigma_0 = (2.326)(1.0) = 2.326 \\ UCL_R &= D_2\sigma_0 = (4.918)(1.0) = 4.918 \\ LCL_R &= D_1\sigma_0 = (0)(1.0) = 0 \end{aligned}$$

Figure 7-6 shows the control chart for the range based on the standard value. Since the control charts were revised in Example 7-1, we plot the 22 in-control samples and exclude samples 3, 22, and 23 because we are assuming that remedial actions have eliminated those causes. Now we can see how close the in-control process comes to meeting the stipulated target values.

The process seems to be out of control with respect to the given standard. Samples 5 and 8 are above the upper control limit, and a majority of the points lie above the center line. Only six of the points plot below the center line. Figure 7-6 thus reveals that the process is not capable of

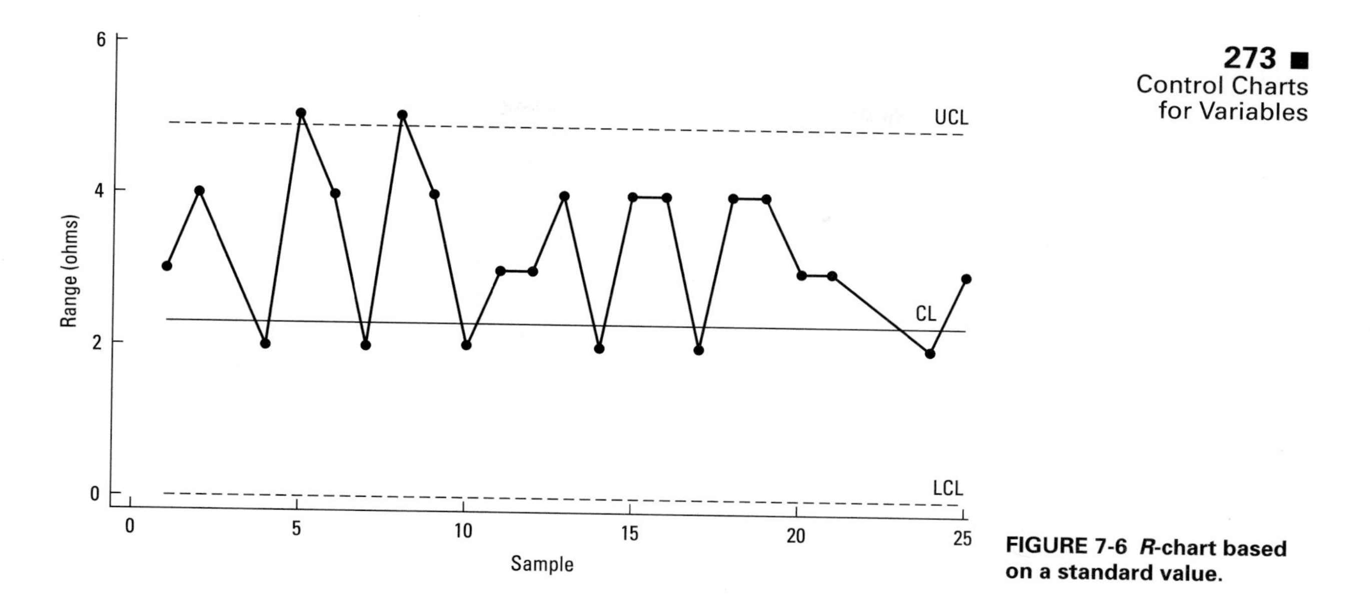

FIGURE 7-6 *R*-chart based on a standard value.

meeting the company guidelines. The target standard deviation σ_0 is 1.0. The estimated process standard deviation from Example 7-1 (calculated after the process was brought to control) is

$$\hat{\sigma} = \frac{\overline{R}}{d_2} = \frac{3.50}{2.326} = 1.505$$

This estimate exceeds the target value of 1.0. Management must look at common causes to reduce the process variability if the standard is to be met. This may require major changes in the methods of operation, the incoming material, or the equipment. Process control will not be sufficient to achieve the desired target.

The $\overline{X}$-chart based on the standard value is shown in Figure 7-7. Several points fall outside the control limits—four points below and two points above. In Example 7-1, the revised center line for the $\overline{X}$-chart was found to be 20.864. Our target center line is now 21.0. Adjusting controllable process parameters could possibly shift the average level up to 21.0. However, the fact that there are points outside both the upper and lower control limits signifies that process variability is the issue here.

Common causes must be examined: That is, reducing variability will only be achieved through process improvement. Figure 7-7 indicates that the target standard deviation of 1.0 is

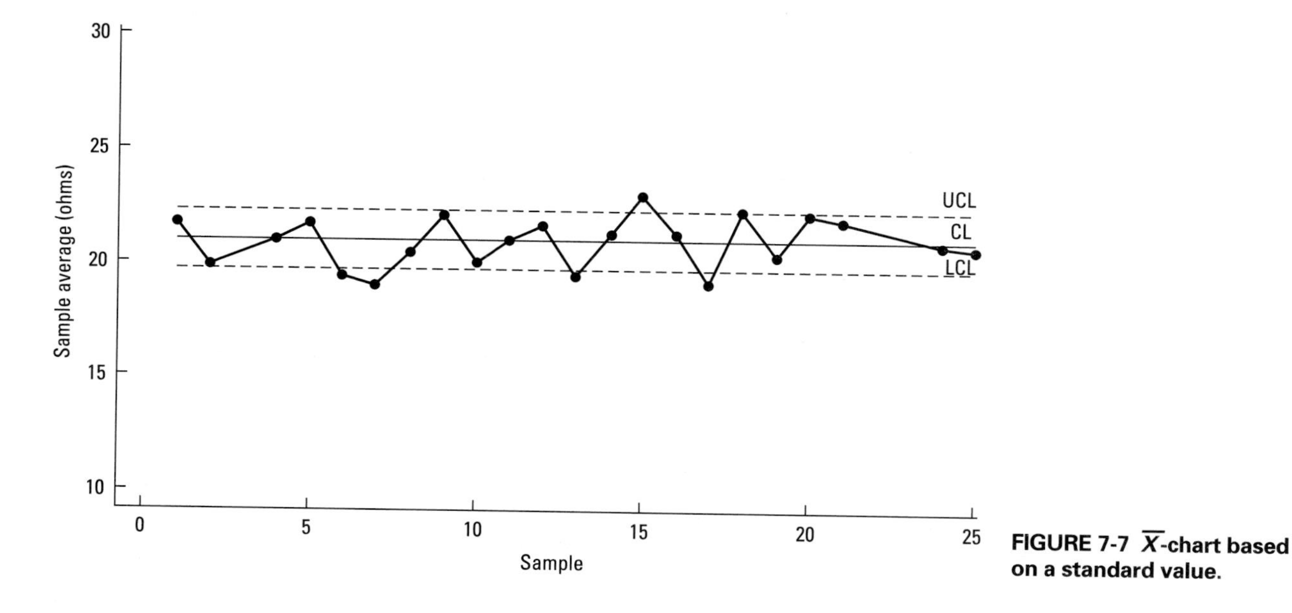

FIGURE 7-7 $\overline{X}$-chart based on a standard value.

not realistic for the current process. Unless management makes major changes in the process, the target value will not be met. Actions on the part of the operators alone are unlikely to cause the necessary reduction in process variability.

Interpretation and Inferences from the Charts

The difficult part of analysis is determining and interpreting the special causes and selecting remedial actions. Effective use of control charts requires operators who are familiar with not only the statistical foundations of control charts but also the process itself. They must thoroughly understand how the different controllable parameters influence the dependent variable of interest. The quality assurance manager or analyst should work closely with the product design engineer and the process designer or analyst to come up with optimal policies.

We discussed five rules in Chapter 6 for determining out-of-control conditions. The presence of a point falling outside the 3σ limits is the most widely used of those rules. Determinations can also be made by interpreting typical plot patterns. Once the special cause is determined, this information plus a knowledge of the plot can lead to appropriate remedial actions.

Often, when the *R*-chart is brought to control, many special causes for the $\overline{X}$-chart are eliminated as well. The $\overline{X}$-chart monitors the centering of the process because $\overline{X}$ is a measure of the center. Thus, a jump on the $\overline{X}$-chart means that the process average has jumped and an increasing trend indicates the process center is gradually increasing. Process centering usually takes place through adjustments in machine settings or such controllable parameters as proper tool, proper depth of cut, or proper feed. On the other hand, reducing process variability in order to allow an *R*-chart to exhibit control is a difficult task that is accomplished through quality improvement.

Once a process is in statistical control, its capability can be estimated by calculating the process standard deviation. This measure can then be used to determine how the process performs with respect to some stated specification limits. The **proportion** of **nonconforming** items can be estimated. Depending on the characteristic being considered, some of the output may be reworked, while some may become scrap. Given the unit cost of rework and scrap, an estimate of the total cost of rework and scrap can be obtained. A more detailed discussion of **process capability** measures is given in Chapter 9. From an *R*-chart that exhibits control, the process standard deviation can be estimated as

$$\hat{\sigma} = \frac{\overline{R}}{d_2}$$

where $\overline{R}$ is the center line and d_2 is a factor tabulated in Appendix A-7. If the distribution of the quality characteristic can be assumed to be normal, then given some specification limits, the standard normal table can be used to determine the proportion of output that is nonconforming.

Example 7-3: Refer to the coil resistance data in Example 7-1. Suppose the specifications are 21 ± 3 ohms.

a. Determine the proportion of the output that is nonconforming, assuming that coil resistance is normally distributed.

Solution. From the revised *R*-chart, we found the center line to be $\overline{R} = 3.50$. The estimated process standard deviation is

$$\hat{\sigma} = \frac{\overline{R}}{d_2} = \frac{3.50}{2.236} = 1.505$$

The revised center line on the $\overline{X}$-chart is $\overline{\overline{X}} = 20.864$, which we use as an estimate of the process mean. Figure 7-8 shows the proportion of the output that is nonconforming. The standardized normal value at the lower specification limit (LSL) is found as

$$z_1 = \frac{(18 - 20.864)}{1.505} = -1.90$$

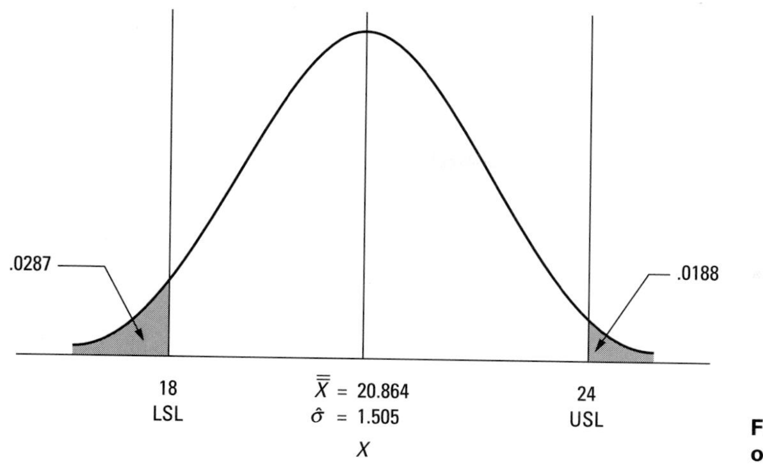

FIGURE 7-8 Proportion of nonconforming output.

The standardized normal value at the upper specification limit (USL) is

$$z_2 = \frac{(24 - 20.864)}{1.505} = 2.08$$

From Appendix A-3, we find that the proportion of the product below the LSL is .0287, and the proportion above the USL is .0188. Thus, the total proportion of nonconforming output is .0475.

b. If the daily production rate is 10,000 coils and if coils with a resistance less than the LSL cannot be used for the desired purpose, what is the loss to the manufacturer if the unit cost of scrap is 50 cents?

Solution. The daily cost of scrap is

$$(10{,}000)(0.0287)(\$0.50) = \$143.50$$

Control Chart Patterns and Corrective Actions

A nonrandom identifiable pattern in the plot of a control chart might provide sufficient reason to look for **special causes** in the system. **Common causes** of variation are inherent to a system; a system operating under only common causes is said to be in a state of statistical control. Special causes, however, could be due to periodic and persistent disturbances that affect the process intermittently. The objective is to identify the special causes and take appropriate remedial action.

Western Electric Company engineers have identified 15 typical patterns in control charts. Your ability to recognize these patterns will enable you to determine *when* action needs to be taken and *what* action to take (AT&T 1984). We discuss nine of these patterns here.

Natural Patterns

A natural pattern is one in which no identifiable arrangement of the plotted points exists. No points fall outside the control limits, the majority of the points are near the center line, and few points are close to the control limits (Gitlow et al. 1989). Natural patterns are indicative of a process that is in control; that is, they demonstrate the presence of a stable system of common causes. A natural pattern is shown in Figure 7-9.

Sudden Shifts in the Level

Many causes can bring about a sudden change (or jump) in pattern level on an $\bar{X}$- or R-chart. Figure 7-10 shows a sudden shift on an $\bar{X}$-chart. Such jumps occur because of changes—intentional or otherwise—in such process settings as temperature, pressure, or depth of cut. A sudden change in the average service level, for example, could

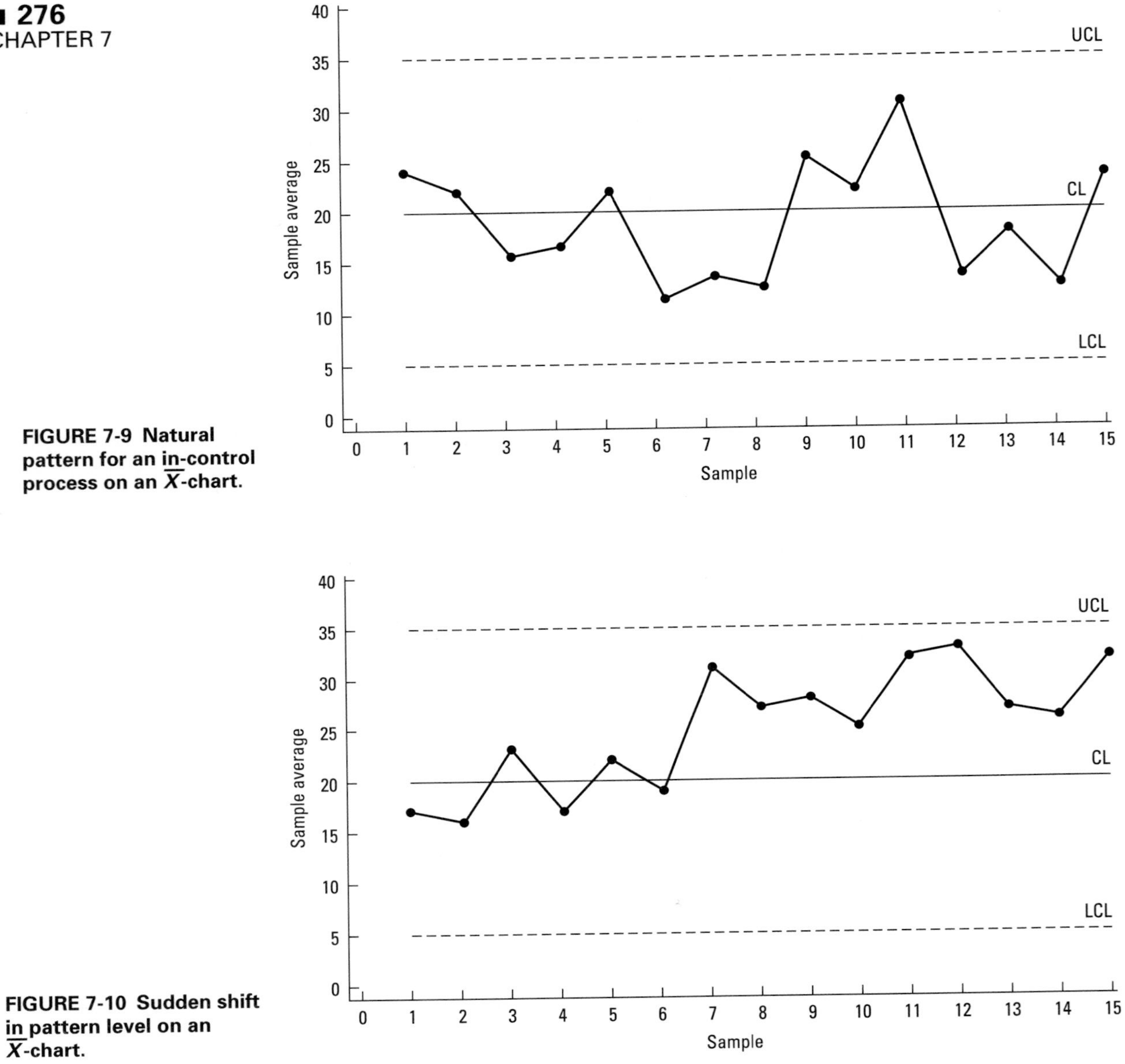

FIGURE 7-9 Natural pattern for an in-control process on an $\overline{X}$-chart.

FIGURE 7-10 Sudden shift in pattern level on an $\overline{X}$-chart.

be a change in customer waiting time at a bank because the number of tellers changed. New operators, new equipment, new measuring instruments, new vendors, and new methods of processing are other reasons for sudden shifts on $\overline{X}$- and R-charts.

Gradual Shifts in the Level

Gradual shifts in level occur when a process parameter changes gradually over a period of time. Afterward, the process stabilizes. An $\overline{X}$-chart might exhibit such a shift because the incoming quality of raw materials or components changed over time, the maintenance program changed, or the style of supervision changed. An R-chart might exhibit such a shift because of a new operator, a decrease in worker skill due to fatigue or monotony, or a gradual improvement in the incoming quality of raw materials because a vendor has implemented a statistical process control system. Figure 7-11 shows an $\overline{X}$-chart exhibiting a gradual shift in the level.

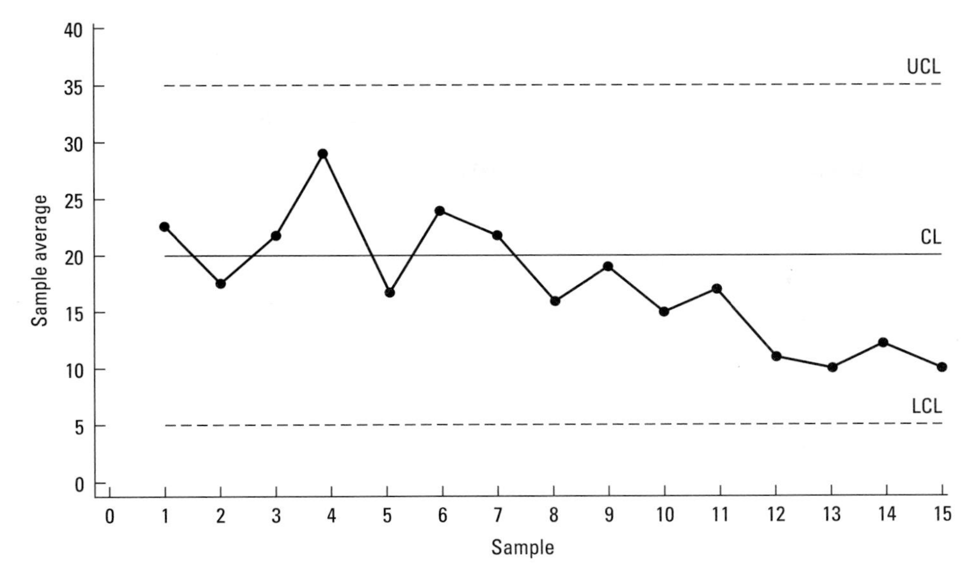

FIGURE 7-11 Gradual shift in pattern level on an $\overline{X}$-chart.

Trending Pattern

Trends differ from gradual shifts in level in that trends do not stabilize or settle down. Trends represent changes that steadily increase or decrease. An $\overline{X}$-chart may exhibit a trend because of tool wear, die wear, gradual deterioration of equipment, buildup of debris in jigs and fixtures, or gradual change in temperature. An R-chart may exhibit a trend because of a gradual improvement in operator skill resulting from on-the-job training, or a decrease in operator skill due to fatigue. Figure 7-12 shows a trending pattern on an $\overline{X}$-chart.

Cyclic Patterns

Cyclic patterns are characterized by a repetitive periodic behavior in the system. Cycles of low and high points will appear on the control chart. An $\overline{X}$-chart may exhibit cyclic behavior because of a rotation of operators, periodic changes in temperature and humidity (such as a cold-morning startup), periodicity in the mechanical or chemical properties of the material, or seasonal variation of incoming components. An

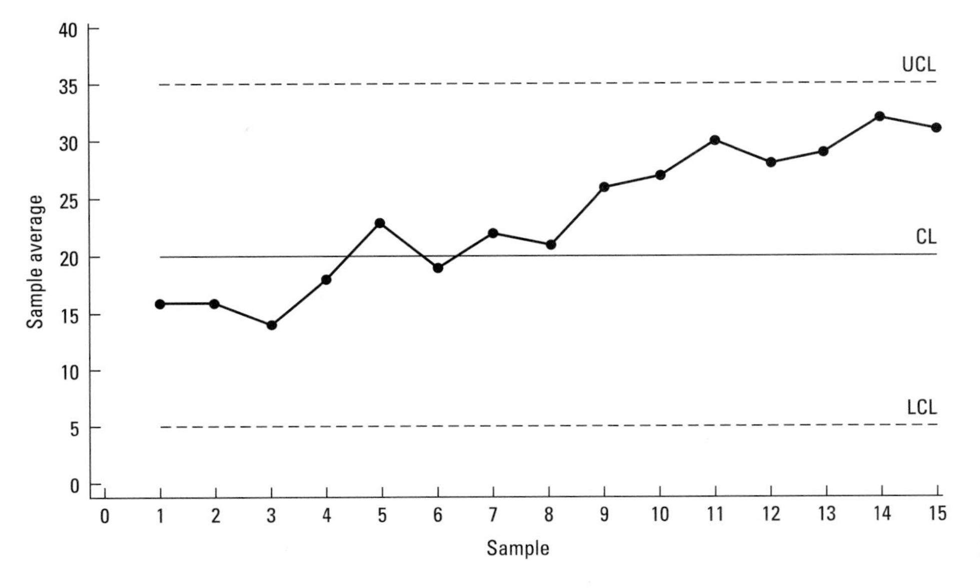

FIGURE 7-12 Trending pattern on an $\overline{X}$-chart.

R-chart may exhibit cyclic patterns because of operator fatigue and subsequent energization following breaks, a difference between shifts, or periodic maintenance of equipment. Figure 7-13 shows a cyclic pattern for an $\overline{X}$-chart. If samples are taken too infrequently, only the high or the low points will be represented, and the graph will not exhibit a cyclic pattern. If control chart users suspect cyclic behavior, then they should take samples frequently to investigate the possibility of a cyclic pattern.

Wild Patterns

Wild patterns are divided into two categories—freaks and bunches (or groups). Control chart points exhibiting either of these two properties are, statistically speaking, significantly different from the other points. Special causes are generally associated with these points.

Freaks are caused by external disturbances that influence one or more samples. Figure 7-14 shows a control chart exhibiting a freak pattern. Freaks are plotted points too small or too large with respect to the control limits. Such points usually fall outside

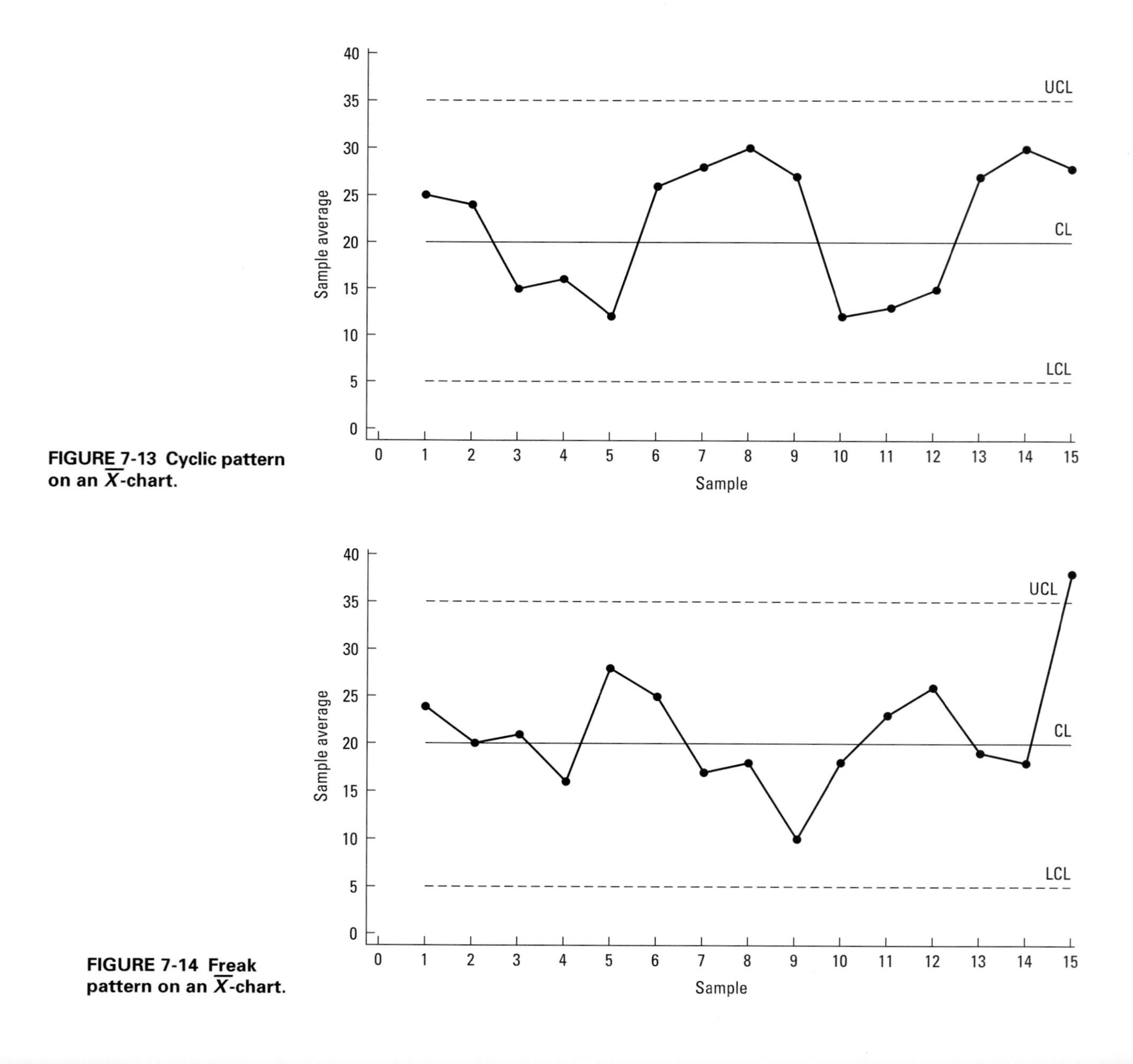

FIGURE 7-13 Cyclic pattern on an $\overline{X}$-chart.

FIGURE 7-14 Freak pattern on an $\overline{X}$-chart.

the control limits and are easily distinguishable from the other points on the chart. It is often not difficult to identify special causes for freaks. You should make sure, however, that there is no measurement or recording error associated with the freak point. Some special causes of freaks include sudden, very short-lived power failures; the use of a new tool for a brief test period; and the failure of a component.

Bunches, or **groups,** are clusters of several observations that are decidedly different from other points on the plot. Figure 7-15 shows a control chart pattern exhibiting bunching behavior. Possible special causes of such behavior include the use of a new vendor for a short period of time, use of a different machine for a brief time period, and a new operator used for a short period.

Mixture Patterns (or the Effect of Two or More Populations)

A mixture pattern is caused by the presence of two or more populations in the sample and is characterized by points that fall near the control limits, with an absence of points near the center line. A mixture pattern can occur when one set of values is too high and another set too low because of differences in the incoming quality of material from two vendors. A remedial action would be to have a separate control chart for each vendor. Figure 7-16 shows a mixture pattern. On an $\overline{X}$-chart, a mixture pattern can also result from overcontrol. If an operator chooses to adjust the machine or process *every* time a point plots near a control limit, the result will be a pattern of large swings. Mixture patterns can also occur on both $\overline{X}$- and R-charts because of two or more machines being represented on the same control chart. Other examples include two or more operators being represented on the same chart, differences in two or more pieces of testing or measuring equipment, and differences in production methods of two or more lines.

Stratification Patterns

A stratification pattern is another possible result when two or more population distributions of the same quality characteristic are present. In this case, the output is combined, or mixed (say, from two shifts), and samples are selected from the mixed output. In this pattern, the majority of the points are very close to the center line, with very few points near the control limits. Thus, the plot can be misinterpreted as indicating unusually good control. A stratification pattern is shown in Figure 7-17. Such a plot could have resulted from plotting data for samples composed of the combined output of two shifts, each different in its performance. It is possible for the sample average (which is really the average of parts chosen from both shifts) to fluctuate very little,

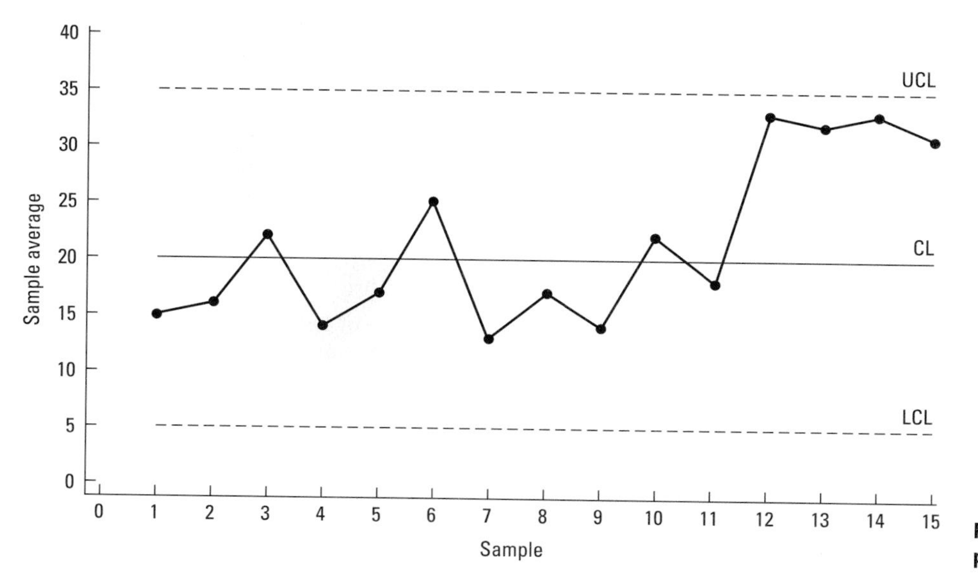

FIGURE 7-15 Bunching pattern on an $\overline{X}$-chart.

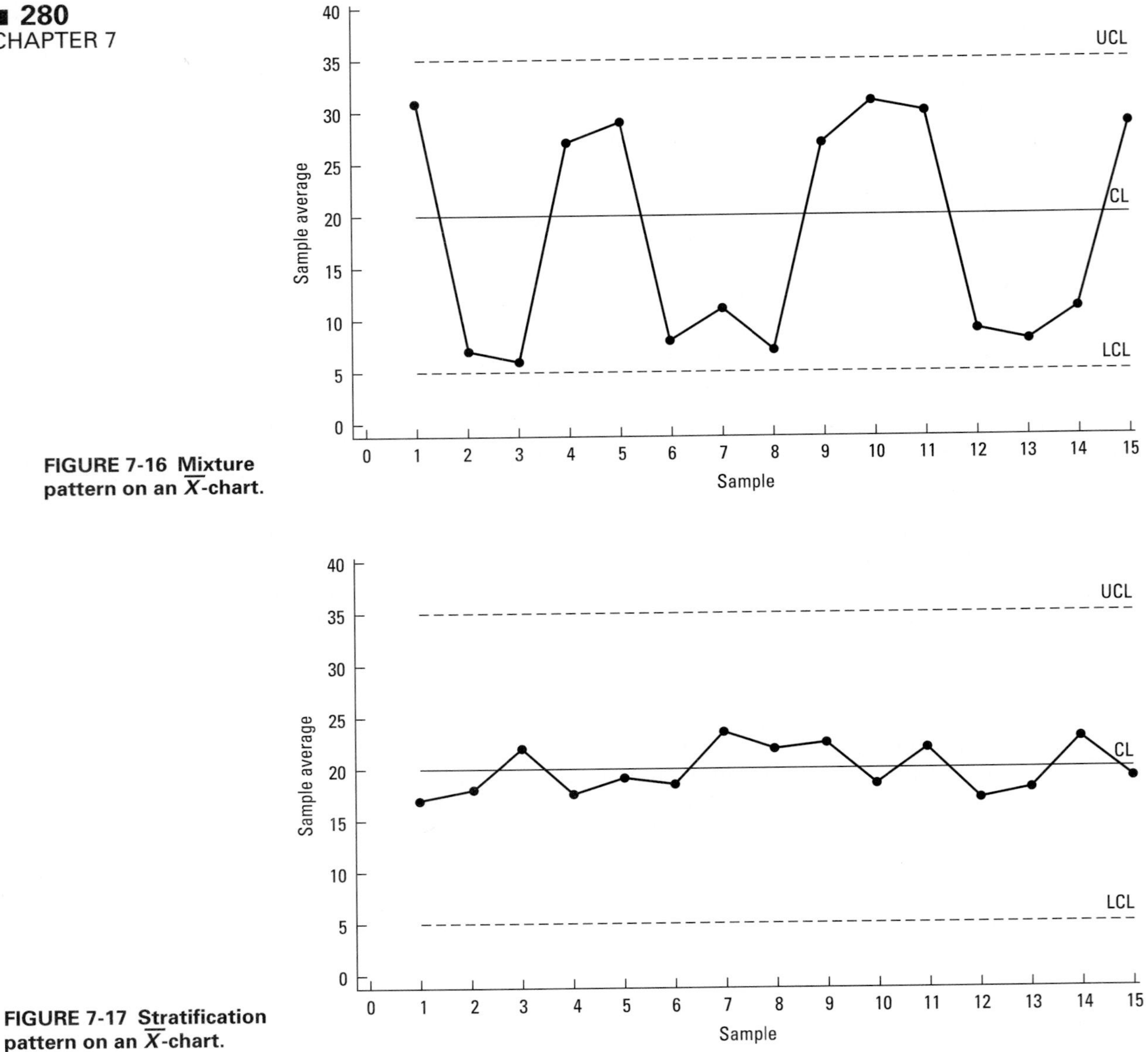

FIGURE 7-16 Mixture pattern on an $\overline{X}$-chart.

FIGURE 7-17 Stratification pattern on an $\overline{X}$-chart.

resulting in a stratification pattern in the plot. Remedial measures in such situations involve having separate control charts for each shift. The method of choosing rational samples should be carefully analyzed so that component distributions are not mixed when samples are selected.

Interaction Patterns

An interaction pattern occurs when the level of one variable affects the behavior of other variables associated with the quality characteristic of interest. Furthermore, the combined effect of two or more variables on the output quality characteristic may be different from the individual effect of each variable. An interaction pattern can be detected by changing the scheme for rational sampling. Suppose that in a chemical process the temperature and pressure are two important controllable variables that affect the output quality characteristic of interest. A low pressure and a high temperature may produce a very desirable effect on the output characteristic, whereas a low pressure by itself may not have that effect. An effective sampling

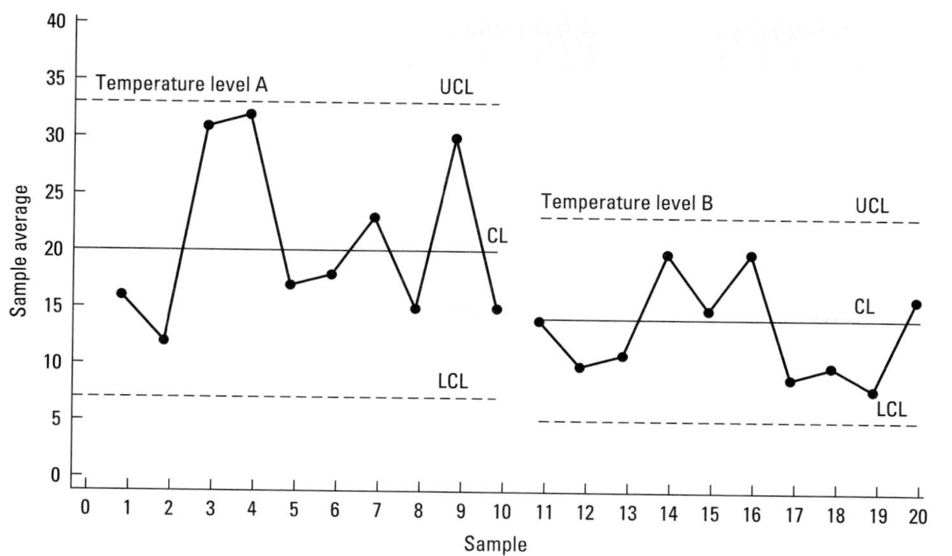

FIGURE 7-18 Interaction pattern between variables on an $\overline{X}$-chart.

method would involve controlling the temperature at several high values and then determining the effect of pressure on the output characteristic for each temperature value. Samples composed of random combinations of temperature and pressure may fail to identify the interactive effect of those variables on the output characteristic. The control chart in Figure 7-18 shows interaction between variables. In the first plot, the temperature was maintained at level A; in the second plot, it was held at level B. Note that the average level and variability of the output characteristic change for the two temperature levels. Also, if the R-chart shows the sample ranges to be small, then information regarding the interaction could be used to establish desirable process parameter settings.

Control Charts for Other Variables

The control chart patterns described in this section also occur in control charts besides $\overline{X}$- and R-charts. When found in other types of control charts, these patterns may indicate different causes than those we discussed in this section, but similar reasoning can be used to determine them. Furthermore, both the preliminary considerations and the steps for constructing control charts described earlier also apply to other control charts.

7-5 CONTROL CHARTS FOR THE MEAN AND STANDARD DEVIATION

Although an R-chart is easy to construct and use, a standard deviation chart (**s-chart**) is preferable for larger sample sizes (greater than 10, usually). As mentioned in Chapter 4, the range accounts for only the maximum and minimum sample values and consequently is less effective for large samples. The sample standard deviation serves as a better measure of process variability in these circumstances. The **sample standard deviation** is given by

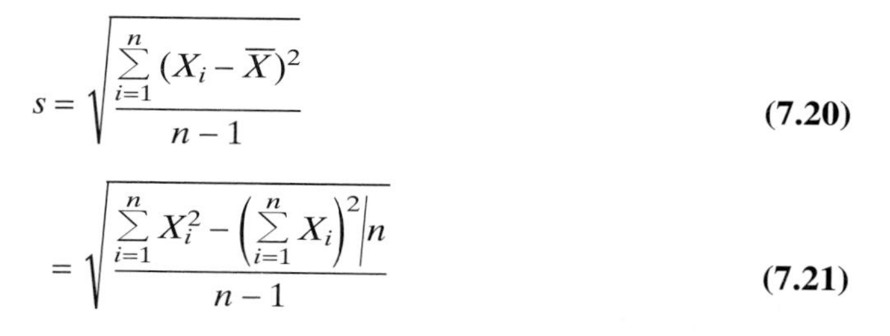

$$s = \sqrt{\frac{\sum_{i=1}^{n} (X_i - \overline{X})^2}{n-1}} \qquad \textbf{(7.20)}$$

$$= \sqrt{\frac{\sum_{i=1}^{n} X_i^2 - \left(\sum_{i=1}^{n} X_i\right)^2 \Big/ n}{n-1}} \qquad \textbf{(7.21)}$$

If the population distribution of a quality characteristic is normal with a **population standard deviation** denoted by σ, the mean and standard deviation of the sample standard deviation are given by

$$E(s) = c_4\sigma \tag{7.22}$$

$$\sigma_s = \sigma\sqrt{1 - c_4^2} \tag{7.23}$$

respectively, where c_4 is a factor that depends on the sample size and is given by

$$c_4 = \left[\frac{2}{(n-1)}\right]^{1/2} \frac{[(n-2)/2]!}{[(n-3)/2]!} \tag{7.24}$$

Values of c_4 are tabulated in Appendix A-7. (See also Montgomery 1991 and Wadsworth et al. 1986.)

No Given Standards

The center line of a standard deviation chart is

$$\text{CL}_s = \bar{s} = \frac{\sum_{i=1}^{g} s_i}{g} \tag{7.25}$$

where g is the number of samples and s_i is the standard deviation of the ith sample. The upper control limit is

$$\text{UCL}_s = \bar{s} + 3\sigma_s = \bar{s} + 3\sigma\sqrt{1 - c_4^2}$$

In accordance with eq. (7.22), an estimate of the population standard deviation σ is

$$\hat{\sigma} = \frac{\bar{s}}{c_4} \tag{7.26}$$

Substituting this estimate of $\hat{\sigma}$ in the preceding expression yields

$$\text{UCL}_s = \bar{s} + \frac{3\bar{s}\sqrt{(1 - c_4^2)}}{c_4} = B_4\bar{s}$$

where $B_4 = 1 + 3\frac{\sqrt{(1 - c_4^2)}}{c_4}$ and is tabulated in Appendix A-7. Similarly,

$$\text{LCL}_s = \bar{s} - \frac{3\bar{s}\sqrt{(1 - c_4^2)}}{c_4} = B_3\bar{s}$$

where $B_3 = \max\left[0, 1 - 3\frac{\sqrt{(1 - c_4^2)}}{c_4}\right]$ and is also tabulated in Appendix A-7. Thus, the 3σ control limits are

$$\begin{aligned} \text{UCL}_s &= B_4\bar{s} \\ \text{LCL}_s &= B_3\bar{s} \end{aligned} \tag{7.27}$$

The center line of the chart for the mean $\overline{X}$ is given by

$$\text{CL}_{\overline{X}} = \overline{\overline{X}} = \frac{\sum_{i=1}^{g} \overline{X}_i}{g} \tag{7.28}$$

The control limits on the $\overline{X}$-chart are

$$\overline{\overline{X}} \pm 3\sigma_{\overline{X}} = \overline{\overline{X}} \pm \frac{3\sigma}{\sqrt{n}}$$

Using eq. (7.26) to obtain $\hat{\sigma}$, we find the control limits to be

$$\begin{aligned} UCL_{\overline{X}} &= \overline{\overline{X}} + \frac{3\overline{s}}{c_4\sqrt{n}} = \overline{\overline{X}} + A_3\overline{s} \\ LCL_{\overline{X}} &= \overline{\overline{X}} - \frac{3\overline{s}}{c_4\sqrt{n}} = \overline{\overline{X}} - A_3\overline{s} \end{aligned} \qquad \textbf{(7.29)}$$

where $A_3 = 3/(c_4\sqrt{n})$ and is tabulated in Appendix A-7.

The process of constructing trial control limits, determining special causes associated with out-of-control points, taking remedial actions, and finding the revised control limits is similar to that explained in the section on $\overline{X}$- and R-charts. The s-chart is constructed first. Only if it is in control should the $\overline{X}$-chart be developed, because the standard deviation of $\overline{X}$ is dependent on $\overline{s}$. If the s-chart is not in control, any estimate of the standard deviation of $\overline{X}$ will be unreliable, which will in turn create unreliable control limits for $\overline{X}$.

Given Standard

If a target standard deviation is specified as σ_0, the center line of the s-chart is found by using eq. (7.22) as

$$CL_s = c_4\sigma_0 \qquad \textbf{(7.30)}$$

The upper control limit for the s-chart is found by using eq. (7.23) as

$$\begin{aligned} UCL_s &= c_4\sigma_0 + 3\sigma_s = c_4\sigma_0 + 3\sigma_0\sqrt{1-c_4^2} \\ &= (c_4 + 3\sqrt{1-c_4^2})\sigma_0 = B_6\sigma_0 \end{aligned}$$

where $B_6 = c_4 + 3\sqrt{1-c_4^2}$ and is tabulated in Appendix A-7. Similarly, the lower control limit for the s-chart is

$$LCL_s = (c_4 - 3\sqrt{1-c_4^2})\sigma_0 = B_5\sigma_0$$

where $B_5 = \max[0, c_4 - 3\sqrt{1-c_4^2}]$ and is tabulated in Appendix A-7. Thus, the control limits for the s-chart are

$$\begin{aligned} UCL_s &= B_6\sigma_0 \\ LCL_s &= B_5\sigma_0 \end{aligned} \qquad \textbf{(7.31)}$$

If a target value for the mean is specified as $\overline{X}_0$, then the center line is given by

$$CL_{\overline{X}} = \overline{X}_0 \qquad \textbf{(7.32)}$$

Equations for the control limits will be the same as those given by eq. (7.16) in the section on $\overline{X}$- and R-charts:

$$\begin{aligned} UCL_{\overline{X}} &= \overline{X}_0 + A\sigma_0 \\ LCL_{\overline{X}} &= \overline{X}_0 - A\sigma_0 \end{aligned} \qquad \textbf{(7.33)}$$

where $A = 3/\sqrt{n}$ and is tabulated in Appendix A-7.

Example 7-4: The thickness of the magnetic coating on audio tapes is an important characteristic. Random samples of size 4 are selected, and the thickness is measured using an optical instrument. Table 7-3 shows the mean $\overline{X}$ and standard deviation s for 20 samples. The specifications are 38 ± 4.5 microns. If a coating thickness is less than the specifications call for, that tape can be used for a different purpose by running it through another coating operation.

a. Find the trial control limits for an $\overline{X}$- and an s-chart.

Solution. The standard deviation chart must first be constructed. The center line of the s-chart is

$$CL_s = \overline{s} = \frac{\sum_{i=1}^{20} s_i}{20} = \frac{95.80}{20} = 4.790$$

TABLE 7-3 Data for Magnetic Coating Thickness (in microns)

Sample	*Sample Mean,* $\overline{X}$	*Sample Standard Deviation, s*	*Sample*	*Sample Mean,* $\overline{X}$	*Sample Standard Deviation, s*
1	36.4	4.6	11	36.7	5.3
2	35.8	3.7	12	35.2	3.5
3	37.3	5.2	13	38.8	4.7
4	33.9	4.3	14	39.0	5.6
5	37.8	4.4	15	35.5	5.0
6	36.1	3.9	16	37.1	4.1
7	38.6	5.0	17	38.3	5.6
8	39.4	6.1	18	39.2	4.8
9	34.4	4.1	19	36.8	4.7
10	39.5	5.8	20	37.7	5.4

The control limits for the s-chart are

$$\text{UCL}_s = B_4\overline{s} = (2.266)(4.790) = 10.854$$
$$\text{LCL}_s = B_3\overline{s} = (0)(4.790) = 0$$

Figure 7-19 shows this standard deviation control chart. None of the points fall outside the control limits, and the process seems to be in a state of control, so the $\overline{X}$-chart is constructed next. The center line of the $\overline{X}$-chart is

$$\text{CL}_{\overline{X}} = \overline{\overline{X}} = \frac{\sum_{i=1}^{20} \overline{X}_i}{20} = \frac{743.5}{20} = 37.175$$

The control limits for the $\overline{X}$-chart are

$$\text{UCL}_{\overline{X}} = \overline{\overline{X}} + A_3\overline{s} = 37.175 + (1.628)(4.790) = 44.973$$
$$\text{LCL}_{\overline{X}} = \overline{\overline{X}} - A_3\overline{s} = 37.175 - (1.628)(4.790) = 29.377$$

Figure 7-20 depicts the $\overline{X}$-chart. All the points are within the control limits, and no unusual nonrandom pattern is visible on the plot.

b. Assuming special causes for the out-of-control points, determine the revised control limits.

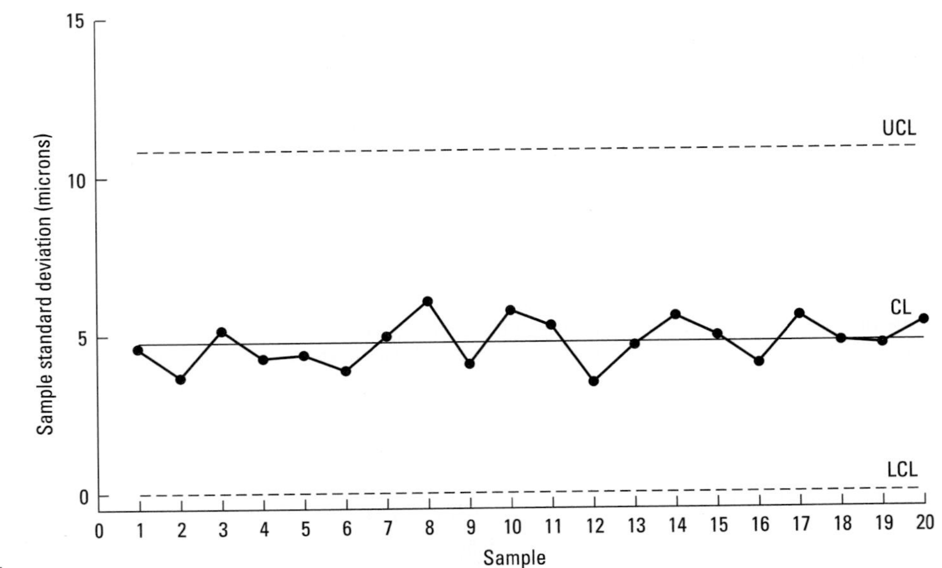

FIGURE 7-19 *s*-Chart for magnetic coating thickness.

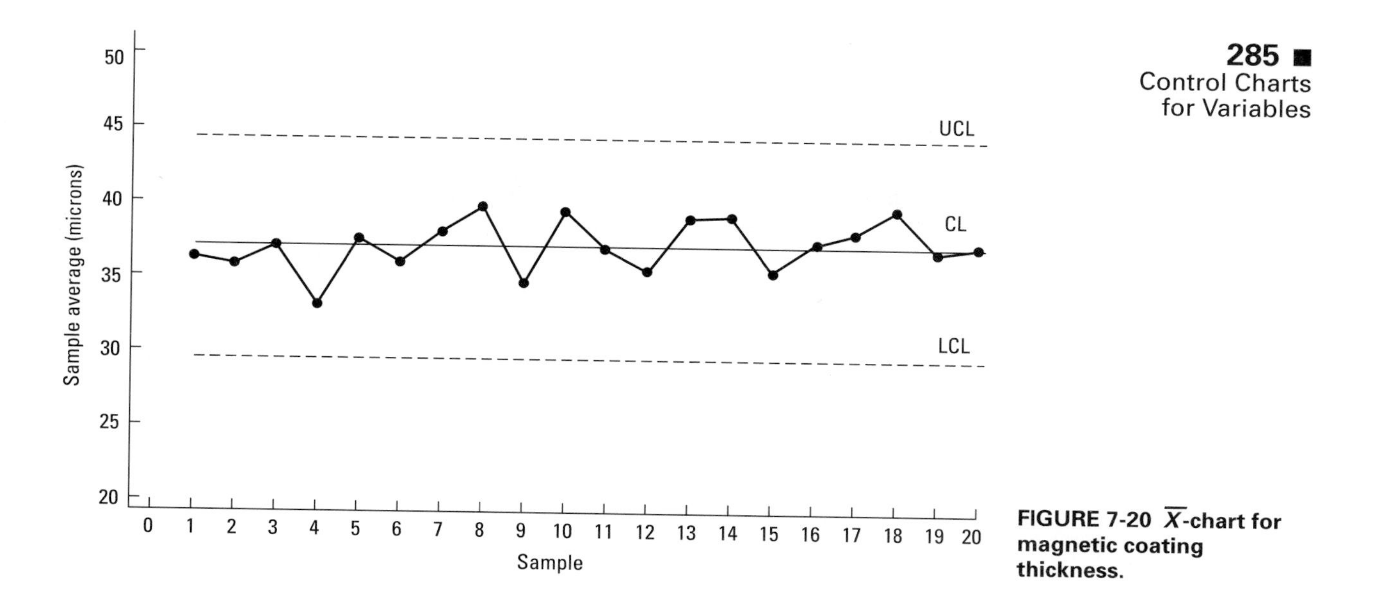

FIGURE 7-20 $\overline{X}$**-chart for magnetic coating thickness.**

Solution. In this case, the revised control limits will be the same as the trial control limits because we believe that no special causes are present in the system.

c. Assuming the thickness of the coating to be normally distributed, what proportion of the product will not meet specifications?

Solution. The process standard deviation may be estimated as

$$\hat{\sigma} = \frac{\bar{s}}{c_4} = \frac{4.790}{0.9213} = 5.199$$

Figure 7-21 shows a normal distribution for the coating thickness, with a process mean of 37.175 microns and an estimated process standard deviation $\hat{\sigma}$ of 5.199 microns. Note that the specification limits are also shown. To find the proportion of the output that does not meet specifications, the standard normal values at the upper and lower specification limits (USL and LSL) must be found. At the lower specification limit we get

$$z_1 = \frac{33.5 - 37.175}{5.199} = -0.71$$

The area below the LSL, found by using the standard normal table in Appendix A-3, is .2389. Similarly, the standard normal value at the upper specification limit is

$$z_2 = \frac{42.5 - 37.175}{5.199} = 1.02$$

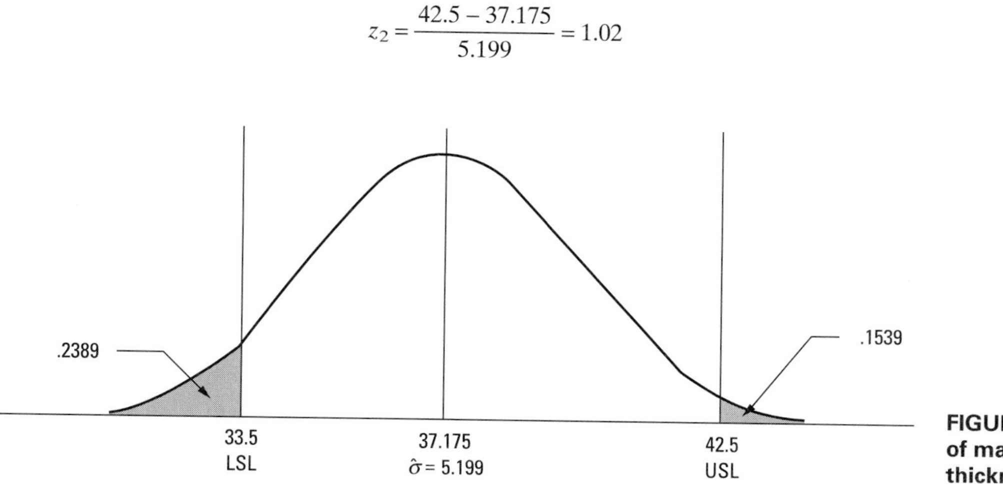

FIGURE 7-21 Distribution of magnetic coating thickness.

From Appendix A-3, the area above the USL is .1539. Hence, the proportion of product not meeting specifications is .2389 + .1539 = .3928.

d. Comment on the ability of the process to produce items that meet specifications.

Solution. A proportion of 39.28% of product not meeting specifications is quite high. On the other hand, we found the process to be in control. This example teaches an important lesson. It is possible for a process to be in control and still not produce conforming items. In such cases, management must look for the prevailing common causes and come up with ideas for process improvement. The existing process is not capable of meeting the stated specifications.

e. If the process average shifts to 37.8 microns, what proportion of the product will be acceptable?

Solution. If the process average shifts to 37.8 microns, the standard normal values must be recalculated. At the LSL,

$$z_1 = \frac{33.5 - 37.8}{5.199} = -0.83$$

From the standard normal table in Appendix A-3, the area below the LSL is .2033. The standard normal value at the USL is

$$z_2 = \frac{42.5 - 37.8}{5.199} = 0.90$$

The area above the USL is .1841. So, the proportion nonconforming is .2033 + .1841 = .3874. Although this change in the process average does reduce the proportion nonconforming, 38.74% nonconforming is still quite significant.

If output that falls below the LSL can be salvaged at a lower expense than that for output above the USL, the company could consider adjusting the process mean in the downward direction to reduce the proportion above the USL. Being aware of the unit costs associated with salvaging output outside the specification limits will enable the company to choose a target value for the process mean. Keep in mind, though, that this approach does not solve the basic problem. The underlying problem concerns the process variability. To make the process more capable, we must find ways of reducing the process standard deviation. This cannot come through process control, because the process is currently in a state of statistical control. It must come through process improvement, some analytical tools of which are discussed at length in Chapter 5.

Example 7-5: The thickness of sheet metal used for making automobile bodies is a characteristic of interest. Random samples of size 4 are taken. The average and the standard deviation are calculated for each sample and are shown in Table 7-4 for 20 samples. Find the control limits for the $\overline{X}$- and s-charts. Determine whether there are any out-of-control points.

TABLE 7-4 Sample Average and Standard Deviation for Sheet Metal Thickness (in mm)

Sample	*Sample Average,* $\overline{X}$	*Sample Standard Deviation,* s	*Sample*	*Sample Average,* $\overline{X}$	*Sample Standard Deviation,* s
1	10.19	0.15	11	10.18	0.16
2	9.80	0.12	12	9.85	0.15
3	10.12	0.18	13	9.82	0.06
4	10.54	0.19	14	10.18	0.34
5	9.86	0.14	15	9.96	0.11
6	9.45	0.09	16	9.57	0.09
7	10.06	0.16	17	10.14	0.12
8	10.13	0.18	18	10.08	0.15
9	9.82	0.14	19	9.82	0.09
10	10.17	0.13	20	10.15	0.12

```
data thick;
 input sample  avex aves;
       aven=4;
       label avex='Average thickness'
       aves='Standard deviation of thickness'
       aven='Sample size';
cards;
1  10.19  0.15
2   9.80  0.12
3  10.12  0.18
4  10.54  0.19
5   9.86  0.14
6   9.45  0.09
7  10.06  0.16
8  10.13  0.18
9   9.82  0.14
10 10.17  0.13
11 10.18  0.16
12  9.85  0.15
13  9.82  0.06
14 10.18  0.34
15  9.96  0.11
16  9.57  0.09
17 10.14  0.12
18 10.08  0.15
19  9.82  0.09
20 10.15  0.12
;
proc shewhart history=thick;
     xschart ave*sample;
```

FIGURE 7-22 Source program for $\overline{X}$- and *s*-charts based on summary data.

Solution. We will use the sample source program using SAS software based on summary data, shown in Figure 7-22, to find the control limits. The name of the characteristic variable is "ave," and "avex" and "aves," respectively, denote the average and standard deviation of the variable "ave." The statement "data thick" defines the name of the data set as "thick." The "input" statement specifies that data for three variables named as "sample," "avex," and "aves" will be input in that order. The variable "sample" represents the sample number. The statement "aven=4" says that the number of observations in each sample taken to measure the variable "ave" is 4. The "label" statement assigns the description to be associated with the corresponding variable in the program output. The "cards" statement indicates that the data to be input follow immediately. Summary data for 20 samples—the sample number, average thickness, and standard deviation of thickness values, in that order—is given next. The "proc shewhart" statement invokes the Shewhart procedure for control charts. In the same statement, "history=thick" indicates that the summary data to be used is found in the data set "thick." The option "xschart ave*sample" dictates that an $\overline{X}$- and *s*-chart option is to be used; the name of the variable is "ave," which is to be plotted versus "sample," the sample number.

Figure 7-23 shows the $\overline{X}$- and *s*-charts for the summary data for the sheet metal thickness. The center line of the $\overline{X}$-chart is 9.994 mm, with upper and lower control limits of 10.228 and 9.761 mm, respectively. On the *s*-chart, the center line is 0.14 mm, with upper and lower control limits of 0.33 and 0 mm, respectively. The standard deviation for sample 14 is above the upper control limit on the *s*-chart. On the $\overline{X}$-chart, sample 4 has an average above the upper control limit, and samples 6 and 16 have averages below the lower control limit. Special causes would need to be investigated for these out-of-control points.

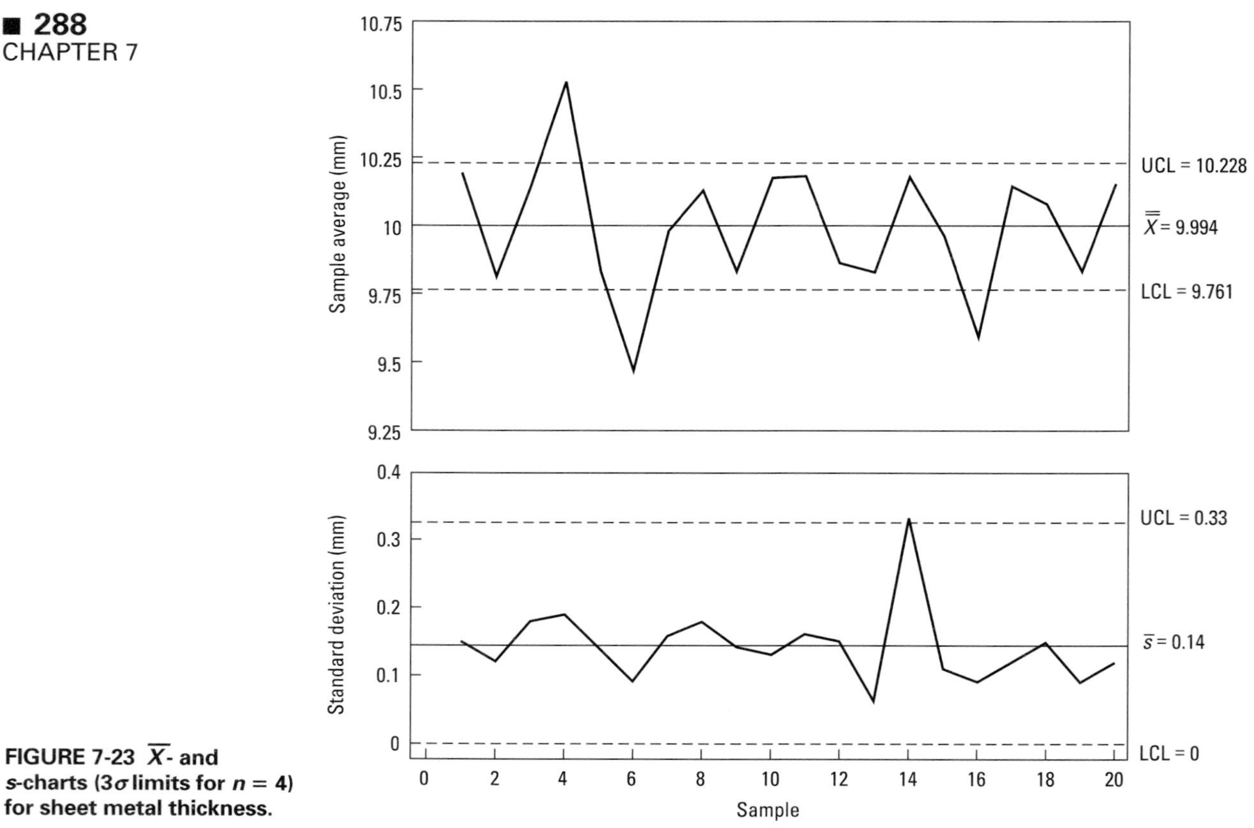

FIGURE 7-23 $\overline{X}$- and s-charts (3σ limits for $n = 4$) for sheet metal thickness.

7-6 CONTROL CHARTS FOR INDIVIDUAL UNITS

For some situations in which the rate of production is low, it is not feasible for a sample size to be greater than 1. Additionally, if the testing process is destructive and the cost of the item is expensive, the sample size might be chosen to be 1. Furthermore, if every manufactured unit from a process is inspected, the sample size is essentially 1. Service applications in marketing and accounting often have sample sizes of 1.

In a control chart for individual units—for which the value of the quality characteristic is represented by X—the variability of the process is estimated from the **moving range** (MR) found from two successive observations. The moving range of two observations is simply the result of subtracting the lesser value. Moving ranges are correlated because they use common rather than independent values in their calculations. That is, the moving range of observations 1 and 2 correlates with the moving range of observations 2 and 3. Because they are correlated, the pattern of the MR-chart must be interpreted carefully. Neither can we assume that X-values in a chart for individuals will be normally distributed like we have in previous control charts. So we must first check the distribution of the individual values. To do this, we might conduct an initial analysis using frequency histograms to identify the shape of the distribution, its skewness, and its kurtosis. This information will tell us whether we can make the assumption of a normal distribution when we establish the control limits.

No Given Standards

An estimate of the process standard deviation is given by

$$\hat{\sigma} = \frac{\overline{MR}}{d_2}$$

where $\overline{\text{MR}}$ is the average of the moving ranges of successive observations. Note that if we have a total of g individual observations, there will be $(g - 1)$ moving ranges. The center line and control limits of the MR-chart are

$$\text{CL}_{\text{MR}} = \overline{MR}$$
$$\text{UCL}_{\text{MR}} = D_4\overline{MR} \qquad \textbf{(7.34)}$$
$$\text{LCL}_{\text{MR}} = D_3\overline{MR}$$

For $n = 2$, $D_4 = 3.267$, and $D_3 = 0$, the control limits become

$$\text{UCL}_{\text{MR}} = 3.267\overline{MR}$$
$$\text{LCL}_{\text{MR}} = 0$$

The center line of the X-chart is

$$\text{CL}_X = \overline{X} \qquad \textbf{(7.35)}$$

The control limits of the ***X*-chart** are

$$\text{UCL}_X = \overline{X} + 3\frac{\overline{MR}}{d_2}$$
$$\text{LCL}_X = \overline{X} - 3\frac{\overline{MR}}{d_2} \qquad \textbf{(7.36)}$$

where (for $n = 2$) Appendix A-7 gives $d_2 = 1.128$.

Given Standard

The preceding derivation is based on the assumption that no standard values are given for either the mean or the process standard deviation. If standard values are specified as $\overline{X}_0$ and σ_0, respectively, the center line and control limits of the X-chart are

$$\text{CL}_X = \overline{X}_0$$
$$\text{UCL}_X = \overline{X}_0 + 3\sigma_0 \qquad \textbf{(7.37)}$$
$$\text{LCL}_X = \overline{X}_0 - 3\sigma_0$$

Assuming $n = 2$, the MR-chart for standard values has the following center line and control limits:

$$\text{CL}_{\text{MR}} = d_2\sigma_0 = (1.128)\sigma_0$$
$$\text{UCL}_{\text{MR}} = D_4 d_2\sigma_0 = (3.267)(1.128)\sigma_0 = (3.685)\sigma_0 \qquad \textbf{(7.38)}$$
$$\text{LCL}_{\text{MR}} = D_3 d_2\sigma_0 = 0$$

One advantage of the X-chart is the ease with which it can be understood. It can also be used to judge capability of a process by plotting the upper and lower specification limits on the chart itself. However, it has several disadvantages compared to an $\overline{X}$-chart. An X-chart is not as sensitive to changes in the process parameters. It typically requires more samples to detect parametric changes of the same magnitude. The main disadvantage of an X-chart, though, is that the control limits can become distorted if the individual items don't fit a normal distribution.

Example 7-6: Table 7-5 shows the Brinell hardness numbers of 20 individual steel fasteners and the moving ranges. The testing process dents the parts so that they cannot be used for their intended purpose. Construct the X-chart and the MR-chart based on two successive observations. Specification limits are 32 ± 7.

Solution. Note that there are 19 moving-range values for 20 observations. The average of the moving ranges is

$$\overline{\text{MR}} = \frac{\sum \text{MR}_i}{19} = \frac{96}{19} = 5.053$$

TABLE 7-5 Brinell Hardness Data for Individual Fasteners

Sample	*Brinell Hardness*	*Moving Range*	*Sample*	*Brinell Hardness*	*Moving Range*
1	36.3	—	11	29.4	1.1
2	28.6	7.7	12	35.2	5.8
3	32.5	3.9	13	37.7	2.5
4	38.7	6.2	14	27.5	10.2
5	35.4	3.3	15	28.4	0.9
6	27.3	8.1	16	33.6	5.2
7	37.2	9.9	17	28.5	5.1
8	36.4	0.8	18	36.2	7.7
9	38.3	1.9	19	32.7	3.5
10	30.5	7.8	20	28.3	4.4

which is also the center line of the MR-chart. The control limits for the MR-chart are

$$\text{UCL}_{\text{MR}} = D_4\overline{\text{MR}} = (3.267)5.053 = 16.508$$
$$\text{LCL}_{\text{MR}} = D_3\overline{\text{MR}} = (0)5.053 = 0$$

We can use Minitab to construct control charts for individual values and moving ranges for the steel fastener hardness data in Table 7-5. Choose **Stat > Control Charts > I-MR.** Figure 7-24 shows the trial control charts. No points plot outside the control limits on the MR-chart. Since the MR-chart exhibits control, we can construct the X-chart for individual data values. The center line of the X-chart is

$$\overline{X} = \frac{\sum X_i}{20} = \frac{658.7}{20} = 32.9$$

The control limits for the X-chart are given by

$$\text{UCL}_X = \overline{X} + \frac{3\overline{\text{MR}}}{d_2} = 32.9 + \frac{3(5.053)}{1.128} = 46.339$$
$$\text{LCL}_X = \overline{X} - \frac{3\overline{\text{MR}}}{d_2} = 32.9 - \frac{3(5.053)}{1.128} = 19.461$$

The X-chart is also shown in Figure 7-24. No out-of-control points are visible. Comparing the individual values with the specification limits, we find no values outside the specification limits. Thus, the observed nonconformance rate is zero and the process is capable.

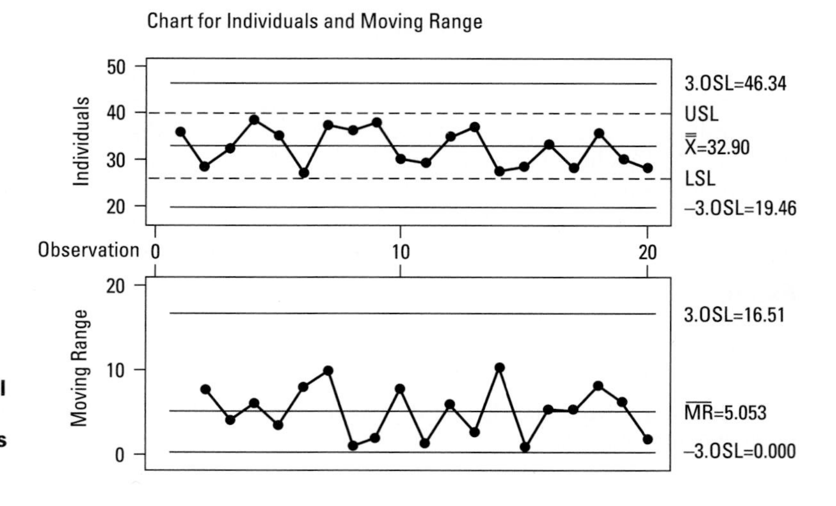

FIGURE 7-24 Control charts for individual values (X-chart) and moving ranges (MR-chart) for the steel fastener hardness data.

7-7 OTHER CONTROL CHARTS

In previous sections we have examined commonly used control charts. Now we will look at several other control charts. These charts are specific to certain situations. Procedures for constructing $\overline{X}$- and R-charts and interpreting their patterns apply to these charts as well, so they are not repeated here.

Cumulative Sum Control Chart for the Process Mean

In the Shewhart control charts, such as the $\overline{X}$- and R-charts, a plotted point represents information corresponding to that observation only. It does not use information from previous observations. On the other hand a **cumulative sum chart,** usually called a **cusum chart,** uses information from all of the prior samples by displaying the cumulative sum of the deviation of the sample values (for instance, the sample mean) from a specified target value.

The cumulative sum at sample number m is given by

$$S_m = \sum_{i=1}^{m} (\overline{X}_i - \mu_0) \tag{7.39}$$

where $\overline{X}_i$ is the sample mean for sample i and μ_0 is the target mean of the process.

Cusum charts are more effective than Shewhart control charts in detecting relatively small shifts in the process mean (of magnitude $0.5\sigma_{\overline{X}}$ to about $2\sigma_{\overline{X}}$). A cusum chart uses information from previous samples, so the effect of a small shift is more pronounced. For situations in which the sample size n is 1 (say, when each part is automatically measured by a machine), the cusum chart is better suited to determining shifts in the process mean than a Shewhart control chart. Because of the magnified effect of small changes, process shifts are easily found by locating the point where the slope of plotted cusum pattern changes.

There are some disadvantages to using cusum charts, however. First, because the cusum chart is designed to detect small changes in the process mean, it can be slow to detect large changes in the process parameters. Because a decision criterion is designed to do well under a specific situation does not mean it will perform equally well under different situations. Details on modifying the decision process for a cusum chart to detect large shifts can be found in Lucus (1976). Second, the cusum chart is not an effective tool in analyzing the historical performance of a process to see whether it is in control or to bring it in control. Thus, these charts are typically used for well-established processes that have a history of being stable.

Recall that for Shewhart control charts, the individual points are assumed to be uncorrelated. Cumulative values, however, *are* related. That is, S_{i-1} and S_i are related because $S_i = S_{i-1} + (\overline{X}_i - \mu_0)$. It is therefore possible for a cusum chart to exhibit runs or other patterns as a result of this relationship. The rules for describing out-of-control conditions based on the plot patterns of Shewhart charts may therefore not be applicable to cusum charts. Finally, training workers to use and maintain cusum charts may be more costly than for Shewhart charts.

Cumulative sum charts can model the proportion of nonconforming items, the number of nonconformities, the sample range, the sample standard deviation, or the process mean. In this section we focus on their ability to detect shifts in the process mean.

Suppose the target value of a process mean when the process is in control is denoted by μ_0. If the process mean shifts upward to a higher value μ_1, an upward drift will be observed in the value of the cusum S_m given by eq. (7.39) because the old lower value μ_0 is still used in the equation even though the X-values are now higher. Similarly, if the process mean shifts to a lower value μ_2, a downward trend will be observed in S_m. The task is to determine whether the trend in S_m is significant so that we can conclude that a change has taken place in the process mean. A template known as a V-mask, proposed by Barnard (1959), is used to make this determination.

Figure 7-25 shows the V-mask, which has two parameters, the lead distance d and the angle θ of each decision line with respect to the horizontal. The V-mask is positioned such that the point P coincides with the last plotted value of the cumulative sum and the line OP is parallel to the horizontal axis. If the previous plotted values are within the two arms of the V-mask—that is, between the upper decision line and the lower decision line—the process is judged to be in control. If any value of the cusum lies outside the arms of the V-mask, the process is considered to be out of control.

In Figure 7-25, notice that strong upward shift in the process mean is visible for sample 5. This shift makes sense given the fact that the cusum value for sample 1 is below the lower decision line, indicating an out-of-control situation. Likewise, the presence of a plotted value above the upper decision line indicates a downward drift in the process mean.

Determination of V-Mask Parameters

The two parameters of a V-mask, d and θ, are determined based on the levels of risk that the decision maker is willing to tolerate. These risks are the Type I and Type II errors described in Chapter 6. The probability of a **Type I error,** α, is the risk of concluding that a process is out of control when it is really in control. The probability of a **Type II error,** β, is the risk of failing to detect a change in the process parameter and concluding that the process is in control when it is really out of control. Let $\Delta\overline{X}$ denote the amount of shift in the process mean that we want to be able to detect, and let $\sigma_{\overline{X}}$ denote the standard deviation of $\overline{X}$. Next, consider the following equation:

$$\delta = \frac{\Delta\overline{X}}{\sigma_{\overline{X}}} \tag{7.40}$$

where δ represents the degree of shift in the process mean, relative to the standard deviation of the mean, that we wish to detect. Then, the **lead distance** for the V-mask is given by

$$d = \frac{2}{\delta^2}\ln\left(\frac{1-\beta}{\alpha}\right) \tag{7.41}$$

If the probability of a Type II error, β, is selected to be small, then eq. (7.41) reduces to

$$d = -\frac{2}{\delta^2}\ln(\alpha) \tag{7.42}$$

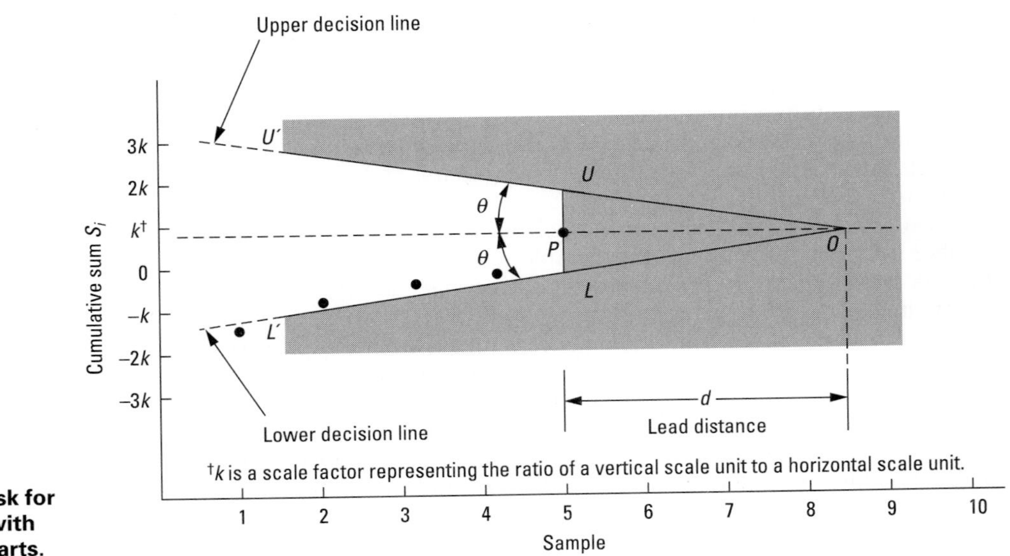

FIGURE 7-25 V-mask for making decisions with cumulative sum charts.

The **angle of decision line** with respect to the horizontal is obtained from

$$\theta = \tan^{-1}\left(\frac{\Delta\overline{X}}{2k}\right) \tag{7.43}$$

where k is a scale factor representing the ratio of a vertical-scale unit to a horizontal-scale unit on the plot. The value of k should be between $\sigma_{\overline{X}}$ and $2\sigma_{\overline{X}}$, with a preferred value of $2\sigma_{\overline{X}}$.

One measure of a control chart's performance is the **average run length (ARL)**. (We discussed ARL in Chapter 6.) This value represents the average number of points that must be plotted before an out-of-control condition is indicated. For a Shewhart control chart, if p represents the probability that a single point will fall outside the control limits, then the average run length is given by

$$\text{ARL} = \frac{1}{p} \tag{7.44}$$

For 3σ limits on a Shewhart $\overline{X}$-chart, the value of p is about .0026 when the process is in control. Hence, the ARL for an $\overline{X}$-chart exhibiting control is

$$\text{ARL} = \frac{1}{.0026} = 385$$

The implication of this is that, on average, if the process is in control, every 385th sample statistic will indicate an out-of-control state. The ARL is usually larger for a cusum chart than for a Shewhart chart. For example, for a cusum chart with comparable risks, the ARL is around 500. Thus, if the process is in control, on average, every 500th sample statistic will indicate an out-of-control situation, so there will be fewer false alarms.

Example 7-7: In the preparation of a drug, the percentage of calcium is a characteristic we want to control. Random samples of size 5 are selected, and the average percentage of calcium is found. The data values from 15 samples are shown in Table 7-6. From historical data, the standard deviation of the percentage of calcium is estimated as 0.2%. The target value for the average percentage of calcium content is 26.5%. We decide to notice shifts in the average percentage of calcium content of 0.1%. Assume an acceptable Type I error level of .05.

Solution. We must first find the deviation of each sample mean $\overline{X}_i$ from the target mean $\mu_0 = 26.5$ and then find the cumulative sum S_i. These values are shown in Table 7-7. From the information given, $\sigma = 0.2$, so $\sigma_{\overline{X}} = \sigma/\sqrt{n} = 0.2/\sqrt{5} = 0.089$. Next,

$$\delta = \frac{\Delta\overline{X}}{\sigma_{\overline{X}}} = \frac{0.1}{0.089} = 1.124$$

The lead distance for the V-mask is

$$d = -\frac{2}{\delta_2}\ln\alpha = -\frac{2}{(1.124)^2}\ln(.05) = 4.742$$

TABLE 7-6 Average Percentage of Calcium

Sample	*Average Percentage of Calcium, $\overline{X}$*	*Sample*	*Average Percentage of Calcium, $\overline{X}$*	*Sample*	*Average Percentage of Calcium, $\overline{X}$*
1	25.5	6	25.9	11	26.9
2	26.0	7	27.0	12	27.8
3	26.6	8	25.4	13	26.2
4	26.8	9	26.4	14	26.8
5	27.5	10	26.3	15	26.6

TABLE 7-7 Cumulative Sum of Data for Calcium Content

Sample i	*Deviation of Sample Mean from Target,* $(\bar{X}_i - \mu_0)$	*Cumulative Sum,* S_i	*Sample* i	*Deviation of Sample Mean from Target,* $(\bar{X}_i - \mu_0)$	*Cumulative Sum,* S_i
1	−1.0	−1.0	9	−0.1	−1.4
2	−0.5	−1.5	10	−0.2	−1.6
3	0.1	−1.4	11	0.4	−1.2
4	0.3	−1.1	12	1.3	0.1
5	1.0	−0.1	13	−0.3	−0.2
6	−0.6	−0.7	14	0.3	0.1
7	0.5	−0.2	15	0.1	0.2
8	−1.1	−1.3			

The units for d are the same as those for the horizontal scale. The angle of the V-mask for $k = 0.125$ is

$$\begin{aligned}\theta &= \tan^{-1}\left(\frac{\Delta\bar{X}}{2k}\right)\\ &= \tan^{-1}\left[\frac{0.1}{2(0.125)}\right]\\ &= \tan^{-1}(0.40) = 21.80^\circ\end{aligned}$$

This cumulative sum plot is shown in Figure 7-26. Using a V-mask with $d = 4.742$ and $\theta = 21.80^\circ$, a downward shift in the process mean is detected rather quickly by the second sample. When the V-mask is positioned on the cumulative sum point for sample 2, the cusum for sample 1 is found to lie above the upper decision line of the V-mask, indicating that the process mean has shifted downward.

Designing a Cumulative Sum Chart for a Specified ARL

The average run length can be used as a design criteria for control charts. If a process is in control, the ARL should be long, whereas if the process is out of control, the ARL should be short. Recall that δ is the degree of shift in the process mean, relative to the **standard deviation of the sample mean,** that we are interested in detecting; that is, $\delta = \Delta\bar{X}/\sigma_{\bar{X}}$. Let $L(\delta)$ denote the desired ARL when a shift in the process mean

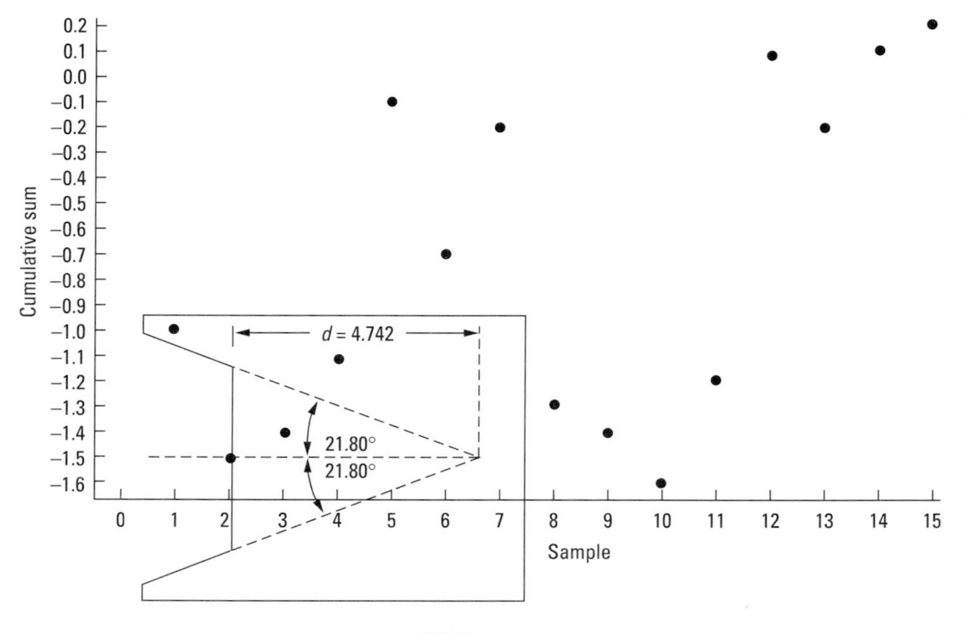

FIGURE 7-26 Cumulative sum chart for average percentage of calcium content.

is on the order of δ. An ARL curve is a plot of δ versus its corresponding average run length, $L(\delta)$. For a process in control, when $\delta = 0$, a large value of $L(0)$ is desirable. For a specified value of δ, we may have a desirable value of $L(\delta)$. Thus, two points on the ARL curve, $[0, L(0)]$ and $[\delta, L(\delta)]$, are specified. The goal is to find the cusum chart parameters d and θ that will satisfy these desirable goals.

Bowker and Lieberman (1972) provide a table (see Table 7-8) for selecting the V-mask parameters d and θ when the objective is to minimize $L(\delta)$ for a given δ. It is assumed that the decision maker has a specified value of $L(0)$ in mind. Table 7-8 gives values for $(k/\sigma_{\overline{X}})\tan\theta$ and d, and the minimum value of $L(\delta)$ for a specified δ. We will use this table in Example 7-8.

Example 7-8: Suppose that for a process in control, we want an ARL of 400. We also decide to detect shifts in the process mean of magnitude $1.5\sigma_{\overline{X}}$—that is, $\Delta\overline{X} = 1.5\sigma_{\overline{X}}$—which means that $\delta = 1.5$. Find the parameters of a V-mask for this process.

Solution. From Table 7-8, for $L(0) = 400$ and $\delta = 1.5$, we have

$$(k/\sigma_{\overline{X}})\tan\theta = 0.75$$
$$d = 4.5$$
$$L(1.5) = 5.2$$

If k, the ratio of the vertical scale to the horizontal scale, is selected to be $2\sigma_{\overline{X}}$, we have

$$2\tan\theta = 0.75 \qquad \text{or} \qquad \tan\theta = 0.375$$

The angle of the V-mask is

$$\theta = \tan^{-1}(0.375) = 20.556°$$

If we feel that 5.2 is too large a value of $L(1.5)$, we then need to reduce the average number of plotted points it takes to first detect a shift in the process mean of magnitude $1.5\sigma_{\overline{X}}$. Currently, it

TABLE 7-8 Selection of Cumulative Sum Control Charts Based on Specified ARL

δ = Deviation from Target Value (in Standard Deviations)		*L(0) = Expected Run Length When Process Is in Control*					
		50	*100*	*200*	*300*	*400*	*500*
0.25	$(k/\sigma_{\overline{x}})\tan\theta$	0.125			0.195		0.248
	d	47.6			46.2		37.4
	$L(0.25)$	28.3			74.0		94.0
0.50	$(k/\sigma_{\overline{x}})\tan\theta$	0.25	0.28	0.29	0.28	0.28	0.27
	d	17.5	18.2	21.4	24.7	27.3	29.6
	$L(0.5)$	15.8	19.0	24.0	26.7	29.0	30.0
0.75	$(k/\sigma_{\overline{x}})\tan\theta$	0.375	0.375	0.375	0.375	0.375	0.375
	d	9.2	11.3	13.8	15.0	16.2	16.8
	$L(0.75)$	8.9	11.0	13.4	14.5	15.7	16.5
1.0	$(k/\sigma_{\overline{x}})\tan\theta$	0.50	0.50	0.50	0.50	0.50	0.50
	d	5.7	6.9	8.2	9.0	9.6	10.0
	$L(1.0)$	6.1	7.4	8.7	9.4	10.0	10.5
1.5	$(k/\sigma_{\overline{x}})\tan\theta$	0.75	0.75	0.75	0.75	0.75	0.75
	d	2.7	3.3	3.9	4.3	4.5	4.7
	$L(1.5)$	3.4	4.0	4.6	5.0	5.2	5.4
2.0	$(k/\sigma_{\overline{x}})\tan\theta$	1.0	1.0	1.0	1.0	1.0	1.0
	d	1.5	1.9	2.2	2.4	2.5	2.7
	$L(2.0)$	2.26	2.63	2.96	3.15	3.3	3.4

Source: A. H. Bowker and G. J. Lieberman, *Engineering Statistics,* 2nd. ed. © 1972, p. 498. Reprinted by permission of Prentice Hall, Upper Saddle River, N.J.

takes about 5.2 points, on average, to detect a shift of this magnitude. Assume that we prefer $L(1.5)$ to be less than 5.0. From Table 7-8, for $\delta = 1.5$ and $L(1.5) < 5.0$, we could choose $L(1.5) = 4.6$, which corresponds to $(k/\sigma_{\overline{X}})\tan\theta = 0.75$, and $d = 3.9$. If k is chosen to be $2\sigma_{\overline{X}}$, we get

$$\tan\theta = \frac{0.75}{2} = 0.375$$

Hence, $\theta = 20.556°$ (the same value as before), and $d = 3.9$ (a reduced value). For $L(1.5) = 4.6$, which is less than 5.0, we have increased the sensitivity of the cusum chart to detect changes of magnitude $1.5\sigma_{\overline{X}}$, but in doing so, we have reduced the ARL for $\delta = 0$—that is, $L(0)$—to 200 from the previous value of 400. So now every 200th observation, on average, will be plotted as an out-of-control point when the process is actually in control.

Moving-Average Control Chart

As mentioned previously, standard Shewhart control charts are quite insensitive to small shifts, and cumulative sum charts are one way to alleviate this problem. A control chart using the moving-average method is another. Such charts are effective for detecting shifts of small magnitude in the process mean. Moving-average control charts can also be used in situations for which the sample size is 1, such as when product characteristics are measured automatically or when the time to produce a unit is long. It should be noted that, by their very nature, moving-average values are correlated.

Suppose samples of size n are collected from the process. Let the first t sample means be denoted by $\overline{X}_1, \overline{X}_2, \overline{X}_3, \ldots, \overline{X}_t$. (One sample is taken for each time step.) The moving average of width w (that is, of w samples) at time step t is given by

$$M_t = \frac{\overline{X}_t + \overline{X}_{t-1} + \cdots + \overline{X}_{t-w+1}}{w} \tag{7.45}$$

At any time step t, the moving average is updated by dropping the oldest mean and adding the newest mean. The variance of each sample mean is

$$\text{Var}(\overline{X}_t) = \frac{\sigma^2}{n}$$

where σ^2 is the population variance of the individual values. The variance of M_t is

$$\begin{aligned}\text{Var}(M_t) &= \frac{1}{w^2}\sum_{i=t-w+1}^{t} \text{Var}(\overline{X}_i) \\ &= \frac{1}{w^2}\sum_{i=t-w+1}^{t} \frac{\sigma^2}{n} \\ &= \frac{\sigma^2}{nw}\end{aligned} \tag{7.46}$$

The center line and control limits for the moving-average chart are given by

$$\begin{aligned}\text{CL} &= \overline{\overline{X}} \\ \text{UCL} &= \overline{\overline{X}} + 3\frac{\sigma}{\sqrt{nw}} \\ \text{LCL} &= \overline{\overline{X}} - 3\frac{\sigma}{\sqrt{nw}}\end{aligned} \tag{7.47}$$

From eq. (7.47), we can see that as w increases, the width of the control limits decreases. So, to detect shifts of smaller magnitudes, larger values of w should be chosen. For the startup period (when $t < w$), the moving average is given by

$$M_t = \frac{\sum_{i=1}^{t} \overline{X}_i}{t}, \qquad t = 1, 2, \ldots, w-1 \tag{7.48}$$

The control limits for this startup period are

$$\text{UCL} = \overline{\overline{X}} + \frac{3\sigma}{\sqrt{nt}}, \qquad t = 1, 2, \ldots, w-1$$
$$\text{LCL} = \overline{\overline{X}} - \frac{3\sigma}{\sqrt{nt}}, \qquad t = 1, 2, \ldots, w-1 \qquad \textbf{(7.49)}$$

Since these control limits change at each sample point during this startup period, an alternative procedure would be to use the ordinary $\overline{X}$-chart for $t < w$ and use the moving-average chart for $t \geq w$.

Example 7-9: The amount of a coloring pigment in polypropylene plastic, produced in batches, is a variable of interest. For 20 random samples of size 5, the average amount of pigment (in kilograms) is shown in Table 7-9. Construct a moving-average control chart of width 6. The process has up to this point been in control with an average range $\overline{R}$ of 0.40 kg.

Solution. Table 7-9 shows the computed values of the moving average M_t based on a width w of 6. For values of $t < 6$, the moving average is calculated using eq. (7.48). For $t \geq 6$, M_t is calculated using eq. (7.45). Also shown in Table 7-9 are the lower and upper control limits for the moving-average chart. To find these limits, eq. (7.49) is used for $t < 6$, and eq. (7.47) is used for $t \geq 6$. The mean of the sample averages is

$$\overline{\overline{X}} = \frac{\sum_{t=1}^{20} \overline{X}_t}{20} = \frac{503.2}{20} = 25.16$$

Since $\overline{R} = 0.40$, an estimate of the process standard deviation is

$$\hat{\sigma} = \frac{\overline{R}}{d_2} = \frac{0.40}{2.326} = 0.172$$

TABLE 7-9 Data and Results for a Moving-Average Control Chart (in kg)

Sample	*Sample Average,* $\overline{X}_t$	*Moving Average* M_t	*Control Limits for* M_t	
			LCL	*UCL*
1	25.0	25.0	24.929	25.391
2	25.4	25.2	24.997	25.323
3	25.2	25.2	25.027	25.293
4	25.0	25.15	25.045	25.275
5	25.2	25.16	25.057	25.263
6	24.9	25.12	25.066	25.254
7	25.0	25.12	25.066	25.254
8	25.4	25.12	25.066	25.254
9	24.9	25.07	25.066	25.254
10	25.2	25.10	25.066	25.254
11	25.0	25.07	25.066	25.254
12	25.7	25.20	25.066	25.254
13	25.0	25.20	25.066	25.254
14	25.1	25.15	25.066	25.254
15	25.0	25.17	25.066	25.254
16	24.9	25.12	25.066	25.254
17	25.0	25.12	25.066	25.254
18	25.1	25.02	25.066	25.254
19	25.4	25.08	25.066	25.254
20	25.8	25.20	25.066	25.254

To calculate the control limits, consider sample 3:

$$\text{UCL} = \overline{\overline{X}} + 3\frac{\sigma}{\sqrt{nt}} = 25.16 + 3\frac{0.172}{\sqrt{(5)(3)}} = 25.293$$

$$\text{LCL} = \overline{\overline{X}} - 3\frac{\sigma}{\sqrt{nt}} = 25.16 - 3\frac{0.172}{\sqrt{(5)(3)}} = 25.027$$

For samples 6–20, the control limits stay the same; the LCL is 25.066 kg and the UCL is 25.254 kg. Figure 7-27 shows a plot of the moving averages and control limits. The moving average for sample 18 plots below the lower control limit, indicating that the process mean has drifted downward. Special causes should be investigated for this out-of-control condition, and appropriate corrective action should be taken.

Exponentially Weighted Moving Average or Geometric Moving-Average Control Chart

The preceding discussion showed that a moving-average chart can be used as an alternative to an ordinary $\overline{X}$-chart to detect small changes in process parameters. The moving-average method is basically a weighted-average scheme. For sample t, the sample means $\overline{X}_t, \overline{X}_{t-1}, \ldots, \overline{X}_{t-w+1}$ are each weighted by $1/w$ [see eq. (7.45)], while the sample means for time steps less than $(t - w + 1)$ are weighted by zero. Along similar lines, a chart can be constructed based on varying weights for the prior observations. More weight can be assigned to the most recent observation, with the weights decreasing for less recent observations. A geometric moving-average control chart, also known as an exponentially weighted moving-average (EWMA) chart, is based on this premise. One of the advantages of a geometric moving-average chart over a moving-average chart is that the former is more effective in detecting small changes in process parameters. The geometric moving average at time step t is given by

$$G_t = r\overline{X}_t + (1 - r)G_{t-1} \tag{7.50}$$

where r is a weighting constant $(0 < r \leq 1)$ and G_0 is $\overline{\overline{X}}$. By using eq. (7.50) repeatedly, we get

$$\begin{aligned} G_t &= r\overline{X}_t + r(1-r)\overline{X}_{t-1} + r(1-r)^2 G_{t-2} \\ &= r\overline{X}_t + r(1-r)\overline{X}_{t-1} + r(1-r)^2\overline{X}_{t-2} \\ &\quad + \cdots + (1-r)^t G_0 \end{aligned} \tag{7.51}$$

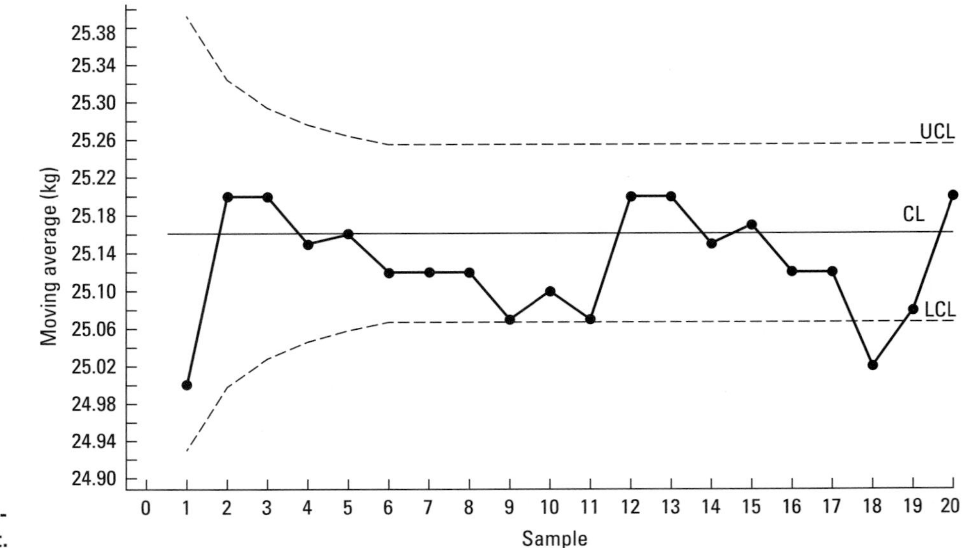

FIGURE 7-27 Moving-average control chart.

Equation (7.51) shows that the weight associated with the ith mean from $t(\overline{X}_{t-i})$ is $r(1-r)^i$. The weights decrease geometrically as the sample mean becomes less recent. The sum of all the weights is 1. Consider, for example, the case for which $r = 0.3$. This implies that in calculating G_t, the most recent sample mean $(\overline{X}_t)$ has a weight of 0.3, the next most recent observation $(\overline{X}_{t-1})$ has a weight of $(0.3)(1-0.3) = 0.21$, the next observation $(\overline{X}_{t-2})$ has a weight of $0.3(1-0.3)^2 = 0.147$, and so on. Here, G_0 has a weight of $(1-0.3)^t$. Since these weights appear to decrease exponentially, eq. (7.51) describes what is known as the exponentially weighted moving-average model (EWMA).

If the sample means $\overline{X}_1, \overline{X}_2, \overline{X}_3, \ldots, \overline{X}_{t-1}$ are assumed to be independent of each other and if the population standard deviation is σ, then the variance of G_t is given by

$$\mathrm{Var}(G_t) = \left(\frac{\sigma^2}{n}\right)\left(\frac{r}{2-r}\right)[1-(1-r)^{2t}] \qquad \textbf{(7.52)}$$

For large values of t, the standard deviation of G_t is

$$\sigma_G = \sqrt{\mathrm{Var}(G_t)} = \sqrt{\frac{\sigma^2}{n}\left(\frac{r}{2-r}\right)}$$

The upper and lower control limits are

$$\begin{aligned} \mathrm{UCL} &= \overline{\overline{X}} + 3\sigma\sqrt{\frac{r}{(2-r)n}} \\ \mathrm{LCL} &= \overline{\overline{X}} - 3\sigma\sqrt{\frac{r}{(2-r)n}} \end{aligned} \qquad \textbf{(7.53)}$$

For small values of t, the control limits are found using eq. (7.52) to be

$$\begin{aligned} \mathrm{UCL} &= \overline{\overline{X}} + 3\sigma\sqrt{\frac{r}{n(2-r)}[1-(1-r)^{2t}]} \\ \mathrm{LCL} &= \overline{\overline{X}} - 3\sigma\sqrt{\frac{r}{n(2-r)}[1-(1-r)^{2t}]} \end{aligned} \qquad \textbf{(7.54)}$$

A geometric moving-average control chart is based on a concept similar to that of a moving-average chart. By choosing an adequate set of weights, however, where recent sample means are more heavily weighted, the ability to detect small changes in process parameters is increased. If the **weighting factor** r is selected as

$$r = \frac{2}{w+1} \qquad \textbf{(7.55)}$$

where w is the **moving-average span,** then the moving-average method and the geometric moving-average method are equivalent. There are guidelines for choosing the value of r. If our goal is to detect small shifts in the process parameters as soon as possible, we use a small value of r (say, 0.1). If we use $r = 1$, the geometric moving-average chart reduces to the standard Shewhart chart for the mean.

Example 7-10: Refer to Example 7-9 regarding the amount of a coloring pigment in polypropylene plastic. Table 7-10 gives the sample averages for 20 samples of size 5. Construct a geometric moving-average control chart using a weighting factor r of 0.2.

Solution. For the data in Table 7-10, the mean of the sample averages is

$$\overline{\overline{X}} = \frac{503.2}{20} = 25.160$$

Since $\overline{R}$ is given as 0.40 in Example 7-9, the estimated process standard deviation is

$$\hat{\sigma} = \frac{\overline{R}}{d_2} = \frac{0.40}{2.326} = 0.172$$

TABLE 7-10 Data and Results for a Geometric Moving-Average Control Chart (in kg)

Sample	*Sample Average,* $\overline{X}_t$	*Geometric Moving Average,* G_t	*Control Limits for Geometric Average* LCL	UCL
1	25.0	25.128	25.114	25.206
2	25.4	25.182	25.101	25.219
3	25.2	25.186	25.094	25.226
4	25.0	25.149	25.090	25.230
5	25.2	25.159	25.087	25.233
6	24.9	25.107	25.086	25.234
7	25.0	25.086	25.085	25.235
8	25.4	25.149	25.084	25.236
9	24.9	25.099	25.084	25.236
10	25.2	25.119	25.084	25.236
11	25.0	25.095	25.083	25.237
12	25.7	25.216	25.083	25.237
13	25.0	25.173	25.083	25.237
14	25.1	25.158	25.083	25.237
15	25.0	25.127	25.083	25.237
16	24.9	25.081	25.083	25.237
17	25.0	25.065	25.083	25.237
18	25.1	25.072	25.083	25.237
19	25.4	25.138	25.083	25.237
20	25.8	25.270	25.083	25.237

The geometric moving average for sample 1 using eq. (7.50) is (for $G_0 = \overline{\overline{X}}$)

$$\begin{aligned} G_1 &= r\overline{X}_1 + (1-r)G_0 \\ &= (0.2)25.0 + (1-0.2)25.16 = 25.128 \end{aligned}$$

The remaining geometric moving averages are calculated similarly. These values are shown in Table 7-10. The control limits for sample 1 are calculated using eq. (7.54):

$$\begin{aligned} \text{UCL} &= 25.160 + 3(0.172)\sqrt{\frac{0.2}{5(2-0.2)}[1-(1-0.2)^2]} \\ &= 25.206 \end{aligned}$$

$$\begin{aligned} \text{LCL} &= 25.160 - 3(0.172)\sqrt{\frac{0.2}{5(2-0.2)}[1-(1-0.2)^2]} \\ &= 25.114 \end{aligned}$$

Similar computations are performed for the remaining samples. For large values of t (say, $t = 15$), the control limits are found by using eq. (7.53):

$$\begin{aligned} \text{UCL} &= 25.160 + 3(0.172)\sqrt{\frac{0.2}{(2-0.2)5}} \\ &= 25.237 \end{aligned}$$

$$\begin{aligned} \text{LCL} &= 25.160 - 3(0.172)\sqrt{\frac{0.2}{(2-0.2)5}} \\ &= 25.083 \end{aligned}$$

Figure 7-28 shows a plot of this geometric moving-average control chart. Notice that samples 16, 17, and 18 are below the lower control limit and that sample 20 plots above the upper control limit. The special causes for these points should be investigated in order to take remedial action. Note that in the moving-average chart in Figure 7-27, sample 18 plotted below the lower control limit, but samples 16, 17, and 20 were within the control limits. The geometric moving-average chart (Figure 7-28), which is a little more sensitive to small shifts in the process parameters than the moving-average chart, identifies these additional points as being out of control.